EXCITOTOXICITY IN NEUROLOGICAL DISEASES

New Therapeutic Challenge

EXCITOTOXICITY IN NEUROLOGICAL DISEASES

New Therapeutic Challenge

edited by

Carlo Ferrarese

Department of Neurology and Neuroscience
University of Milano-Bicocca, Ospedale San Gerardo
Monza (Milano), Italy

and

M. Flint Beal

Department of Neurology and Neuroscience
Weill Medical College of Cornell University
New York, NY, U.S.A.

KLUWER ACADEMIC PUBLISHERS
Boston / Dordrecht / London

Distributors for North, Central and South America:
Kluwer Academic Publishers
101 Philip Drive
Assinippi Park
Norwell, Massachusetts 02061 USA
Telephone (781) 871-6600
Fax (781) 681-9045
E-Mail <kluwer@wkap.com>

Distributors for all other countries:
Kluwer Academic Publishers Group
Post Office Box 322
3300 AH Dordrecht, THE NETHERLANDS
Telephone 31 786 576 000
Fax 31 786 576 254
E-Mail <services@wkap.nl>

Electronic Services <http://www.wkap.nl>

Library of Congress Cataloging-in-Publication Data

Excitotoxicity in neurological diseases : new therapeutic challenge / edited by Carlo Ferrarese and M. Flint Beal.
p. cm.
Includes bibliographical references and index.
ISBN 1-4020-7680-0 (alk. paper)
1. Nervous system—Diseases—Treatment. 2. Neurotoxic agents. 3. Nervous system—Degeneration—Treatment. 4. Neurotoxicology. I. Ferrarese, Carlo, 1955- II. Beal, M. Flint.

RC346.5.E97 2004
616.8'046—dc22 2003060524

Printed on acid-free paper.
Printed in the United States of America

Table of contents

Preface ix

Acknowledgement xi

Introduction and historical notes 1
M. Flint Beal
Department of Neurology and Neuroscience
Weill Medical College, Cornell University
New York Presbyterian Hospital, NY 10021, New York, USA.

Part I: Basic mechanisms

1. Contribution of astrocyte glutamate release to excitotoxicity 13
Daniela Rossi, Paola Bezzi, Maria Domercq, Liliana Brambilla, Jacopo Meldolesi and Andrea Volterra
Department of Pharmacological Sciences
University of Milan
20133 Milan, Italy.

2. Glutamate uptake and transporters 27
Danbolt N.C., Holmseth S., Skår A., Lehre K.P., Furness D.N.
Department of Anatomy, Institute of Basic Medical Sciences
University of Oslo
Blindern N-0317, Oslo, Norway.

3. NMDA receptors 51
Fabrizio Gardoni, Flaminio Cattabeni and Monica Di Luca
Department of Pharmacological Sciences
University of Milan
20133 Milan, Italy.

4. AMPA receptor and synaptic plasticity 65
Maria Passafaro and Carlo Sala
CNR Institute of Neuroscience, Cellular and Molecular Pharmacology
20129 Milan, Italy.

5. Metabotropic glutamate receptors and neurodegeneration 79
Valeria Bruno, Giuseppe Battaglia and Ferdinando Nicoletti
Department of Pharmacological Sciences
University of Rome "La Sapienza"
00185 Rome, Italy.

6. Mechanisms of excitotoxicity and excitoprotection 103
Mark P. Mattson
Laboratory of Neurosciences, National Institute of Aging
Gerontology Research Centre
MD 21224 Baltimore, USA.

Part II: Acute and chronic neurological disorders

7. Epilepsy and seizures: excitotoxicity or excitotrophicity? 137
Karen Gale
Department of Pharmacology
Georgetown University Medical Center
DC 20007 Washington, USA.

8. Excitotoxicity in cerebral ischemia 171
Alberto Chiarugi, Elena Meli, Andrea Cozzi, Flavio Moroni and Domenico E. Pellegrini-Giampietro
Department of Preclinical and Clinical Pharmacology
University of Florence
50139 Florence, Italy.

9. Excitotoxicity and traumatic brain injury 189
Niklas Marklund, John F. Stover and Tracy K. McIntosh
Department of Neurosurgery
University of Pennsylvania and Veterans Administration Medical Center
Philadelphia, Pennsylvania 19104, USA.

10. Excitotoxicity in amyotrophic lateral sclerosis: selective vulnerability of motor neurons 217
Tiziana Mennini and Caterina Bendotti
Istituto di Ricerche Framacologiche Mario Negri
20157 Milan, Italy.

11. Glutamate transmission in the pathogenesis of Parkinson's disease 229
Antonio Pisani and Paolo Calabresi
Department of Neurology
University of Rome "Tor Vergata"
00133 Rome, Italy.

12. Excitotoxicity in Huntington's Disease 243
M. Flint Beal
Department of Neurology and Neuroscience
Weill Medical College, Cornell University
New York Presbyterian Hospital, NY 10021, USA.

13. The glutamatergic system in Alzheimer's disease brain: dysfunction associated with amyloid β-peptide and oxidative stress 251
Allan D. Butterfield
Department of Chemistry, Center of Membrane Sciences
University of Kentucky
Lexington, KY 40506-0055, USA.

14. Neurotoxicity and prion disease 263
Judyth Sassoon, Filmore Banks and David R. Brown
Department of Biology and Biochemistry
University of Bath
BA2 7AY, Bath, UK.

15. Glutamate excitotoxicity in multiple sclerosis 285
David Pitt and Cedric S. Raine
Department of Neurology
Albert Einstein College of Medicine, Yeshive University
Bronx, NY 10461, New York, USA.

16. HIV-1 associated dementia 305
Marcus Kaul and Stuart A. Lipton
Center for Neuroscience and Aging
The Burnham Institute
La Jolla, CA 92037, USA.

17. Markers of excitotoxicity in patients: tool for diagnosis or therapy? 317
Simone Beretta, Angelo Aliprandi and Carlo Ferrarese
Department of Neurology
University of Milano-Bicocca
Ospedale San Gerardo, 20052 Monza (Milan), Italy.

18. Anti-excitotoxic therapies in neurological diseases: current evidences and new perspectives 339
Laura Brighina, Angelo Aliprandi, Simone Beretta and Carlo Ferrarese
Department of Neurology
University of Milano-Bicocca
Ospedale San Gerardo, 20052 Monza (Milan), Italy.

PREFACE

It can be reasonably anticipated that, over the next generations, the proportion of elderly people will remarkably increase and, with this, the number of persons suffering from acute (e.g. cerebral ischemia) or chronic neurodegenerative disorders. To date, approved drugs only alleviate the symptoms of these diseases (for instance, acetylcholinesterase inhibitors in Alzheimer disease and L-dopa and dopamine-agonists in Parkinson disease), while none seems to stop the progression of the degenerative processes underlying them.
The development of effective preventive or protective therapies has been impeded by the limitations of our knowledge of the causes and the mechanisms by which neurons die in neurodegenerative disorders.

Evidence accumulated in the past 20 years indicated that the major excitatory neurotransmitter glutamate may play a role as neurotoxin in several conditions. In particular, the glutamatergic system dysfunction seems to be an early event working as a common pathway in the pathogenesis of a large number of acute and chronic neurological disorders, in strict conjunction with other important mechanisms, such as oxidative stress and energetic failure, and probably triggered by different mechanisms in various diseases.
In consideration of that, drug discovery efforts over the last decade have been focused on the search for drugs that either reduce glutamate synaptic levels or block its postsynaptic effects.

Despite numerous reviews on basic mechanisms and clinical aspects of the excitotoxic phenomenon, so far no comprehensive book has covered the topic in all its complexity, starting from basic pharmacological mechanisms, to animal models of diseases and finally to clinical pathogenic and therapeutic implications.
This book is intended both for scientists already involved in the field, who may want a broader knowledge of all the aspects related to this complex mechanism, as well as for PhD or MD students, residents and neurologists wishing to learn more about a common mechanism involved in neuronal damage, driven by scientific curiosity or by the need to understand how new drugs, already on the market, may or may not work in various neurological disorders.

With this aim, the book is basically divided into two parts, designed to be mutually integrated:

- The first part will cover all basic mechanisms of excitotoxicity, from glutamate metabolism, release, activity on various pre- and post-synaptic receptors, reuptake and their regulations
- The second part will focus on each neurological disorder where this mechanism seem to play a major role, by examining in details pre-clinical and clinical evidences as well as treatment strategies in animal models and in patients.

  Although the targets of excitotoxicity are neurons, its players and actors are working also in cells outside the central nervous system. Thus, peripheral *ex vivo* markers from patients might bridge the gap between preclinical studies in animal models and patient care, and may help to verify the excitotoxic hypothesis at the bedside.

To accomplish this complex task, world leaders in various fields, who did major contributions to define the problem and to open new avenues, and are also actively working at present time, agreed to put their expertise to launch the first complete and advanced level book on the role of excitotoxicity in neurological disorders.

Carlo Ferrarese, MD, PhD
Professor and Chairman of Neurology
University of Milano-Bicocca
Ospedale San Gerardo, Monza, Italy

ACKNOWLEDGEMENT

We are indebted to Dr. Simone Beretta, M.D. and Ph.D. student at the Department of Neuroscience of the University of Milano-Bicocca, for his essential contribution in the various phases of the manuscript preparation, including careful revisions of each chapter of this book.

INTRODUCTION AND HISTORICAL NOTES
EXCITOTOXICITY IN NEUROLOGIC DISEASES

M. Flint Beal, M.D.
Department of Neurology and Neuroscience, Weill Medical College of Cornell University, New York Presbyterian Hospital, New York, NY USA

INDEX

1. Introduction
2. Cellular mechanisms of excitotoxicity
3. Free radicals
4. Nitric oxide
5. Slow excitotoxicity
6. Conclusions

Excitotoxicity refers to neuronal death caused by activation of excitatory amino acid receptors. Several lines of evidence have linked excitotoxic cell death to the pathogenesis of both acute and chronic neurologic diseases. The initial observation that glutamate was neurotoxic was that of Lucas and Newhouse, who found that administration of glutamate to mice resulted in retinal degeneration (Lucas and Newhouse, 1957). Subsequent studies of Olney and colleagues linked neurotoxicity to the activation of excitatory amino acid receptors, and the term "excitotoxin" was coined (Olney, 1969). Further advances were those of Rothman linking release of excitatory amino acids to anoxic cell death in hippocampal cultures (Rothman , 1984), and of Choi linking calcium influx to delayed cell death caused by excitatory amino acids (Choi, 1987). More work has linked activation of excitatory amino acid receptors to free radical generation and nitric oxide, both of which may lead to oxidative stress (Dawson et al., 1991: Lafon-Cazal et al., 1993).

2. Cellular mechanisms of excitotoxicity

An important advance in clarifying mechanisms of excitotoxicity was that of Choi that delayed glutamate neurotoxicity was calcium dependent (Choi, 1987). Subsequent studies showed that calcium load in cultured cortical neurons correlates with subsequent neuronal degeneration (Hartley et al., 1993; Eimerl et al., 1994), whereas intracellular calcium concentrations do not. This supports the idea that much of the glutamate-induces Ca^{2+} load is sequestered into mitochondria rather than free in the cytoplasm. Evidence has shown that mitochondria and Na^{+} /Ca^{2+} exchange, which is the major means of efflux of mitochondrial Ca^{2+}, buffer glutamate-induced calcium loads in cultured cortical neurons (White et al 1995; Wang et al 1994). The increases in mitochondrial calcium also lead to metabolic dysfunction as shown by a lowering in intracellular pH (Wang et al., 1994). The release of calcium from intracellular stores may also contribute to excitotoxicity (Frandsen and Schousboe, 1991; Lei et al., 1992). Blocking calcium uptake into mitochondria abolishes excitotoxicity (Stout et al., 1998; Nicholls and Budd, 2000).

Randall and Thayer showed that there are three phases of changes in intracellular calcium preceding cell death in cultured hippocampal neurons (Randall and Thayer, 1992). There is an initial phase of increased intracellular lasting 5-10 min followed by a latent phase of approximately 2 h in which calcium returns to normal. The third phase consists of a graduated sustained rise in intracellular sodium that reaches a plateau associated with cell death. Tymianski *et al.,* showed that cell-permanent Ca^{2+} chelators reduced excitotoxic cell injury (Tymianski et al., 1993). These authors also demonstrated "source specificity" of the Ca^{2+} load, showing the Ca^{2+} entering through the NMDA channel was more toxic than that entering through other sources (Tymianski et al., 1993). This observation is consistent with the earlier work of Choi and colleagues demonstrating that the NMDA receptor mediates most of the excitotoxic effects of glutamate (Choi, 1987). It has been suggested that Ca^{2+} influx by the NMDA receptor may have access to proteins or compartments which make this Ca^{2+} more efficacious in producing cell death. It was shown that NMDA receptors are specifically coupled to nitric oxide production by the scaffolding protein PSD-95 (Sattlet et al., 1999). Ischemic brain damage is *in vivo* and excitotoxic cell damage *in vitro* is blocked by disrupting the NMDA-receptor PSD 95 interaction (Aarts et al., 2002).

The means by which increased intracellular calcium leads to cell death may involve several mechanisms. These include activation of protein kinases, phospholipases, nitric oxide synthase, proteases, endonucleases, inhibition of protein synthesis, mitochondrial damage, and free radical generation (Beal, 1992). Evidence to support or refute these various possibilities is limited, particularly *in vivo.* Kainic acid neurotoxicity is associated with activation of calpain *in vivo* (Siman and Noszek, 1988). Furthermore, calpain inhibitors reduce AMPA-induced neurotoxicty *in vitro* although they are not effective against glutamate toxicity (Caner et al., 1993; Manev et al., 1991). Calpain inhibitors show efficacy in a gerbil model of global ischemia and in models of focal ischemia in rats (Bartus et al., 1994; Hong et al., 1994; Lee et al., 1991).

3. Free radicals

The initial report linking free radicals to excitotoxicity was that of Dykens, who showed that kainate-induced damage to cerebellar neurons could be attenuated by superoxide dismutase, allopurinol, and hydroxy radical scavengers such as mannitol (Dykens et al., 1987). Other studies showed that glutathione depletion exacerbates excitotoxcity while the free radical scavengers α-tocopherol, ascorbic acid, and ubiquinone show neuroprotective effects (Bridges et al., 1991; Favit et al., 1992; Majewska and Bell, 1990). The 21-aminosteroids and α-phenyl-*N-tert*-butylnitrone, which scavenge free radicals also protect against excitotoxicity *in vitro.* The vitamin E analogue trolox protects cultured neurons from AMPA toxicity (Chow et al., 1994). Cultured cortical neurons which overexpress superoxide dismutase are resistant to both glutamate and ischemia induced neurotoxicity (Chan et al., 1990).

Direct evidence linking excitotoxicty to free radical generation comes from studies using electron paramagnetic resonance which show that NMDA dose-dependently increases superoxide formation in cultured cerebellar neurons (Lafon-Cazal et al., 1993). The effects are blocked by NMDA antagonists or removing extracellular Ca^{2+}. This is consistent with the findings of Dykens that exposure of isolated cortical mitochondria to 2.5 uM Ca^{2+}, which is similar to concentrations which occur in the setting of excitotoxicity, leads to free radical generation (Dykens, 1994). In synaptosomes NMDA, kainic acid and AMPA all stimulate free radical generation (Bondy and Lee, 1993). Electron paramagnetic resonance also showed generation of free radicals *in vivo* following systemic administration of kainic acid (Sun et al., 1992).

Evidence linking increases in intracellular calcium to mitochondrial production of reactive oxygen species has been obtained *in vitro* (Dugan et al 1995; Reynolds and Hastings, 1995). Dugan and colleagues used the oxidation-sensitive dye dihydrorhodamine-123 with confocal microscopy to demonstrate that exposure to NMDA, but not kainate, ionomycin or elevated potassium led to oxygen radical production in cultured neurons (Dugan et al., 1995). This was confirmed by studies using electron paramagnetic resonance. The increase in oxygen radical production was blocked by inhibitors of mitochondrial electron transport and mimicked by an uncoupler of electron transport. In contrast, inhibitors of nitric oxide synthase and arachidonic acid metabolism had no effect.

The study of Reynolds and Hastings used the oxidation-sensitive dye dichlorodihydrofluorescein to study the effects of glutamate in neuronal cultures (Reynolds and Hastings, 1995). Glutamate at excitotoxic concentrations caused localized areas of increased fluorescence at the margins of the cell body which were dependent on NMDA receptor activation and calcium entry, and which were blocked by an uncoupler of mitochondrial electron transport. These two studies therefore suggest a critical role of Ca^{2+} dependent uncoupling of neuronal mitochondrial electron transport in the production of reactive oxygen species following glutamate exposure.

Beal and co-workers examined the relationship of excitotoxicity to free radical production *in vivo* (Schulz et al., 1995). They showed that malonate, which produces excitotoxic lesions, leads to increased hydroxy radical generation as assessed by the salicylate trapping method. The free radical spin trap *N-tert*-butyl-α (2-sulfophenyl)-nitrone (S-PBN) attenuated both hydroxy radical generation and neurotoxicity. It also attenuated striatal lesions produced by NMDA, AMPA, and kainate. These findings provide direct *in vivo* evidence for a role of free radicals in excitotoxicity.

4. Nitric oxide

There is a prominent role of nitric oxide in excitotoxicity. Dawson and colleagues originally demonstrated that nitric oxide synthase inhibitors and hemoglobin, which scavenges nitric oxide, block glutamate neurotoxicity *in vitro* (Dawson et al., 1991). They subsequently showed that pretreatment of cultures with quisqualate, which preferentially kills nitric oxide synthase neurons, blocks glutamate neurotoxicity in the cultures (Dawson et al., 1993). Excitotoxicity can occur in the absence of nitric oxide synthase since cultured kidney neurons which lack the enzyme show excitotoxicity when transfected with NMDA receptors (Anegawa et al., 1995).

Initial studies of the effects of inhibition of nitric oxide synthase on NMDA-induced excitotoxicity *in vivo* were also conflicting (Moncada et al., 1992; Lenrer-Natoli et al., 1992). This appears to be due to the nonspecificity of the nitric oxide synthase inhibitors utilized, which have effects on both the neuronal and endothelial isoforms, leading to vascular effects (Huang et al., 1994). Evidence strongly favoring a role of neuronal nitric oxide synthase in focal ischemic lesions has come from studies showing that lesions are attenuated in mice with a knockout of the enzyme (Huang et al., 1994).

Several studies showed that 7-nitroindazole is a relatively specific inhibitor of the neuronal isofom of nitric oxide synthase *in vivo*. It has no effects on blood pressure or on acetylcholine-induced vasorelaxation (Babbedge et al., 1993; Moore et al., 1993; Yoshida et al., 1994). 7-Nitroindazole reduces focal ischemic lesions (Yoshida et al., 1994). Beal and co-workers found that it significantly attenuated excitotoxicity produced by NMDA but not by AMPA or kainate (Schulz et al., 1994). This is consistent with the observations of Dawson and colleagues *in vitro* (Dawson et al., 1991; Dawson et al., 1993), and suggests that Ca^{2+} influx via the NMDA receptor leads to activation of neuronal nitric oxide synthase.

5. Slow excitotoxicity

The role of excitotoxicity in neurodegenerative diseases is speculative. In these diseases, there is no evidence for an increase in glutamate concentrations, with the exception of glutamate in the CSF in ALS patients. Furthermore, increases in glutamate concentrations by themselves may not be sufficient to cause excitotoxicity (Massieu et al., 1995). A search for increases in concentrations of other endogenous excitotoxins such as quinolinic acid has been unsuccessful (Heyes et al., 1991). The concept of slow or weak excitotoxicity has therefore been proposed (Beal, 1992; Albin and Greenamyre, 1992). One possibility to account for this would be a receptor abnormality which could lead to increased calcium influx.

Another possibility is that slow excitotoxicity could occur as a consequence of an impairment in energy metabolism. This could occur by a variety of mechanisms including genetic mutations in mitochondrial electron transport or Krebs cycle enzymes. The possibility that impaired energy metabolism could result in excitotoxicity was originally demonstrated by the work of Novelli and co-workers (Novelli et al., 1988). They showed that inhibitors of oxidative phosphorylation of Na^+ -K^+ ATPase allowed glutamate to become neurotoxic at concentrations which ordinarily exhibited no neurotoxicity. This was felt to be due to a reduction in ATP leading to partial neuronal depolarizatiion. This may then lead to relief of the voltage-dependent Mg^{2+} block of the NMDA receptor leading to persistent receptor activation by ambient levels of glutamate.

Consistent with this possibility Zeevalk and Nicklas showed that partial energy impairment in cultured chick retina with either iodoacetate (a glycolysis inhibitor) or with cyanide (an inhibitor of oxidative phosphorylation) leads to NMDA receptor activation and excitotoxicity in the absence of any increase in extracellular concentrations of glutamate (Zeevalk and Nicklas, 1990). Furthermore graded titration of membrane potential with potassium mimicked the toxicity produced by graded metabolic inhibition (Zeevalk and Nicklas, 1991). Potassium channel activators, which hyperpolarize the cell membrane, can block excitotoxicity *in vitro* (Abele and Miller, 1990). Inhibitors of the Na^+ -K^+ ATPase produce lesions in rat substantia nigra and striatum (Lees and Leong, 1996). There is therefore a complex interrelationship between excitotoxicity and mitochondrial function, in that mitochondrial dysfunction can lead to excitotoxicity which then further impairs mitochondrial function.

6. Conclusions

There is a large body of evidence implicating excitotoxicity in the pathogenesis of neurologic diseases. In the setting of acute cerebral ischemia, epilepsy or trauma, the primary mechanism of excitotoxic cell death appears to be necrosis. With milder insults, which are associated with preservation of ATP levels, the primary mechanism of cell death appears to be apoptosis (Ankarcrona et al., 1995). This is likely to be the case in neurodegenerative diseases and other illnesses, which are more slowly evolving. Excitotoxicity remains an important disease mechanism, which is implicated in many neurologic illnesses and it is an important target in the development of therapies for these diseases.

REFERENCES

Aarts M., Liu Y., Liu L., Besshoh S., Arundine M., Gurd J.W., Wang Y.T., Salter M.W., Tymianski M. Treatment of ischemic brain damage by perturbing NMDA receptor-PSD-95 protein interactions. *Science* 2000, 298:846-850

Abele A.E. and Miller R.J. Potassium channel activators abolish excitotoxicity in cultured hippocampal pyramidal neurons. *Neurosci Lett* 1990, 115:195-200

Albin R.L. and Greenamyre J.T. Alternative excitotoxic hypotheses. *Neurology* 1992, 42:733-738

Anegawa N.J., Lynch D.R., Verdoorn T.A., et al. Transfection of N-methyl-D-aspartate receptors in a nonneuronal cell line leads to cell death. *J Neurochem* 1995, 64:2004-2012

Ankarcrona M., Dypbukt J.M., Bonfoco E., Zhivotovsky B., Orrenius S., Lipton SA., Nicotera P. Glutamate-induced neuronal death: a succession of necrosis or apoptosis depending on mitochondrial function. *Neuron* 1995, 15:961-973

Babbedge R. C., Bland-Ward P.A., Hart S.L., et al. Inhibition of rat cerebellar nitric oxide synthase by 7-nitro indazole and related substituted indazoles. *Br J Pharmacol* 1993, 110, 225-228

Bartus R.T., Baker K.L., Heiser A.D., et al. Postischemic administration of AK275, a calpain inhibitor, provides substantial protection against focal ischemic brain damage. *J Cereb Blood Flow Metab*, 1994, 14:537-544

Beal M. F. Does impairment of energy metabolism result in excitotoxic neuronal death in neurodegenerative illnesses? *Ann Neurol* 1992, 31:119-130

Bondy S.C. and Lee D. K. Oxidative stress induced by glutamate receptor agonists. *Brain Res* 1993, 610:229 – 233

Bridges R.J., Koh J.Y., Hatalski C.G., et al. Increased excitotoxic vulnerability of cortical cultures with reduced levels of glutathione. *Eur J Pharmacol* 1991, 192:199-200

Caner H., Collins J. L. Harris S.M. , et al. Attenuation of AMPA-induced neurotoxicity by a calpain inhibitor. *Brain Res* 1993, 607:354-356.

Chan P.H., Chu L., Chen S.F., et al. Reduced neurotoxicity in transgenic mice overexpressing human copper-zinc superoxide dismutase. *Stroke* 1990, 21:III80

Choi D.W. Ionic dependence of glutamate neurotoxicity. J Neurosci 1987, 7: 369-379

Chow H.S., Lynch I. J.J., Rose K. et al Trolox attenuates cortical neuronal injury induced by iron, ultraviolet light, glucose deprivation or AMPA. *Brain Res* 1994, 639: 102-108.

Dawson V.L., Dawson T.M., Bartley D.A. et al. Mechanisms of nitric oxide mediated neurotoxicity in primary brain cultures. *J Neurosci* 1993, 13:2651-2661

Dawson V.L.. Dawson T.M., London E.D. et al. Nitric oxide mediates glutamate neurotoxicity in primary cortical cultures. *Proc Natl Acad Sci U*SA 1991, 88:6368-6371

Dugan L. L., Sensi S.L. Canzoniero L. M.T. Mitochondrial production of reactive oxygen species in cortical neurons following exposure to N-methyl-D-aspartate. *J Neurosci* 1995, 15:6377-6388

Dykens J.A. Isolated cerebral and cerebellar mitochondria produce free radicals when exposed to elevated Ca^{2+} and Na^+: implications for neurodegeneration. *J Neurochem* 1994, 63:584-591

Dykens J.A., Stern A., Trenkner E. Mechanisms of kainate toxicity to cerebellar neurons *in vitro* is analogous to reperfusion tissue injury. *J Neurochem* 1987, 49:1222-1228

Eimerl S. and Schramm M. The quantity of calcium that appears to induce neuronal death. *J Neurochem.*, 1994, 62:1223-1226

Favit A., Nicoletti F., Scapagnini U. et al. Ubiquinone protects cultured neurons agonist spontaneous and excitotoxin-induced degeneration. *J Cereb Blood Flow Metab* 1992, 12:638-645

Frandsen A., and Schousboe A. Dantrolene prevents glutamate cytotoxicity and Ca^{2+} release from intracellular stores in cultured cerebral cortical neurons. *J Neurochem* 1991,56:1075-1078.

Hartley D.M., Kurth M.C., Bjerkness L., et al. Glutamate receptor-induced Ca^{2+} accumulation in cortical cell culture correlates with subsequent neuronal degeneration. *J Neurosci* 1993, 13: 1993-2000.

Heyes M.P., Swartz K.J., Markey S.P., et al. Regional brain and cerebrospinal fluid quinolinic acid concentrations in Huntington's Disease. *Neurosci Lett* 1991, 122:265-269

Hong S-C, Goto Y, Lanzino G., et al., Neuroprotection with a calpain inhibitor in a model of focal cerebral ischemia. *Stroke* 1994, 25, 663-669

Huang Z., Huang P.L., Panahian N., et al. Effects of cerebral ischemia in mice deficient in neuronal nitric oxide synthase. *Science* 1994, 265:1883-1885

Lafon-Cazal M., Pietri S., Culcasi M., et al. NMDA-dependent superoxide production and neurotoxicity. *Nature* 1993, 364:535-537

Lee K.S., Frank S., Vanderklish P., et al. Inhibition of proteolysis protects hippocampal neurons from ischemia. *Proc Natl Acad Sci USA* 1991, 88:7233-7237

Lees G.T. and Leong W. The sodium-potassium ATPase inhibitor ouabain is neurotoxic in the rat substantia nigra and striatum. *Neurosci Lett* 1995, 188:113-116

Lei S.Z., Zhang D., Abele A.E. et al. Blockade of NMDA receptor-mediated mobilization of intracellular Ca^{2+} prevents neurotoxicity. *Brain Res* 1992,598:196-202

Lerner-Natoli M., Rondouin G., de Block F., et al. Chronic NO synthase inhibition fails to protect hippocampal neurons against NMDA toxicity. *Neuroreport* 1992, 3:1109-1112

Lucas D.R. and Newhouse J. P. The toxic effect of sodium L-glutamate on the inner layers of the retina. *Arch. Ophthalmol.*, 1957, 58:193-201.

Majewska M.D. and Bell J.A. Ascorbic acid protects neurons from injury induced by glutamate and NMDA. *NeuroReport* 1990, 1:194-196

Manev H., Favaron M., Siman R., et al. Glutamate neurotoxicity is independent of calpain 1 inhibition in primary cultures of cerebellar granule cells. *J Neurochem* 1991, 57:1288-1295

Massieu L., Morales-Villagran A., Tapia R. Accumulation of extracellular glutamate by inhibition of its uptake is not sufficient for inducing neuronal damage: an *in vivo* microdialysis study. *J. Neurochem*, 1995, 64:2262-2272

Moncada C., Lekieffre D., Arvin B., et al. Effect of NO synthase inhibition on NMDA- and ischaemia-induced hippocampal lesions. *Neuroreport* 1992, 3:530-532

Moore P.K., Wallace P., Gaffen Z., et al. Characterization of the novel nitric oxide synthase inhibitor 7-nitroindazole and related indazoles. Antinociceptive and cardiovascular effects. *Br J Pharmacol* 1993,110:219-224

Nicholls D.G. and Budd S.L. Mitochondria and neuronal survival. *Physiol Rev* 2000, 80:315-360

Novelli A., Reilly J.A., Lysko P.G., et al. Glutamate becomes neurotoxic via the N-methyl-D-aspartate receptor, when intracellular energy levels are reduced. *Brain Res* 1988, 451:205-212

Olney J.W. Brain lesions, obesity and other disturbances in mice treated with monosodium glutamate. *Science*, 1969, 164: 719-721.

Randall R.D. and Thayer S.A. Glutamate-induced calcium transient triggers delayed calcium overload and neurotoxicity in rat hippocampal neurons. *J Neurosci* 1992, 12:1882-1895

Reynolds I.J. and Hastings, T.G. Glutamate induces the production of reactive oxygen species in cultured forebrain neurons following NMDA receptor activation. *J Neurosc* 1995, 15:3318-3327

Rothman S.R. Synaptic release of excitatory amino acid neurotransmitter mediates anoxic neuronal death. *J Neurosci* 1984, 4: 1884-1891

Sattler R., Xiong Z., Lu W.Y., Hafner M., MacDonald J.F., Tymianski M. Specific coupling of NMDA receptor activation to nitric oxide neurotoxicity by PSD-95 protein. *Science* 1999, 284:1845-1848

Schulz J.B., Henshaw D.R., Siwek D. et al. Involvement of free radicals in excitotoxicity *in vivo*. *J Neurochem* 1995, 64:2239-2247

Schulz J.B., Matthews R.T., Henshaw D.R., et al. Inhibition of neuronal nitric oxide synthase (NOS) protects against neurotoxicity produced by 3-nitropropionic acid, malonate and MPTP. *Soc Neurosci Abst* 1994, 20:1661

Siman R. and Noszek J. C. Excitatory amino acids activate calpain I and induce structural protein breakdown in vivo. *Neuron* 1988, 1:279-287

Stout A.K., Raphael H.M., Kanterewicz B.I., Klann E., Reynolds I.J. Glutamate-induced neuron death requires mitochondrial calcium uptake. *Nat Neurosci* 1998,1:366-373

Sun A.Y., Cheng Y., Bu Q., et al. The biochemical mechanisms of the excitotoxicity of kainic acid. Free radical formation. *Mol Chem Neuropathol* 1992,17:51-63

Tymianski M., Charlton M.P., Carlen P. L. et al. Source specificity of early calcium neurotoxicity in cultured embryonic spinal neurons. *J. Neurosci* 1993, 13:2085-2104

Tymianski M., Wallace M.C., Spigelman I., et al. Cell-permanent Ca^{2+} chelators reduce early excitotoxic and ischemic neuronal injury *in vitro* and *in vivo*. *Neuron* 1993, 11:221-235

Wang G.J., Randall R.D., and Thayer S. A. Glutamate-induced intracellular acidification of cultured hippocampal neurons demonstrates altered energy metabolism resulting from Ca^{2+} loads. *J Neurophysiol* 1994, 72:2563-2569

White R. J. and Reynolds I. J. Mitochondria and Na^+/Ca^{2+} exchange buffer glutamate-induced calcium loads in cultured cortical neurons. *J Neurosci* 1995, 15:1318-1328

Yoshida T., Limmroth V., Irikura K., et al. The NOS inhibitor, 7-nitroindazole, decreases focal infarct volume but not the response to topical acetylcholine in pial vessels. *J Cereb Blood Flow Metab* 1994,14:924-929

Zeevalk G.D. and Nicklas W.J. Chemically induced hypoglycemia and anoxia: relationship to glutamate receptor-mediated toxicity in retina. *J Pharmacol Exp Ther* 1990, 253:1285-1292

Zeevalk G.D. and Nicklas W.J. Mechanisms underlying initiation of excitotoxicity associated with metabolic inhibition. *J Pharmacol Exp Ther* 1991, 257:870-878

Part I

Basic Mechanisms

Chapter 1
CONTRIBUTION OF ASTROCYTE GLUTAMATE RELEASE TO EXCITOTOXICITY

Daniela Rossi[1], Paola Bezzi[1,2], Maria Domercq[1*], Liliana Brambilla[1], Jacopo Meldolesi[3], and Andrea Volterra[1,2]

(1) Department of Pharmacological Sciences and Center of Excellence on Neurodegenerative Diseases , University of Milan, Via Balzaretti 9, 20133 Milan, Italy; (2) Institute of Cell Biology and Morphology, University of Lausanne, Rue du Bugnon 9, 1005 Lausanne, Switzerland (3) Department of Neuroscience, Vita-Salute San Raffaele University and Excellence Centre in Cell Differentiation, via Olgettina 58, 20132 Milan, Italy.

*Present Address: Departamento de Neurociencias, Universidad del Pais Vasco, 48940 Leida, Spain.

Abstract: Although glial cells have been traditionally viewed as supportive partners of neurons, recent studies demonstrated that astrocytes possess functional receptors and are able to release transmitters by regulated pathways. Astrocytes were found to react to synaptically released neurotransmitters by undergoing intracellular calcium elevation which subsequently triggers an exocytosis-like glial transmitter release. These findings led to a new concept of neuron-glia intercommunication where astrocytes play an unsuspected dynamic role by integrating neuronal inputs and modulating synaptic activity. The discovery that glial release of the excitatory amino acid glutamate is controlled by molecules linked to inflammatory functions, such as cytokines and prostaglandins, suggested that glia-to-neuron signalling may be implicated in physiological processes but also in pathological situations. Indeed, a local and parenchimal inflammatory reaction characterised by astroglia and microglia activation has been reported in several brain pathologies including prion diseases and various dementias like Alzheimer's disease and the AIDS dementia complex. In agreement, stimulation of the calcium-dependent glial glutamate release process via activation of the chemokine receptor, CXCR4, by its natural ligand, SDF1α, is crucial for normal brain communication. However, the interaction of the same receptor with the HIV-1 coat protein gp120 in pathological conditions caused de-regulation of the glutamate system and excitotoxic neuronal cell death. The findings herein reported suggest that a better comprehension of the glial-neuron glutamatergic interplay may provide information about normal brain functions and may highlight possible molecular targets for therapeutical interventions in pathology.

Keywords: astrocytes, glutamate release, excitotoxicity, glial signalling, calcium.

INDEX

1. Introduction.
2. Towards understanding the active roles of astrocytes.
3. Astrocyte glutamate release in physiology and in pathology.
    3.1 Calcium-independent glutamate release from astrocytes.
    3.2 Calcium-dependent glutamate release from astrocytes.
4. Contribution of astrocyte calcium-dependent glutamate release to neurodegeneration: the case of AIDS neuropathology.
5. Conclusions.

1. Introduction.

Glia are a group of cells largely represented in the central nervous system (CNS). Based on their diverse morphology and function they can be distinguished in various subclasses, including oligodendrocytes, responsible for myelination of the CNS; microglia, comparable to cerebral macrophages; and astrocytes, which are thought to perform several functions owing to their strategic location between neurons and capillaries. For decades astrocytes have been regarded solely as passive supportive elements responsible for maintaining the optimal environment for neurons, an assumption supported by the evidence that they respond to neuronal metabolic necessities and regulate the composition of extracellular brain fluid. However, this vision was recently challenged by studies showing that astroglial cells modulate synaptic activity and actively participate to the CNS function and formation. Astrocytes were found to express a range of neurotransmitter receptors, often mirroring that of the neighbouring synapses. There is now evidence that such receptors can be activated by the spillover of neurotransmitters during synaptic activity and are able to induce elevation of astrocyte internal calcium ($[Ca^{2+}]_i$). This in turn can start release of transmitters (e.g. glutamate) from astrocytes and intercellular communication, including astrocyte-to-neuron signalling. Astrocyte-released transmitters can activate neuronal receptors and thereby modify electrical excitability properties as well as synaptic activity.

Our group recently described a novel calcium-dependent glutamate release process from astrocytes, mediated by prostaglandins (PGs) and tumour necrosis factor α (TNFα) (Bezzi et al., 1998; Bezzi et al., 2001). In normal conditions, this regulated glial process is apparently part of a physiological modulatory interplay between neurons and astrocytes. However, specific pathological events may alter the glial pathway leading to amplification of the glutamatergic input to neurons and to excitotoxic neuronal death (Bezzi et al., 2001). Importantly, the discovery that "pro-inflammatory" mediators control glia-to-neuron signalling suggests that brain inflammation may interfere with the circuits between astrocytes and neurons and cause their de-regulation.

In this chapter, we will examine recent advances made in understanding the active role of astrocytes in glutamatergic functions, starting from the physiological processes, and then considering the glial contribution to excitotoxic neurodegeneration.

2. Towards understanding the active roles of astrocytes.

Astrocytes, the most numerous glial cells in the CNS, are intimately associated with neurons, for instance in hippocampus their processes tightly enwrap about 50% of the synapses (Pfrieger and Barres, 1996; Ventura and Harris, 1999). This close interaction between the two cell types allows astrocytes to provide nerve cells with structural, metabolic and trophic support. For example, glucose, the main source of energy for brain, enters the CNS via uptake from astroglial cells, where it is either glycolytically processed to produce metabolic substrates for neurons or stored in the form of glycogen (Tsacopoulos and Magistretti, 1996) (Figure 1). Astrocytes play a critical role also in regulating the homeostasis of ions, in particular potassium (Karwoski et al., 1989), by means of permeable channels or active pumps located on their membrane. Indeed, the maintenance of a physiological extracellular K^+ concentration is critical for normal propagation of neuronal action potentials, while unbalanced fluxes may result in neuronal depolarisation and eventually action potential blockage. Moreover, astroglia control the extracellular concentration of synaptically released neurotransmitters by means of specific transporter proteins (Coco et al., 1997; Dehnes et al., 1998) located on their plasma membrane. Astrocyte glutamate uptake is apparently the preponderant system for removal of synaptic glutamate.

Once glutamate is taken up by the astrocytes, one of its known fates is conversion to glutamine by the astrocyte-specific enzyme, glutamine synthetase, followed by shuttling back to neurons, to replenish the glutamate synaptic pool (Waniewski and Martin, 1986) (Figure 1).

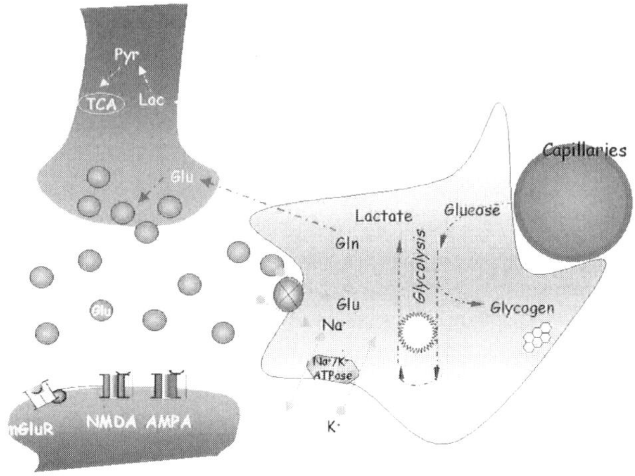

Figure 1. Schematic representation of neuron-astrocyte metabolic coupling. Synaptically released glutamate (Glu) is taken up by specific transporter proteins (in blue) located on the astrocyte plasma membrane. Once the transmitter has entered the glial cells, one of its metabolic fates is transformation to glutamine (Gln) and shuttling back to neurons, so to replenish the glutamate synaptic pool (glutamate-glutamine cycle). In addition, astrocyte glutamate uptake is involved in processes of energy supply from astrocytes to neurons. Indeed, during the uptake process sodium ions are co-transported with glutamate into the astrocytes. As a consequence, Na^+ activates the Na^+/K^+ ATPase pump leading to hydrolysis of ATP, increased glycolysis and uptake of glucose from the blood vessels. During the glycolytic process, lactate is produced and delivered to neurons where it enters the tricarboxylic acid cycle (TCA). Glutamate uptake and lactate release appear to respect a 1:1 stoichiometry, an evidence which would tightly link the level of synaptic activity to the metabolic supply from astrocytes.

In addition to the above supportive functions, astrocytes have been shown to synthesise and release a number of compounds (e.g. eicosanoids, nitric oxide, cytokines, nucleotides, neurosteroids, neurotrophins, amino acids and peptides) through which they can actively exchange signals with various cell types of the CNS, including neurons and neighbouring astrocytes. Moreover, they express a variety of ion channels and neurotransmitter receptors (Porter and McCarthy, 1997; Verkhratsky and Kettenmann, 1996), including ionotropic (s)-α-amino-3-hydroxy-5-methyl-4-isoxazole propionic acid (AMPA)/kainate and metabotropic glutamate receptors (mGluRs), that allow them to actively react to neuronal and other environmental inputs (Steinhauser and Gallo, 1996). Stimulation of such astrocyte receptors induces rises in their internal calcium levels (Dani et al., 1992; Pasti et al., 1997; Porter and McCarthy, 1997) followed by glutamate release, which may act as a back-signalling molecule, inducing neuromodulation (Hassinger et al., 1995; Nedergaard, 1994; Parpura et al., 1994) (see next section).

The discovery of these unsuspected active properties of astrocytes highlights new aspects of their cellular and functional organisation as well as the existence of an authentic network of chemical signalling in brain where astrocytes represent an intermediary and integrative station.

3. Astrocyte glutamate release in physiology and in pathology.

Several mediators, including excitatory and inhibitory amino acids, are synthesised and released via different mechanisms from glia. The amino acid glutamate is released from astrocytes through multiple pathways, that are activated under different conditions, at different loci and/or with different modalities. Among the different mechanisms involved we count the Ca^{2+}-dependent process outlined above, and several Ca^{2+}-independent processes involving the reversed operation of re-uptake carriers, exchange via the cystine-glutamate antiporter or volume-sensitive organic anion channels.

From here on we will focus on these different pathways by which glutamate is released from glia, in particular discussing the role of the Ca^{2+}-dependent process in physiological and pathological circumstances.

3.1 Ca^{2+}-Independent glutamate release from astrocytes.

The first direct evidence that astrocytes release endogenous glutamate was reported in 1990 by Attwell and colleagues (Szatkowski et al., 1990), who showed that inversion of the ion gradients driving the glutamate uptake systems in cultured glia induced a non-vesicular release of the transmitter. The Authors pointed out that this condition is unlikely to occur during normal synaptic transmission, but is probably the major source of excitotoxic extracellular glutamate accumulation during brain ischemia (Rossi et al., 2000). A second reported mechanism taking place via glial carriers is the so-called 'exchange' process which leads to glutamate release independently of ion gradient changes. Under normal conditions this process is imperceptible being an homoexchange of glutamate, which is simultaneously released and taken up by the transporters. However, when substrates other than glutamate [e.g. t-pyrrolidine-2,4-dicarboxylate (t-PDC)] compete with the amino acid for the transport, an heteroexchange occurs which results in a net release of glutamate (Volterra et al., 1996). Similar processes may be mediated also by other systems, like the cystine/glutamate antiporter. This latter normally takes up cystine in exchange for glutamate, and uses cystine as precursor for the synthesis of glutathione. When cystine is applied to brain slices it induces glutamate-dependent membrane currents in neurons, suggesting that this system could potentially be implicated in neurotransmission or/and excitotoxicity (Warr et al., 1999).

A different glutamate release process, known as 'hypoosmotic release', was identified by Kimelberg and colleagues. They described that astrocyte swelling, induced by exposure to a hypotonic buffer, leads to release of osmolites such as the amino acids glutamate, aspartate and taurine (Kimelberg et al., 1990). The molecular mechanism by which this process occurs is still unclear although some evidence suggests the implication of volume-activated anion channels of undefined molecular composition. The process is probably activated during physiological osmoregulation, but also under a number of pathological conditions such as traumatic injuries, ischemia, and hepatic encephalopathy, where it contributes to excitotoxic brain damage.

3.2 Ca^{2+}-dependent glutamate release from astrocytes.

The initial demonstration that astrocytes signal to neurons in response to physiological stimuli by Ca^{2+}-dependent glutamate release came in 1994 by the group of Philip Haydon. The authors described that stimulation of mixed neuron-glia cultures with bradykinin triggered astrocyte $[Ca^{2+}]_i$ rises and glutamate release. Surprisingly, $[Ca^{2+}]_i$ elevations in neurons followed those in astrocytes and were mediated by glutamate released from glia (Parpura et al., 1994).

Support to the existence of a bi-directional glutamatergic signalling between glia and neurons *in situ* was provided by Pasti and co-workers (Pasti et al., 1997). They showed that stimulation of neuronal afferents in acute cortical and hippocampal slices induced $[Ca^{2+}]_i$ oscillations in surrounding astrocytes. Most likely such events were consequent to stimulation of metabotropic receptors (mGluRs) by synaptically released glutamate. $[Ca^{2+}]_i$ oscillations in astrocytes were followed by $[Ca^{2+}]_i$ rises in the surrounding neurons, possibly the result of glutamate release from astrocytes.

Indeed, parallel studies by our group demonstrated that co-activation of mGluRs and AMPA/kainate receptors resulted in glutamate release from astrocytes via a calcium-dependent event. Interestingly, such release was occlusive with the Ca^{2+}-dependent mechanism activated by bradykinin, while it was additive to the Ca^{2+}-independent mechanism mediated by glutamate transporters (Bezzi et al., 1998). As a consequence, we proposed that: i) different astrocyte receptors might be coupled to a common Ca^{2+}-dependent transduction pathway ultimately leading to glutamate release; and ii) Ca^{2+}-dependent and -independent release mechanisms might use different glial glutamate pools.

Since it was previously reported that co-stimulation of mGluRs and AMPA receptors in striatal neurons induced Ca^{2+}-dependent activation of phospholipase A_2 (PLA_2) and arachidonic acid release (Dumuis et al., 1990), we addressed the question as whether a similar cascade might be involved in the GluR-mediated glutamate release from glia. We found that co-activation of mGluRs and AMPA receptors triggers Ca^{2+}-dependent release of radioactively labelled arachidonate from the cell membrane. In addition, specific PLA_2 and cyclooxygenase (COX) inhibitors blocked the glutamate release suggesting that arachidonic acid metabolism plays a crucial role in the process. Interestingly, formation of the COX metabolite prostaglandin (PG)E_2 was observed as a consequence of co-activation of mGluRs and AMPARs and its direct application to astrocyte cultures produced potent $[Ca^{2+}]_i$ elevations.

In view of these observations, the question arose as whether astrocytes in intact brain tissue released glutamate by the same mechanism identified in culture. In agreement with this possibility, we found that co-activation of mGluRs and AMPA receptors as well as direct PGE_2 application induced Ca^{2+}-dependent release of glutamate in acute hippocampal slices in which synaptic release had been inhibited. Moreover, application of PGE_2 to the slices induced $[Ca^{2+}]_i$ rises in both astrocytes and pyramidal neurons. Neuronal $[Ca^{2+}]_i$ elevations followed the glial ones and resulted to be only indirectly caused by PGE_2, via glutamate release from astrocytes, as they were sensitive to GluR antagonists (Bezzi et al., 1998).

Overall, these data provided the first demonstration *in situ* that eicosanoids, particularly PGE_2, are implicated in Ca^{2+}-dependent glutamate release from astrocytes and astrocyte-neuron communication. In addition to its physiological relevance, the discovery of such mechanism offered new perspectives on the pathogenesis of various brain diseases.

Indeed, increased amounts of PGs have been reported in a number of brain pathologies including cerebral ischemia, epilepsy, spinal hyperalgesia, prion diseases and various dementias like the AIDS dementia complex (ADC) (Collaco-Moraes et al., 1996; Griffin et al., 1994; Malmberg and Yaksh, 1992; Naffah-Mazzacoratti et al., 1995; Williams et al., 1997; Williams et al., 1994). Since the Ca^{2+}-dependent process of glutamate release from astrocytes appears under the control of PGs, it is conceivable that acute or chronic alterations in prostaglandin production might lead to de-regulation of this glutamate system and to excitotoxic neuronal damage.

4. Contribution of astrocyte calcium-dependent glutamate release to neurodegeneration: the case of AIDS neuropathology.

Several lines of evidence suggest that a local, parenchimal inflammatory response is often present in brain during chronic neurodegenerative disorders such as Alzheimer's disease. Such condition is characterised by specific morphological and functional changes of astroglia and microglia, broadly defined as reactive gliosis. The signals exchanged between the two glial cell types during these events are largely unknown, yet their transition from the resting to the activated condition appears to be associated with a marked up-regulation of several genes and the secretion of factors like cytokines, eicosanoids, reactive oxygen species, nitric oxide and excitatory amino acids (Perry et al., 1995). Although the inflammatory reaction was originally thought to be beneficial for tissue-repairing processes, several observations now suggest that it may actually contribute to the exacerbation of the neurodamaging processes via the secretion of toxic factors. For instance, overproduction of prostaglandins in astrocytes might favour an increased and deleterious glutamatergic input to neurons. Other inflammatory mediators such as cytokines may also contribute to neurodegeneration.

Cytokines are a family of polypeptides, including interleukins (IL), interferons (IFN), tumour necrosis factors (TNF) and structurally associated chemokines and growth factors, which are known to perform diverse actions. Some of these agents are produced as inactive precursors and achieve the secreted mature form only after cleavage by specific enzymes.

A.

B.

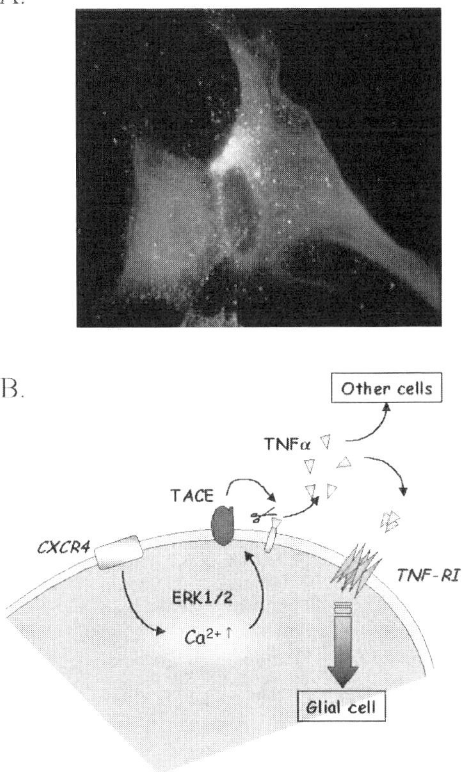

Figure 2. CXCR4-mediated shedding of soluble TNFα and its role in glial signalling. (A) Constitutive surface expression of TNFα (green) in astrocytes (GFAP-positive cells-red). (B) Schematic representation of the molecular events linking CXCR4 activation to TNFα release. Once released, the cytokine acts as an autocrine/paracrine factor and stimulates its receptor, TNF-RI, on the same cell and/or on the surrounding ones, providing additional signaling events (Bezzi et al., 2001) (see Figure 3).

Two recent studies, focused on TNFα (Bezzi et al., 2001) and IL-1β (Blasi et al., 1999) respectively, reported that such molecules are constitutively expressed in glial cells (Figure 2A) and are rapidly (seconds) released in response to physiological stimuli, the transmitter glutamate and the chemokine, SDF-1α (Bezzi et al., 2001; Noda et al., 2000). In particular, investigations performed by our group discovered that activation of the chemokine receptor CXCR4 by its endogenous ligand SDF-1α leads to stimulation of the ERK-MAPK pathway followed by TNFα-converting enzyme (TACE)-dependent shedding of the cytokine. Once released from astrocytes, TNFα acts autocrinally on its TNF-RI receptor to induce PGE_2 formation, followed by $[Ca^{2+}]i$ elevation and glutamate release (Figure 2B). Interestingly, CXCR4 is expressed also in microglia and its stimulation results in potent TNFα release, particularly when the microglia are in a reactive state. In the pathological brain, astrocytes and microglia often form local foci of reactive cells around the sites of infection or injury.

Simultaneous stimulation of CXCR4 in contiguous microglia and astrocytes results in a synergistic amplification of TNFα release that, independently of its cellular source, converges to strongly potentiate Ca^{2+}-dependent glutamate release from astrocytes. We have provided evidence that this pathway contributes to the slow excitotoxic apoptotic death of surrounding neurons (Figure 3).

Not only identification of the above CXCR4-dependent signalling cascade defines for the first time a role for TNFα in rapid glial communications, but also (Bezzi et al., 2001) implicates glial glutamate release specifically in the development of AIDS-associated neuropathology. In particular, previous evidence suggested that an increasing number of neurons die by apoptosis (Bagetta et al., 1996; Shi et al., 1996) during disease progression. The neurons are not directly infected by the HIV-1 virus and die by NMDA-mediated excitotoxicity (Kaul et al., 2001; Meucci and Miller, 1996; Toggas et al., 1996). An important role in the pathogenesis of ADC is played by gp120, the coat protein of HIV-1. Thus, expression of this protein in transgenic mice per se reproduces several features of the ADC neuropathology (Toggas et al., 1994). Based on these premises, we tried to clarify the sequence of events that leads from brain infection by HIV-1 to neurodegeneration. In particular, we found that gp120, which during the pathology is probably shedded from HIV-infected microglia, can act as a CXCR4 agonist similar to the endogeneous ligand, SDF-1α, in both astrocytes and microglia, triggering potent astrocyte glutamate release. Importantly, the level of CXCR4 stimulation by the viral protein is not under physiological control as in the case of the endogenous chemokine, which may eventually lead to glutamate-dependent excitotoxic neuronal cell death (Bezzi et al., 2001) (Figure 3). In our experimental model, we can obtain neuroprotection by blocking the astrocyte signalling cascade at any of the identified steps: CXCR4 receptor activation, TNFα release, prostaglandin formation or, finally, by scavenging released glutamate or by blocking NMDA receptors.

Antiretroviral therapy does not fully control AIDS-associated neuropathology while significantly prolonging the life expectations of AIDS-affected subjects. Therefore, there is a need for drugs limiting the development of ADC and ensuring a better quality of life. Clinical trials are currently underway with memantine, an inhibitor of the NMDA receptor channels. Our work suggests that additional molecules, some of which already on the market (eg, the COX antagonists), could be also beneficial.

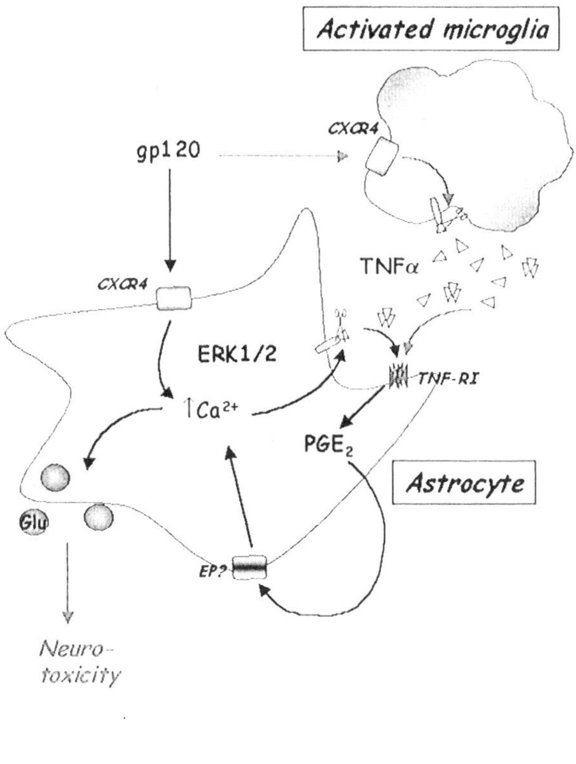

Figure 3. Schematic representation of the CXCR4-mediated astrocyte-glutamate release process in pathological conditions. The interaction of the HIV-1 coat protein, gp120 (T tropic strain IIIB), with the astrocyte receptor CXCR4 can trigger a sequence of intracellular events similar to the endogenous ligand, SDF-1α, including $[Ca^{2+}]i$ rises, stimulation of the ERK-MAPK pathway, activation of the metalloproteinase TACE and shedding of TNFα from its membrane-bound precursor. The interaction of TNFα with its receptor, TNF-RI, starts a second sequence of intracellular signalling events leading to formation of PGE_2 followed by Ca^{2+}-dependent glutamate release. If activated microglia is present, e.g. during ADC progression, gp120 activates CXCR4 on both astrocytes and microglia at the same time. As a consequence, the two glial cell types contribute synergistically to the release of high levels of TNFα which leads to potentiation of the downstream signalling and ultimately to excessive glutamate release, which causes slow neuronal apoptosis.

5. Conclusions.

The discovery that astrocytes are active partners of neurons in brain communications and possess regulated forms of transmitter release is dense of implications for understanding brain processes in health and disease. In particular, some of the findings briefly summarized in this review, indicate that release of transmitter (glutamate) from astrocytes is implicated in physiological modulation of synaptic functions. However, de-regulation of the glial pathway can cause apoptotic neuronal cell death. Therefore, Ca^{2+}-dependent glutamate release from astrocytes as well as additional emerging glial signalling pathways have to be considered as important contributors to normal and pathological brain functions. This indicates that glial and neuronal physiologies must be studied in integration, and glial derangements in pathological conditions should not be considered only as epiphenomena or late reactions to neuronal injury, but rather as intrinsic components of the neurodamaging processes.

Acknowledgements. This work was supported by the Ministero dell'Università e della Ricerca Scientifica e Tecnologica of Italy (Cofin 2000 and 2002), European Community (QLK6-CT1999-02203), Telethon Italy (GGP02052) to A.V., Istituto Superiore di Sanità of Italy (IV National Program on AIDS Research; National Program on Alzheimer's Disease) to J.M. D.R. is recipient of an EMBO Long Term Fellowship (279-2002). The authors wish to thank Chiara Valori for experimental help and for drawings and artwork.

REFERENCES

Bagetta, G., Corasaniti, M.T., Malorni, W., Rainaldi, G., Berliocchi, L., Finazzi-Agro, A. and Nistico, G. (1996) The HIV-1 gp120 causes ultrastructural changes typical of apoptosis in the rat cerebral cortex. *Neuroreport*, **7**, 1722-1724.

Bezzi, P., Carmignoto, G., Pasti, L., Vesce, S., Rossi, D., Rizzini, B.L., Pozzan, T. and Volterra, A. (1998) Prostaglandins stimulate calcium-dependent glutamate release in astrocytes. *Nature*, **391**, 281-285.

Bezzi, P., Domercq, M., Brambilla, L., Galli, R., Schols, D., De Clercq, E., Vescovi, A., Bagetta, G., Kollias, G., Meldolesi, J. and Volterra, A. (2001) CXCR4-activated astrocyte glutamate release via TNFalpha: amplification by microglia triggers neurotoxicity. *Nat Neurosci*, **4**, 702-710.

Blasi, F., Riccio, M., Brogi, A., Strazza, M., Taddei, M.L., Romagnoli, S., Luddi, A., D'Angelo, R., Santi, S., Costantino-Ceccarini, E. and Melli, M. (1999) Constitutive expression of interleukin-1beta (IL-1beta) in rat oligodendrocytes. *Biol Chem*, **380**, 259-264.

Coco, S., Verderio, C., Trotti, D., Rothstein, J.D., Volterra, A. and Matteoli, M. (1997) Non-synaptic localization of the glutamate transporter EAAC1 in cultured hippocampal neurons. *Eur J Neurosci*, **9**, 1902-1910.

Collaco-Moraes, Y., Aspey, B., Harrison, M. and de Belleroche, J. (1996) Cyclo-oxygenase-2 messenger RNA induction in focal cerebral ischemia. *J Cereb Blood Flow Metab*, **16**, 1366-1372.

Dani, J.W., Chernjavsky, A. and Smith, S.J. (1992) Neuronal activity triggers calcium waves in hippocampal astrocyte networks. *Neuron*, **8**, 429-440.

Dehnes, Y., Chaudhry, F.A., Ullensvang, K., Lehre, K.P., Storm-Mathisen, J. and Danbolt, N.C. (1998) The glutamate transporter EAAT4 in rat cerebellar Purkinje cells: a glutamate-gated chloride channel concentrated near the synapse in parts of the dendritic membrane facing astroglia. *J Neurosci*, **18**, 3606-3619.

Dumuis, A., Pin, J.P., Oomagari, K., Sebben, M. and Bockaert, J. (1990) Arachidonic acid released from striatal neurons by joint stimulation of ionotropic and metabotropic quisqualate receptors. *Nature*, **347**, 182-184.

Griffin, D.E., Wesselingh, S.L. and McArthur, J.C. (1994) Elevated central nervous system prostaglandins in human immunodeficiency virus-associated dementia. *Ann Neurol*, **35**, 592-597.

Hassinger, T.D., Atkinson, P.B., Strecker, G.J., Whalen, L.R., Dudek, F.E., Kossel, A.H. and Kater, S.B. (1995) Evidence for glutamate-mediated activation of hippocampal neurons by glial calcium waves. *J Neurobiol*, **28**, 159-170.

Karwoski, C.J., Lu, H.K. and Newman, E.A. (1989) Spatial buffering of light-evoked potassium increases by retinal Muller (glial) cells. *Science*, **244**, 578-580.

Kaul, M., Garden, G.A. and Lipton, S.A. (2001) Pathways to neuronal injury and apoptosis in HIV-associated dementia. *Nature*, **410**, 988-994.

Kimelberg, H.K., Goderie, S.K., Higman, S., Pang, S. and Waniewski, R.A. (1990) Swelling-induced release of glutamate, aspartate, and taurine from astrocyte cultures. *J Neurosci*, **10**, 1583-1591.

Malmberg, A.B. and Yaksh, T.L. (1992) Hyperalgesia mediated by spinal glutamate or substance P receptor blocked by spinal cyclooxygenase inhibition. *Science*, **257**, 1276-1279.

Meucci, O. and Miller, R.J. (1996) gp120-induced neurotoxicity in hippocampal pyramidal neuron cultures: protective action of TGF-beta1. *J Neurosci*, **16**, 4080-4088.

Naffah-Mazzacoratti, M.G., Bellissimo, M.I. and Cavalheiro, E.A. (1995) Profile of prostaglandin levels in the rat hippocampus in pilocarpine model of epilepsy. *Neurochem Int*, **27**, 461-466.

Nedergaard, M. (1994) Direct signaling from astrocytes to neurons in cultures of mammalian brain cells. *Science*, **263**, 1768-1771.

Noda, M., Nakanishi, H., Nabekura, J. and Akaike, N. (2000) AMPA-kainate subtypes of glutamate receptor in rat cerebral microglia. *J Neurosci*, **20**, 251-258.

Parpura, V., Basarsky, T.A., Liu, F., Jeftinija, K., Jeftinija, S. and Haydon, P.G. (1994) Glutamate-mediated astrocyte-neuron signalling. *Nature*, **369**, 744-747.

Pasti, L., Volterra, A., Pozzan, T. and Carmignoto, G. (1997) Intracellular calcium oscillations in astrocytes: a highly plastic, bidirectional form of communication between neurons and astrocytes in situ. *J Neurosci*, **17**, 7817-7830.

Perry, V.H., Bell, M.D., Brown, H.C. and Matyszak, M.K. (1995) Inflammation in the nervous system. *Curr Opin Neurobiol*, **5**, 636-641.

Pfrieger, F.W. and Barres, B.A. (1996) New views on synapse-glia interactions. *Curr Opin Neurobiol*, **6**, 615-621.

Porter, J.T. and McCarthy, K.D. (1997) Astrocytic neurotransmitter receptors in situ and in vivo. *Prog Neurobiol*, **51**, 439-455.

Rossi, D.J., Oshima, T. and Attwell, D. (2000) Glutamate release in severe brain ischaemia is mainly by reversed uptake. *Nature*, **403**, 316-321.

Shi, B., De Girolami, U., He, J., Wang, S., Lorenzo, A., Busciglio, J. and Gabuzda, D. (1996) Apoptosis induced by HIV-1 infection of the central nervous system. *J Clin Invest*, **98**, 1979-1990.

Steinhauser, C. and Gallo, V. (1996) News on glutamate receptors in glial cells. *Trends Neurosci*, **19**, 339-345.

Szatkowski, M., Barbour, B. and Attwell, D. (1990) Non-vesicular release of glutamate from glial cells by reversed electrogenic glutamate uptake. *Nature*, **348**, 443-446.

Toggas, S.M., Masliah, E. and Mucke, L. (1996) Prevention of HIV-1 gp120-induced neuronal damage in the central nervous system of transgenic mice by the NMDA receptor antagonist memantine. *Brain Res*, **706**, 303-307.

Toggas, S.M., Masliah, E., Rockenstein, E.M., Rall, G.F., Abraham, C.R. and Mucke, L. (1994) Central nervous system damage produced by expression of the HIV-1 coat protein gp120 in transgenic mice. *Nature*, **367**, 188-193.

Tsacopoulos, M. and Magistretti, P.J. (1996) Metabolic coupling between glia and neurons. *J Neurosci*, **16**, 877-885.

Ventura, R. and Harris, K.M. (1999) Three-dimensional relationships between hippocampal synapses and astrocytes. *J Neurosci*, **19**, 6897-6906.

Verkhratsky, A. and Kettenmann, H. (1996) Calcium signalling in glial cells. *Trends Neurosci*, **19**, 346-352.

Volterra, A., Bezzi, P., Rizzini, B.L., Trotti, D., Ullensvang, K., Danbolt, N.C. and Racagni, G. (1996) The competitive transport inhibitor L-trans-pyrrolidine-2, 4-dicarboxylate triggers excitotoxicity in rat cortical neuron-astrocyte co-cultures via glutamate release rather than uptake inhibition. *Eur J Neurosci*, **8**, 2019-2028.

Waniewski, R.A. and Martin, D.L. (1986) Exogenous glutamate is metabolized to glutamine and exported by rat primary astrocyte cultures. *J Neurochem*, **47**, 304-313.

Warr, O., Takahashi, M. and Attwell, D. (1999) Modulation of extracellular glutamate concentration in rat brain slices by cystine-glutamate exchange. *J Physiol*, **514 (Pt 3)**, 783-793.

Williams, A., Van Dam, A.M., Ritchie, D., Eikelenboom, P. and Fraser, H. (1997) Immunocytochemical appearance of cytokines, prostaglandin E2 and lipocortin-1 in the CNS during the incubation period of murine scrapie correlates with progressive PrP accumulations. *Brain Res*, **754**, 171-180.

Williams, A.E., van Dam, A.M., Man, A.H.W.K., Berkenbosch, F., Eikelenboom, P. and Fraser, H. (1994) Cytokines, prostaglandins and lipocortin-1 are present in the brains of scrapie-infected mice. *Brain Res*, **654**, 200-206.

Chapter 2
GLUTAMATE UPTAKE AND TRANSPORTERS

Danbolt NC[1]; Holmseth, S[1], Skår, A[1], Lehre KP[1], Furness DN[2]
[1] Department of Anatomy, Centre for Molecular Biology and Neuroscience, Institute of Basic Medical Sciences, University of Oslo, P.O.Box 1105 Blindern, N-0317 Oslo, Norway . [2] MacKay Institute of Communication and Neuroscience, Keele University, Keele, Staffs, ST5 5BG, England.

Abstract: Glutamate uptake is mediated by a family of glutamate transporter proteins present in the plasma membranes of glial cells and neurons. These proteins enable the nervous system to maintain low extracellular concentrations of glutamate. This is important to avoid harmful overactivation of glutamate receptors and to secure a high signal to noise ratio in synaptic transmission. In addition to this simple, but essential glutamate removal role, the glutamate transporters also appear to have more sophisticated functions in the modulation of neurotransmission. They are present at very high densities and may modify the time course of synaptic events, the extent and pattern of activation and desensitization of receptors outside the synaptic cleft and at neighboring synapses (intersynaptic crosstalk). Disturbed control of glutamate transporters appears to play a role in the pathogenesis of most neurological diseases.

Keywords: glutamate transport, excitatory amino acid transporters, excitotoxicity, intersynaptic crosstalk.

INDEX

1. Introduction.
2. A variety of transporters.
3. Properties of EAAT-type transporters.
    3.1 Transport mechanism.
    3.2 Structure
    3.3 Attachment
4. Distribution
    4.1 EAAT1
    4.2 EAAT2
    4.3 EAAT3
    4.4 EAAT4

5. Physiological roles.
6. Roles of glutamate transporters in disease.
7. Concluding remarks.

1. Introduction.

The amino acid glutamate is considered to be the major excitatory neurotransmitter in the mammalian central nervous system (CNS) and is involved, directly or indirectly, in more or less all normal and abnormal brain functions, including memory and cognition as well as regulation of cell migration, differentiation and elimination. Because glutamate modulates fundamental nervous tissue processes, and because glutamate acts on its receptors from the extracellular fluid, it follows that the extracellular concentration of glutamate must be tightly controlled. Both too little and too much activation of glutamate receptors are harmful. A major mechanism for securing optimal receptor activation is removal from the extracellular fluid by means of glutamate uptake. This uptake is mediated by the glutamate transporter proteins present in both neurons and glial cells. Considering that glutamate is constantly being released (via several different mechanisms) and that brain tissue contains huge amounts (10 mmol/kg) of glutamate despite being highly sensitive to it, it seems logical that CNS tissue displays a very high glutamate uptake activity (for review and references see: Danbolt, 2001).

2. A variety of transporters.

A number of transporters able to transport glutamate are found in brain tissue (Fig. 1; for review see: Danbolt, 2001). These transporters belong to five different gene families, but only one of them is normally called the "glutamate transporter family". The latter family of proteins is also referred to as the sodium- and potassium-coupled "Excitatory Amino Acid Transporters (EAATs)" and has seven members, five of which are glutamate transporters: EAAT1 (GLAST), EAAT2 (GLT), EAAT3 (EAAC) , EAAT4 and EAAT5.

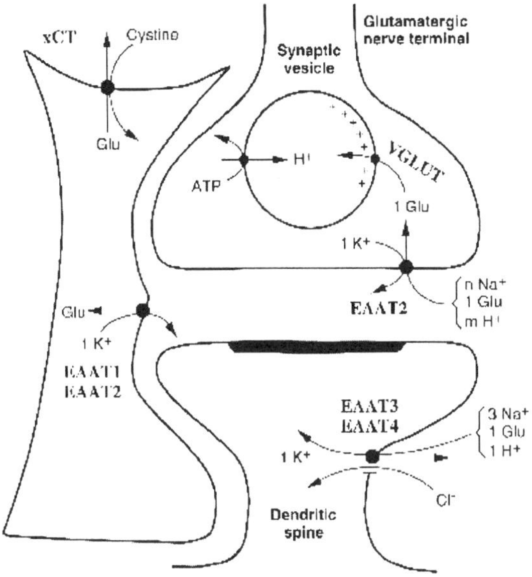

Fig. 1. Brain tissue expresses a variety of transporters able to transport glutamate. Glutamate transporters (EAAT1 and EAAT2) in glial cell plasma membranes remove glutamate from the extracellular fluid. A few percent of the EAAT2-molecules are found in glutamatergic terminals. The glutamate transporters EAAT3 and EAAT4 are found post synaptically in the plasma membranes of dendritic spines. These transporters, which include EAAT4 (only in cerebellar Purkinje cells) and EAAC (throughout the brain), have a relatively high associated Cl^--channel activity. Synaptic vesicles are loaded with glutamate from cytosol by means of a glutamate transporter located in the plasma membrane of synaptic vesicles (VGLUT). Cystine, which is essential for the synthesis of glutathione (an important antioxidant, for review see: Dringen, 2000), is transported into cells in exchange with glutamate by xCT. Modified from Danbolt 2001 with permission (Copyright 2001 by Elsevier Science Ltd).

These are the transporters which are responsible for the clearance of glutamate from the extracellular fluid. They are, as described in more detail below, expressed in the plasma membranes of both glial cells and neurons.

Brain plasma membranes also express the glutamate-cystine exchanger (Sato et al., 1999; Flynn and McBean, 2000) and the sodium dicarboxylate transporter number two (SDCT2; Chen et al., 1999). These two proteins have neither significant sequence homology with each other nor with the EAAT-type of transporters. The glutamate-cystine exchanger may be a major mediator of glutamate release (Baker et al., 2002) because it uses the transmembrane gradient of glutamate to transport cystine into cells by exchanging glutamate and cystine in a 1:1 relationship (Bannai, 1986). SDCT2 plays a role in the metabolism of dicarboxylates such as succinate, but there are currently no data suggesting that it is of importance for excitatory amino acid transport and it will not be discussed here.

Brain cells also contain intracellular transporters for glutamate. These include mitochondrial transporters (Fiermonte et al., 2002) and a family of "vesicular glutamate transporters (VGLUTs)" which are responsible for the filling of synaptic vesicles with glutamate. The latter family contains seven genes. Three of the proteins encoded in these genes (VGLUT1-3) have been characterized and shown to transport glutamate (Bellocchio et al., 2000; Takamori et al., 2000; Fremeau et al., 2001; Herzog et al., 2001; Takamori et al., 2001; Fremeau et al., 2002; Gras et al., 2002; Takamori et al., 2002).

3. Properties of the EAAT-type transporters.

3.1 Transport mechanism

All of the five glutamate transporter proteins (EAAT1-5) mediate Na^+- and K^+-coupled transport of excitatory amino acids (e.g. L-glutamate, L- and D-aspartate). They operate as shuttles which catalyze the stoichiometric (coupled) transport of sodium and hydrogen ions and excitatory amino acids in one direction and potassium ions in the other (Fig. 2).

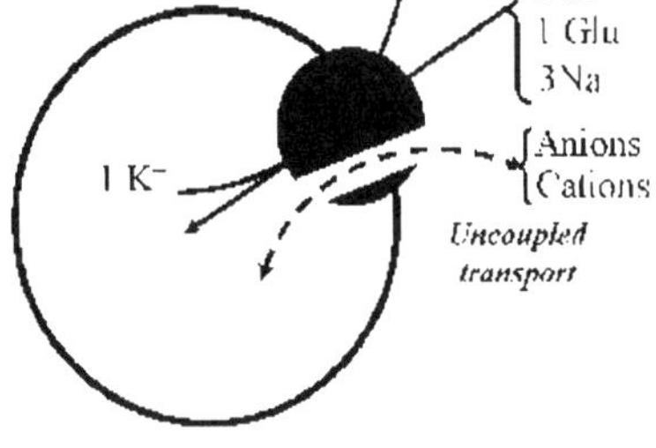

Fig. 2. The glutamate transporters catalyze the influx of 1 glutamate, 3 sodium ions and 1 hydrogen ion in the exchange of 1 potassium ion (Zerangue and Kavanaugh, 1996; Levy et al., 1998). This transport is accompanied (at least in the case of EAAT1) of 400 molecules of water (MacAulay et al., 2001). In addition to this, the transporters operate as ion channels (Wadiche et al., 1995; Fairman et al., 1995; Arriza et al., 1997; Fairman et al., 1998).

Sodium is required for glutamate binding while potassium is required for net transport (for review see: Kanner and Schuldiner, 1987; Danbolt, 2001). According to recent studies of human EAAT3 (Zerangue and Kavanaugh, 1996) and rat EAAT2 (Levy et al., 1998) net glutamate transport has the following stoichiometry: 1 glutamate is taken up together with 3 Na^+ and 1 H^+ in exchange for 1 K^+. It is important to realize that this process is reversible at most of the individual steps in the transport cycle. This implies both that the transporters can operate in reversed mode (Fig. 3) releasing glutamate (a phenomenon believed to be important in disease: see below) and that the transporters are able to perform "exchange" in a 1:1 relationship between substrate molecules on the outside and substrate molecules on the inside. For instance, if an alternative substrate (e.g. L-[^{3}H]glutamate, DL-aspartate, hydroxyaspartate or L-trans-pyrrolidine-2,4-dicarboxylate (PDC)) is added on the outside, it may exchange with internal glutamate giving rise to both accumulation of added substrate and release of endogenous glutamate. Exchange requires sodium because sodium is required for substrate binding, but does not require potassium. In fact, if potassium is absent, the transporters are locked in exchange mode (Kanner and Schuldiner, 1987; Pines and Kanner, 1990). It is also important to realize that exchange and net transport can occur at the same time, particularly at high substrate concentrations, probably because of competition between internal potassium binding and internal substrate binding. When a high concentration of PDC is added to the outside, PDC goes into the cells, but a glutamate leak is also observed (Volterra et al., 1996; for a detailed description and references see: Danbolt, 2001). Consequently, a transportable inhibitor does not "block" transport. It competitively inhibits uptake, but it also causes release. The term "blockers" should therefore be restricted to non-transportable inhibitors (Danbolt, 2001) like TBOA (Shimamoto et al., 1998) and dihydrokainate. The latter is selective for EAAT2 (Arriza et al., 1994), while the former inhibits all the five subtypes (for review see: Bridges et al., 1999).

In addition to stoichiometric transport, these proteins also function as ion channels (for review see: Seal and Amara, 1999) as well as water channels (at least in the case of EAAT1) (MacAulay et al., 2001). Sodium-dependent glutamate binding to the transporters triggers anion conductance (Wadiche et al., 1995; Billups et al., 1996; Wadiche and Kavanaugh, 1998). This anion conductance is thermodynamically independent of the transport process. The transport rate does not depend on the anion concentrations (as long as they do not change the pH or membrane potential), and the transport occurs in the absence of these anions. Consequently, the anions pass through glutamate transporters down their electrochemical gradients as if the anions were passing through a glutamate-gated channel. This property is particularly prominent in EAAT4 and EAAT5 and almost non-existent in EAAT2 (at least in the presence of physiologically relevant amino acids).

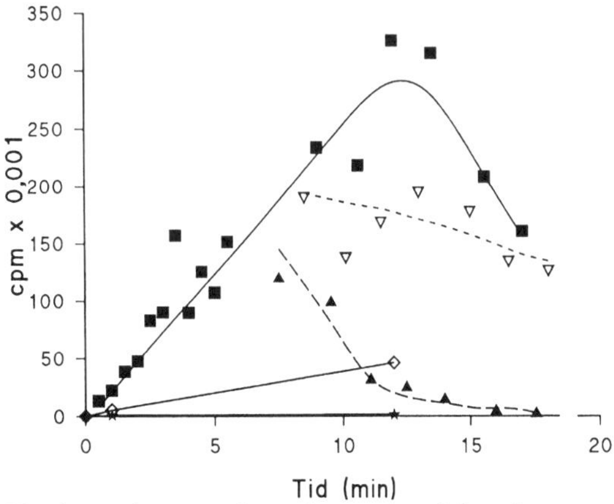

Fig. 3. Uptake and reversed uptake catalyzed by glutamate transporters. The electrochemical gradients across plasma membranes drives excitatory amino acid transport and thereby also determines the direction of transport. The figure shows the time course of glutamate uptake in a reconstituted system consisting of artificial cell membranes (liposomes) containing glutamate transporters and neither $Na^{+}K^{+}$ATPase nor ion channels (as described: Danbolt et al., 1990; Trotti et al., 1995). The liposomes are loaded with K^{+} and the uptake reaction is started by diluting the liposomes in a solution of Na^{+} and tritiated L-glutamate. Glutamate accumulates inside the liposomes (solid squares) until the ion gradients are dissipated. Then glutamate leaks out due to reversal of the transporters giving rise to the bell-shaped time course. Addition of ionophores during the uptake reaction abolishes ion gradients and induces reversal (solid triangles). No uptake is observed when ionophores are present from the start of the uptake reaction (asterisks). Dihydrokainate inhibits uptake (open squares), but is also able to block release due to reversal because it is a non-transportable substrate as illustrated by the addition of both ionophores and dihydrokainate (open triangles).

3.2 Structure

The EAAT1-5 are glycoproteins consisting of 543, 573, 523, 564 and 560 amino acid residues, respectively (for review see: Seal and Amara, 1999; Danbolt, 2001). On SDS-PAGE gels they run with apparent molecular masses in the range of 60-75 kDa. Their topologies are still debated and seem to be unusual with one or two re-entrant loops, a hydrophobic membrane associated linker region and 8 transmembrane domains (Grunewald and Kanner, 2000; Seal et al., 2000). Both the termini are intracellular. The transporters appear to exist in brain membranes as oligomers, probably homo-oligomers (Haugeto et al., 1996; Dehnes et al., 1998; Eskandari et al., 2000). The subunits are non-covalently connected, but covalent bonds may form (at least in the case of EAAT1 and EAAT2) if the sulfhydryl groups are oxidized. This causes dithiotreitol-sensitive oligomer bands to appear on SDS-PAGE. Another different cause of high molecular mass bands in the electrophoresis gels is SDS-resistant aggregation observed after solubilization in mild detergents, especially at elevated temperatures (see section 7 in: Danbolt, 2001).

3.3 Attachment

The differentiated subcellular distribution of the transporter proteins (see below) as well as rapid changes in cell surface expression imply that these proteins are kept in position by transporter-binding proteins forming attachment to the cytoskeleton (for review see Danbolt, 2001). Four such proteins have so far been reported: Ajuba which binds EAAT2 (Marie et al., 2002), GTRAP3-18 which binds EAAT3 (Lin et al., 2001) and the two EAAT4-binding proteins GTRAP41 and GTRAP48 (Jackson et al., 2001).

4. Distribution.

4.1 EAAT1

EAAT1 (GLAST) protein is expressed in glial cells (Fig. 4) throughout the central nervous system (CNS), and the major glutamate transporter in the following CNS regions: the cerebellum (Lehre and Danbolt, 1998), the circumventricular organs (Berger and Hediger, 2000) and the retina (for review see: Rauen et al., 1999). It is also the major transporter in the inner ear (Furness and Lehre, 1997; Takumi et al., 1997).

EAAT1 is present in the CNS from early developmental stages onwards, but at low concentrations initially. In the rat, the EAAT1 level at birth is in the order of 5 % of the adult level (Ullensvang et al., 1997). The sharpest concentration increase is observed during the most active period of synaptogenesis (in the rat from the end of the second postnatal week to the end of the fourth week) reaching near adult levels by P35. The concentration of EAAT1 in the young adult rat hippocampus is approximately 0.3 mg per gram tissue or about one-quarter of that of EAAT2, while the concentration of EAAT1 in the cerebellar molecular layer is approximately 1.8 mg per gram tissue or about six times higher than EAAT2 (Lehre and Danbolt, 1998). Neuronal expression of EAAT1 has not been detected anywhere in the nervous system in vivo, neither in the adult nor during development (Lehre et al., 1993; Chaudhry et al., 1995; Lehre et al., 1995; Rothstein et al., 1995; Schmitt et al., 1997; for review see: Danbolt, 2001). EAAT1 immunoreactive astrocytes are seen light microscopically throughout gray as well as white matter. Densely labeled processes extend from the cell bodies. Electron microscopical studies confirm labeling of astrocyte cell bodies as well as of their processes. Also astrocytic processes forming perivascular end-feet are labeled. It should, however, be mentioned that a possible expression of EAAT1 mRNA in granule cells of the mouse hippocampus (gyrus dentatus) at birth (P0) has been described (Sutherland et al., 1996) and it has also been reported that EAAT1 is transiently expressed in a subpopulation of cultured embryonic hippocampal neurons (Plachez et al., 2000).

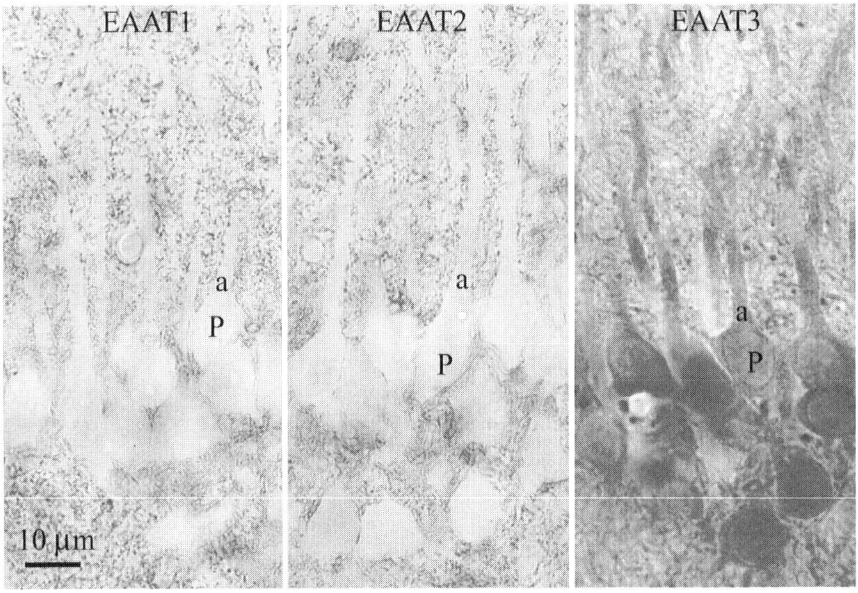

Fig. 4. Light microscopic immunocytochemical localization of EAAT1, EAAT2 and EAAT3 in the pyramidal layer of the CA1 region of the hippocampus, illustrating the comparative distributions of these transporter subtypes. The distributions of EAAT1 and EAAT2 are remarkably similar in this region, these transporters being localized to the tiny astroglial processes associated with the dendritic fields of the pyramidal cells (P), but completely sparing the latter=s somata and apical dendrites (a). EAAT3 shows a completely different distribution as it is found within the cytoplasm of the pyramidal cell somata and dendrites.

4.2 EAAT2

EAAT2 (GLT) is, like EAAT1, expressed in astroglial cells (Fig. 4) throughout the brain and spinal cord, and is present in the CNS from early developmental stages. But in contrast to EAAT1, EAAT2 is initially expressed in neurons (e.g. Northington et al., 1999). This neuronal expression is retained in the retina (for review see: Rauen et al., 1999), but not in the brain and spinal cord where astrocytes take over as the major sites of EAAT2-expression. This apparent switch has already taken place at P11 (the youngest animals which have been studied by electron microscopical immunocytochemistry: Ullensvang et al., 1997) and is also observed in cell cultures (e.g. Plachez et al., 2000). It is not clear if the neurons reduce their EAAT2 expression or if it just becomes relatively minor because of the massive increase in glial expression which is one or two orders of magnitude higher. Most neurons in adult rat brain contain mRNA encoding EAAT2 (Torp et al., 1994; Schmitt et al., 1996; Torp et al., 1997; Berger and Hediger, 1998) and up-regulation of neuronal expression is frequently observed in diseased tissue (e.g. Martin et al., 1997).

In the newborn rat forebrain, EAAT2 levels are less than 1 % of adult levels (Ullensvang et al., 1997). Like EAAT1, the sharpest increase in expression level occurs during the most active phase of synaptogenesis. The increase in transport activity (from rat forebrain) parallels that of the EAAT2 protein expression (Ullensvang et al., 1997) in agreement with the notion that EAAT2 is the major glutamate transporter in the forebrain, brainstem and spinal cord (all parts of the CNS except those where EAAT1 is dominant; see above). The quantitative importance of EAAT2 is further illustrated by the facts that EAAT2 was the glutamate transporter which was isolated when reconstitution of transport activity was used to monitor the purification process (Danbolt et al., 1990), that antibodies to EAAT2 are able to precipitate most of the reconstitutable glutamate transport activity from a crude rat forebrain brain detergent extract (Danbolt et al., 1992; Haugeto et al., 1996) and that EAAT2 knockout mice have a very low intact glutamate uptake activity (Tanaka et al., 1997).

Almost all of the EAAT2 protein in normal juvenile and adult brain and spinal cord is present in astroglial cell membranes. This has been clear from the publication of the first immunocytochemical study on EAAT2 (Danbolt et al., 1992) and has been confirmed in all later studies (Hees et al., 1992; Levy et al., 1993; Rothstein et al., 1994; Chaudhry et al., 1995; Lehre et al., 1995; Schmitt et al., 1996; Milton et al., 1997; Berger and Hediger, 2000; for review see: Danbolt, 2001).

The concentration of the EAAT2 protein in young normal adult rat hippocampus appears to be as high as 1.3 mg per gram tissue (Lehre and Danbolt, 1998), or about 12000 molecules per μm^3 of tissue (stratum radiatum, CA1), in accordance with the functional importance of this protein. In the cerebellar molecular layer the concentration is considerably lower at 0.3 mg/g tissue, or about 2800 molecules per μm^3 (Lehre and Danbolt, 1998). Because essentially all the EAAT2 is present in the astroglial plasma membrane and because the total astroglial plasma membrane surface areas in hippocampus are 1.4 $\mu m^2/\mu m^3$, and in cerebellum 3.8 $\mu m^2/\mu m^3$, (Lehre and Danbolt, 1998), these concentrations translate into densities of about 8500 and 740 molecules per μm^2 astroglial cell membrane, respectively (Lehre and Danbolt, 1998). It is, however, important to realize that these values are from adult rat brain and that the concentrations change dramatically during development (Ullensvang et al., 1997), that the expression is regulated via a number of different mechanisms and that it changes during disease (see below).

Although most of the EAAT2 protein is found in glial cells, there is also, as mentioned above, some EAAT2 in neurons (Chen et al., 2002; Schmitt et al., 2002), in particular in the glutamatergic nerve terminals (Y. Dehnes, D.N. Furness and N.C. Danbolt, unpublished). This may explain the glutamate uptake activity in glutamatergic nerve terminals (Gundersen et al., 1993; for review see Danbolt, 2001). Recently, a study (Suchak et al., 2003) based on subcellular fractionation, immunoblotting and measurements of radioactive amino acids strongly supports the notion that nerve terminals take up glutamate by a process predominantly mediated by EAAT2. Chen and coworkers (2002) reported nerve terminal localization of a variant of EAAT2 (EAAT2b). This study, however, is weakened by the facts that the only part of EAAT2b which differs from EAAT2a, is the extreme C-terminal (11 amino acids) which has significant sequence identity with a number of other proteins, including synaptotagmin 6 and the sodium ascorbate transporter 2 as well as a number of enzymes and receptor proteins. Thus, we do not know which variant of EAAT2 is expressed in the terminals. What we do know is that the levels are normally very low as compared to those in astroglia.

4.3 EAAT3

EAAT3 (EAAC) is also widely expressed in neurons (Fig. 4) throughout the central nervous system and is found in highest concentrations in the hippocampus, cerebellum (Fig. 2) and basal ganglia (Rothstein et al., 1994; Conti et al., 1998; Kugler and Schmitt, 1999), but the total tissue concentration (adult rat forebrain) is only around 1 % of that of EAAT2 in the hippocampus and less than 10 % of both EAAT2 and EAAT4 in the cerebellum (S. Holmseth, Y. Dehnes, K.P. Lehre and N.C. Danbolt, unpublished). This is in agreement with the low uptake activity in EAAT2 knockout mice (Tanaka et al., 1997) and the seemingly normal uptake activities in EAAT3 knockout mice (Peghini et al., 1997).

EAAT3 protein and mRNA have a wide distribution in the brain and may be found in most if not all glutamatergic neurons, as well as in GABAergic neurons like the cerebellar Purkinje cells and hippocampal interneurons (Rothstein et al., 1994); Conti et al., 1998; Kugler and Schmitt, 1999; Berger and Hediger, 1998; Kugler and Schmitt, 1999). EAAT3 is also detected (Kugler and Schmitt, 1999) in peripheral neurons (spinal ganglia), in oligodendrocytes in various white matter regions of the CNS, in ependymal cells, and in epithelial cells of the plexus choroideus of the four ventricle, as well as in satellite cells of spinal ganglia.

In contrast to EAAT1, EAAT2 and EAAT4 (e.g. Chaudhry et al., 1995; Lehre et al., 1995; Dehnes et al., 1998), a large part of the EAAT3-immunoreactivity localizes to cytoplasmic structures (Conti et al., 1998; Kugler and Schmitt, 1999). The cytoplasmic localization of EAAT3 is believed to imply that EAAT3 can be rapidly mobilized from the cytoplasm to the plasma membrane (see below).

EAAT3 is, at least in the cerebral cortex, present throughout the dendritic ramifications, including the spines and is therefore close to the synapses (Conti et al., 1998). Recent studies (He et al., 2000; He et al., 2001) have concluded that EAAT3 is present in the non-synaptic parts of the spine membrane and not in the synaptic area.

4.4 EAAT4

In contrast to the wide distribution of EAAT1-3, EAAT4 is in the adult restricted to the Purkinje cells of the cerebellar molecular layer (Fig. 5; Dehnes et al., 1998) apart from a small number of neurons in the neocortex (Dehnes et al., 1998; Massie et al., 2001). A transient expression is observed in the forebrain, but the even maximum levels reached at around P10 (Furuta et al., 1997) are low compared to those in the adult cerebellum.

The highest concentration of EAAT4 is found in the parts of the Purkinje cell spine membranes which faces astroglia, but EAAT4 is also detectable in all parts of the Purkinje-cell plasma membrane with the exception of the axon and the part of the spine membrane which surrounds the post synaptic density and faces nerve terminal membranes (Dehnes et al., 1998). The levels in cytoplasmatic structures are close to background (Dehnes et al., 1998). The average concentration of EAAT4 in the molecular layer of the adult rat is about 0.2 mg/g tissue. Because the surface area of the spines is about 1.1 $\mu m^2/\mu m^3$, it follows that the average density of EAAT4 is about 1800 transporter molecules per μm^2 spine membrane (Dehnes et al., 1998).

5. Physiological roles.

The obvious role of the EAAT-type transporters is to remove glutamate from the extracellular fluid. This is essential for nervous system function in order to protect the brain against excitotoxicity, secure a high signal to noise ratio and recycle glutamate. This fundamental role of glutamate uptake was already recognized at the time when the excitatory action of glutamate was discovered (Hayashi, 1954; Curtis et al., 1960; Johnston, 1981; for review see: Roberts et al., 1981; Fonnum, 1984; Balcar et al., 2001; Danbolt, 2001). The extremely high glutamate uptake activity in brain tissue was noted even earlier (Stern et al., 1949).

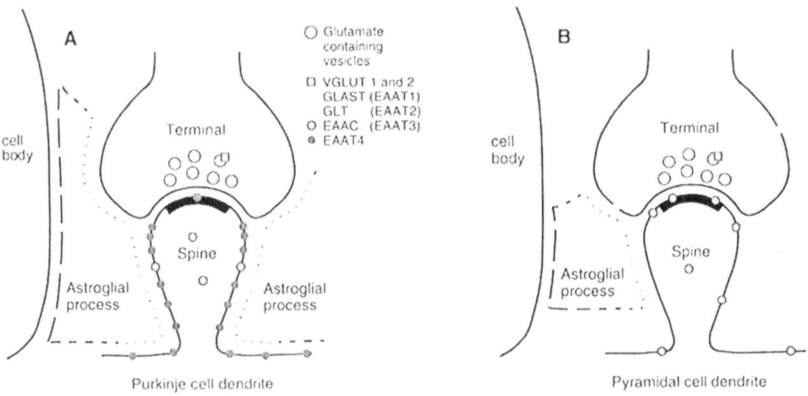

Fig. 5. Schematic representation of the detailed localization of glutamate transporters in the vicinity of the glutamatergic synapses. A. Parallel fiber to Purkinje cell dendritic spine synapses in the cerebellar molecular layer. These synapses are completely surrounded by astroglial membranes containing EAAT1 and EAAT2 at average densities of 4700 and 740 μm^{-2} membrane, respectively (Lehre and Danbolt, 1998). The concentrations are highest towards neuropil and lower towards other structures like cell bodies, vascular endothelium, large dendrites and pia mater. EAAT4 is concentrated in the glia-covered parts of the membranes of Purkinje cell dendrites, highest at the spines and thinner dendrites. There is a very low concentration of EAAT4 at the postsynaptic density. EAAT4 is unevenly distributed in the molecular layer and is expressed at different densities in different parasagittal zones, average density at around 1800 molecules μm^{-2} (Dehnes et al., 1998). EAAT3 is present in Purkinje cell membrane as well as cytoplasm (Conti et al., 1998; Kugler and Schmitt, 1999), but quantitative data and information on the precise subcellular distribution of this are currently unavailable. B. Glutamatergic synapse in the hippocampus (e.g. Schaffer collateral to pyramidal cell synapse). The synapses are only contacted and not surrounded by astroglia. EAAT2 and EAAT1 are present in the astroglial membranes at average densities of 8500 and 2300 μm^{-2} membrane, respectively (Lehre and Danbolt, 1998), with highest concentration towards neuropil (Chaudhry et al., 1995). Quantitative information is neither available on EAAT3 nor on EAAT2 in the glutamatergic nerve endings (for review see Danbolt, 2001). Modified from Danbolt 2001 with permission (Copyright 2001 by Elsevier Science Ltd).

This explains why the brain needs glutamate transporters, but not why it needs five of them, nor why they display uncoupled ion channel activities, nor why the regulation is so complex and the distribution so differentiated.

A predominant expression in astroglial cells may be evolutionarily beneficial because this transfers the burden of glutamate uptake from neurons to glial cells. This may be important during periods of neuronal stress considering that the transport requires energy and is electrogenic (net positive charge moving inwards).

A careful regulation (for review see: Gegelashvili and Schousboe, 1998; Sims and Robinson, 1999; Gegelashvili et al., 2000; Danbolt, 2001) of the global brain glutamate uptake activity is in agreement with the notion that both hyper- and hypo-activation of glutamate receptors are harmful. Likewise, local regulation also makes sense considering that the metabolic need varies between regions both during normal brain activity and during disease. However, uptake into glial cells only comes into play after glutamate has escaped from the synaptic clefts. Consequently, regulation of glial glutamate uptake may imply that glial cells are taking part in the control of transmitter reentering the clefts, in the control of activation of extra-synaptic receptors (spillover) and in the control of transmitter diffusing from one synapse and onto receptors in a neighboring synapse (crosstalk). The literature on spillover and crosstalk is now too extensive to be listed here (e.g. Bergles et al., 1999; Kullmann et al., 1999; Rusakov et al., 1999; Mitchell and Silver, 2000; Semyanov and Kullmann, 2000; Diamond, 2001; Oliet et al., 2001; Arnth-Jensen et al., 2002; Lehre and Rusakov, 2002; Voutsinosporche et al., 2003). A book with the title "The Tripartite Synapse" has just been published to honor the important role of astroglia (Volterra et al., 2002).

The above arguments more or less completely explain the function, distribution and regulation of EAAT2. This transporter subtype is, as already mentioned, the most abundant of the transporters and it displays a dynamically regulated expression mostly in astroglial cells. This subtype is also the one which seems to be primarily a transporter because it has very low ion channel conductancy. EAAT2-deficient mice show neurodegeneration, spontaneous epilepsy and die before adulthood (Tanaka et al., 1997). Interestingly, none of the other subtypes are upregulated and none of them can compensate for the loss of EAAT2. Similarly, EAAT1-deficient mice have impaired cerebellar and retinal function (Watase et al., 1998). Consequently, the primary roles of EAAT1 and EAAT2 seem obvious and the overall picture is therefore in reasonable agreement with the above understanding of their physiological roles.

The above described understanding, however, gives little room for the other transporter subtypes and gives no clue as to the importance of the ion channel activities. In fact, mice lacking EAAT3 appear fairly normal apart from amino aciduria (Peghini et al., 1997). Also the EAAT4 expression in Purkinje cells (dehnes 1998 #5555) seems redundant in the parts of the spine membrane which face glial membranes containing extremely high EAAT1 and

EAAT2 levels. Consequently, it seems legitimate to ask whether these transporters do something else. EAAT4 has a 10-fold higher affinity for glutamate (Fairman et al., 1995) than EAAT1 and EAAT2 and should therefore be able to compete with them for glutamate binding. Because the chloride conductance of EAAT4 is much higher than those of the glial transporters (Fairman et al., 1995), it might be speculated if this is the clue to its function. Perhaps it is more important as an inhibitory glutamate receptor than as a glutamate transporter.

6. Roles of glutamate transporters in disease.

Perturbations of glutamate transport have now been reported in all major human neurological disorders (for review see: Danbolt, 2001; Maragakis and Rothstein, 2001). Time will show how many of these reports will withstand further investigations, but considering the complex network of signaling pathways (Sims and Robinson, 1999; Gegelashvili and Schousboe, 1998; Gegelashvili et al., 2000; Danbolt, 2001; Robinson, 2002) controlling the expression, targeting and activities of the glutamate transporters, perturbation in glutamate uptake (directly or indirectly) is to be expected whenever brain tissue is disturbed. Glutamate transporter function is modulated by fatty acids (Trotti et al., 1995; Zerangue et al., 1995; Dunlop et al., 1999), by red-ox mechanisms (Volterra et al., 1994; Trotti et al., 1997), by phosphorylation (Dowd and Robinson, 1996; Davis et al., 1998; Sims et al., 2000; Duan et al., 1999; Gegelashvili et al., 2000; Munir et al., 2000), by growth factors, cytokines and unknown soluble factors released from neurons (Gegelashvili et al., 1996; Gegelashvili et al., 1997; Schlag et al., 1998).

In ischemia and hypoglycemia, glutamate transporters reverse and catalyze massive efflux of glutamate when the transmembrane electrochemical gradients are sufficiently weakened (Nicholls and Attwell, 1990; Rossi et al., 2000). Down regulation of EAAT2 is observed in several conditions, including ischemia-hypoxia, traumatic brain injury, amyotrophic lateral sclerosis, brain cancer, viral infections, liver failure and possibly Alzheimers disease and epilepsy (for review see: Ye and Sontheimer, 1999; Danbolt, 2001; Maragakis and Rothstein, 2001; Jones, 2002). Several drugs (including some antiepileptic drugs and some antipsychotics) modulate EAAT2-expression (e.g. Schneider et al., 1998; Hassel et al., 2001; Melone et al., 2001; McCullumsmith and Meador-Woodruff, 2002; Ralph-Williams et al., 2003). Regardless of the initiating cause, changes in glutamate uptake may have profound consequences and may thereby represent important events in the development of disease and tissue damage.

7. Concluding remarks.

The major line of defense against excitotoxicity appears to be astroglial glutamate uptake mediated by EAAT2 in most of the CNS and by EAAT1 in the retina and the cerebellum. The role of the other subtypes remain a mystery which is well worth a clarification, considering their high densities at specific locations. Finally, disturbed control of glutamate transporters appears to play a role in the pathogenesis of most neurological diseases and pharmacological tools to modulate their expression are needed.

Acknowledgements: This work was supported by the Norwegian Elite Research Programme (Toppforskningsprogrammet), the Norwegian Research Council, EU BIOMED (contract QLG3-CT-2001-02004). Correspondence should be addressed to: Dr. Niels C. Danbolt, the Department of Anatomy, the Institute of Basic Medical Sciences, University of Oslo, P.O.Box 1103 Blindern, N-0317 Oslo, Norway. Fax: (+47) 22 85 12 78.

REFERENCES

Arnth-Jensen, N., Jabaudon, D., and Scanziani, M. (2002) Cooperation between independent hippocampal synapses is controlled by glutamate uptake. Nat Neurosci 5, 325-331.

Arriza, J.L., Eliasof, S., Kavanaugh, M.P., and Amara, S.G. (1997) Excitatory amino acid transporter 5, a retinal glutamate transporter coupled to a chloride conductance. Proc Natl Acad Sci USA 94, 4155-4160.

Arriza, J.L., Fairman, W.A., Wadiche, J.I., Murdoch, G.H., Kavanaugh, M.P., and Amara, S.G. (1994) Functional comparisons of three glutamate transporter subtypes cloned from human motor cortex. J Neurosci 14, 5559-5569.

Baker, D.A., Xi, Z.X., Shen, H., Swanson, C.J., and Kalivas, P.W. (2002) The origin and neuronal function of in vivo nonsynaptic glutamate. J Neurosci 22, 9134-9141.

Balcar, V.J., Takamoto, A., and Yoneda, Y. (2001) Neurochemistry of L-glutamate transport in the CNS: a review of thirty years of progress. Collect. Czech. Chem. Commun. 66, 1315-1340.

Bannai, S. (1986) Exchange of cystine and glutamate across plasma membrane of human fibroblasts. J Biol Chem 261, 2256-2263.

Bellocchio, E.E., Reimer, R.J., Fremeau, R.T.jr. , and Edwards, R.H. (2000) Uptake of glutamate into synaptic vesicles by an inorganic phosphate transporter. Science 289, 957-960.

Berger, U.V. and Hediger, M.A. (1998) Comparative analysis of glutamate transporter expression in rat brain using differential double in situ hybridization. Anat Embryol (Berl) 198, 13-30.

Berger, U.V. and Hediger, M.A. (2000) Distribution of the glutamate transporters GLAST and GLT-1 in rat circumventricular organs, meninges and dorsal root ganglia. J Comp Neurol 421, 385-399.

385-399.

Bergles, D.E., Diamond, J.S., and Jahr, C.E. (1999) Clearance of glutamate inside the synapse and beyond. Curr Opin Neurobiol 9, 293-298.

Billups, B., Rossi, D., and Attwell, D. (1996) Anion conductance behavior of the glutamate uptake carrier in salamander retinal glial cells. J Neurosci 16, 6722-6731.

Bridges, R.J., Kavanaugh, M.P., and Chamberlin, A.R. (1999) A pharmacological review of competitive inhibitors and substrates of high-affinity, sodium-dependent glutamate transport in the central nervous system. Curr Pharm Des 5, 363-379.

Chaudhry, F.A., Lehre, K.P., Campagne, M.V., Ottersen, O.P., Danbolt, N.C., and Storm-Mathisen, J. (1995) Glutamate transporters in glial plasma membranes: highly differentiated localizations revealed by quantitative ultrastructural immunocytochemistry. Neuron 15, 711-720.

Chen, W., Aoki, C., Mahadomrongkul, V., Gruber, C.E., Wang, G.J., Blitzblau, R., Irwin, N., and Rosenberg, P.A. (2002) Expression of a variant form of the glutamate transporter GLT1 in neuronal cultures and in neurons and astrocytes in the rat brain. J Neurosci 22, 2142-2152.

Chen, X., Tsukaguchi, H., Chen, X.Z., Berger, U.V., and Hediger, M.A. (1999) Molecular and functional analysis of SDCT2, a novel rat sodium- dependent dicarboxylate transporter. J Clin Invest 103, 1159-1168.

Conti, F., DeBiasi, S., Minelli, A., Rothstein, J.D., and Melone, M. (1998) EAAC1, a high-affinity glutamate transporter, is localized to astrocytes and GABAergic neurons besides pyramidal cells in the rat cerebral cortex. Cereb Cortex 8, 108-116.

Curtis, D.R., Phillis, J.W., and Watkins, J.C. (1960) The chemical excitation of spinal neurons by certain acidic amino acids . J. Physiol. 150, 656-682.

Danbolt, N.C. (2001) Glutamate uptake. Prog Neurobiol 65, 1-105.

Danbolt, N.C., Pines, G., and Kanner, B.I. (1990) Purification and reconstitution of the sodium- and potassium-coupled glutamate transport glycoprotein from rat brain. Biochemistry US 29, 6734-6740.

Danbolt, N.C., Storm-Mathisen, J., and Kanner, B.I. (1992) An [Na^+ + K^+]coupled L-glutamate transporter purified from rat brain is located in glial cell processes. Neuroscience 51, 295-310.

Davis, K.E., Straff, D.J., Weinstein, E.A., Bannerman, P.G., Correale, D.M., Rothstein, J.D., and Robinson, M.B. (1998) Multiple signaling pathways regulate cell surface expression and activity of the excitatory amino acid carrier 1 subtype of Glu transporter in C6 glioma. J Neurosci 18, 2475-2485.

Dehnes, Y., Chaudhry, F.A., Ullensvang, K., Lehre, K.P., Storm-Mathisen, J., and Danbolt, N.C. (1998) The glutamate transporter EAAT4 in rat cerebellar Purkinje cells: a glutamate-gated chloride channel concentrated near the synapse in parts of the dendritic membrane facing astroglia. J Neurosci 18, 3606-3619.

Diamond, J.S. (2001) Neuronal glutamate transporters limit activation of NMDA receptors by neurotransmitter spillover on CA1 pyramidal cells. J Neurosci 21, 8328-8338.

Dowd, L.A. and Robinson, M.B. (1996) Rapid stimulation of EAAC1-mediated Na^+-dependent L-glutamate transport activity in C6 glioma cells by phorbol ester. J Neurochem 67, 508-516.

Dringen, R. (2000) Metabolism and functions of glutathione in brain. Prog Neurobiol 62, 649-671.

Duan, S.M., Anderson, C.M., Stein, B.A., and Swanson, R.A. (1999) Glutamate induces rapid upregulation of astrocyte glutamate transport and cell-surface expression of GLAST. J Neurosci 19, 10193-10200.

Dunlop, J., Lou, Z., Zhang, Y., and McIlvain, H.B. (1999) Inducible expression and pharmacology of the human excitatory amino acid transporter 2 subtype of L-glutamate transporter. Br J Pharmacol 128, 1485-1490.

Eskandari, S., Kreman, M., Kavanaugh, M.P., Wright, E.M., and Zampighi, G.A. (2000) Pentameric assembly of a neuronal glutamate transporter. Proc Natl Acad Sci USA 97, 8641-8646.

Fairman, W.A., Sonders, M.S., Murdoch, G.H., and Amara, S.G. (1998) Arachidonic acid elicits a substrate-gated proton current associated with the glutamate transporter EAAT4. Nat Neurosci 1, 105-113.

Fairman, W.A., Vandenberg, R.J., Arriza, J.L., Kavanaugh, M.P., and Amara, S.G. (1995) An excitatory amino-acid transporter with properties of a ligand-gated chloride channel. Nature 375, 599-603.

Fiermonte, G., Palmieri, L., Todisco, S., Agrimi, G., Palmieri, F., and Walker, J.E. (2002) Identification of the mitochondrial glutamate transporter - Bacterial expression, reconstitution, functional characterization, and tissue distribution of two human isoforms. J Biol Chem 277, 19289-19294.

Flynn, J. and McBean, G.J. (2000) Kinetic and pharmacological analysis of L-[^{35}S]cystine transport into rat brain synaptosomes. Neurochem Int 36, 513-521.

Fonnum, F. (1984) Glutamate: a neurotransmitter in mammalian brain. J Neurochem 42, 1-11.

Fremeau, R.T. Jr, Burman, J., Qureshi, T., Tran, C.H., Proctor, J., Johnson, J., Zhang, H., Sulzer, D., Copenhagen, D.R., Storm-Mathisen, J., Reimer, R.J., Chaudhry, F.A., and Edwards, R.H. (2002) The identification of vesicular glutamate transporter 3 suggests novel modes of signaling by glutamate. Proc Natl Acad Sci U S A 99, 14488-14493.

Fremeau, R.T.Jr, Troyer, M.D., Pahner, I., Nygaard, G.O., Tran, C.H., Reimer, R.J., Bellocchio, E.E., Fortin, D., Storm-Mathisen, J., and Edwards, R.H. (2001) The expression of vesicular glutamate transporters defines two classes of excitatory synapse. Neuron 31, 247-260.

Furness, D.N. and Lehre, K.P. (1997) Immunocytochemical localization of a high-affinity glutamate-aspartate transporter, GLAST, in the rat and guinea-pig cochlea. Eur J Neurosci 9, 1961-1969.

Furuta, A., Rothstein, J.D., and Martin, L.J. (1997) Glutamate transporter protein subtypes are expressed differentially during rat CNS development. J Neurosci 17, 8363-8375.

Gegelashvili, G. and Schousboe, A. (1998) Cellular distribution and kinetic properties of high-affinity glutamate transporters. Brain Res Bull 45, 233-238.

Gegelashvili, G., Civenni, G., Racagni, G., Danbolt, N.C., Schousboe, I., and Schousboe, A. (1996) Glutamate receptor agonists up-regulate glutamate transporter GLAST in astrocytes. Neuroreport 8, 261-265.

Gegelashvili, G., Danbolt, N.C., and Schousboe, A. (1997) Neuronal soluble factors differentially regulate the expression of the GLT1 and GLAST glutamate transporters in cultured astroglia. J Neurochem 69, 2612-2615.

Gegelashvili, G., Dehnes, Y., Danbolt, N.C., and Schousboe, A. (2000) The high-affinity glutamate transporters GLT1, GLAST and EAAT4 are regulated via different signalling mechanisms. Neurochem Int 37, 163-170.

Gras, C., Herzog, E., Bellenchi, G.C., Bernard, V., Ravassard, P., Pohl, M., Gasnier, B., Giros, B., and Elmestikawy, S. (2002) A third vesicular glutamate transporter expressed by cholinergic and serotoninergic neurons. J Neurosci 22, 5442-5451.

Grunewald, M. and Kanner, B.I. (2000) The accessibility of a novel reentrant loop of the glutamate transporter GLT-1 is restricted by its substrate. J Biol Chem 275, 9684-9689.

Gundersen, V., Danbolt, N.C., Ottersen, O.P., and Storm-Mathisen, J. (1993) Demonstration of glutamate/aspartate uptake activity in nerve endings by use of antibodies recognizing exogenous D-aspartate. Neuroscience 57, 97-111.

Hassel, B., Iversen, E.G., Gjerstad, L., and Taubøll, E. (2001) Up-regulation of hippocampal glutamate transport during chronic treatment with sodium valproate. J Neurochem 77, 1285-1292.

Haugeto, Ø., Ullensvang, K., Levy, L.M., Chaudhry, F.A., Honoré, T., Nielsen, M., Lehre, K.P., and Danbolt, N.C. (1996) Brain glutamate transporter proteins form homomultimers. J Biol Chem 271, 27715-27722.

Hayashi, T. (1954) Effects of sodium glutamate on the nervous system. Keio J. Med. 3, 183-192.

He, Y., Hof, P.R., Janssen, W.G., Rothstein, J.D., and Morrison, J.H. (2001) Differential synaptic localization of GluR2 and EAAC1 in the macaque monkey entorhinal cortex: a postembedding immunogold study. Neurosci Lett 311, 161-164.

He, Y., Janssen, W.G.M., Rothstein, J.D., and Morrison, J.H. (2000) Differential synaptic localization of the glutamate transporter EAAC1 and glutamate receptor subunit GluR2 in the rat hippocampus. J Comp Neurol 418, 255-269.

Hees, B., Danbolt, N.C., Kanner, B.I., Haase, W., Heitmann, K., and Koepsell, H. (1992) A monoclonal antibody against a Na^+-L-glutamate cotransporter from rat brain. J Biol Chem 267, 23275-23281.

Herzog, E., Bellenchi, G.C., Gras, C., Bernard, V., Ravassard, P., Bedet, C., Gasnier, B., Giros, B., and El Mestikawy, S. (2001) The existence of a second vesicular glutamate transporter specifies subpopulations of glutamatergic neurons. J Neurosci 21, U1-U6.

Jackson, M., Song, W., Liu, M.Y., Jin, L., Dykes-Hoberg, M., Lin, C.I., Bowers, W.J., Federoff, H.J., Sternweis, P.C., and Rothstein, J.D. (2001) Modulation of the neuronal glutamate transporter EAAT4 by two interacting proteins. Nature 410, 89-93.

Johnston, G.A.R. (1981) Glutamate uptake and its possible role in neurotransmitter inactivation. In: Glutamate: Transmitter in the Central Nervous System, 77-87. Roberts, P J, Storm-Mathisen, J, and Johnston, G A R. John Wiley & Sons, Chichester, New York, Brisbane, Toronto.

Jones, N. (2002) Bigger is better when it comes to the G spot . New Scientist, http://www.newscientist.com/news/news.jsp?id=ns99992495.

Kanner, B.I. and Schuldiner, S. (1987) Mechanism of transport and storage of neurotransmitters. CRC Crit Rev Biochem 22, 1-38.

Kugler, P. and Schmitt, A. (1999) Glutamate transporter EAAC1 is expressed in neurons and glial cells in the rat nervous system. Glia 27 , 129-142.

Kullmann, D.M., Min, M.Y., Asztely, F., and Rusakov, D.A. (1999) Extracellular glutamate diffusion determines the occupancy of glutamate receptors at CA1 synapses in the hippocampus. Philos Trans R Soc Lond B Biol Sci 354, 395-402.

Lehre, K., Levy, L., Chaudhry, F., Storm-Mathisen, J., Ottersen, O., and Danbolt, N. (1993) Localization of glutamate transporters in brain by site-directed immunocytochemistry. J Neurochem 61, S251.

Lehre, K.P. and Danbolt, N.C. (1998) The number of glutamate transporter subtype molecules at glutamatergic synapses: chemical and stereological quantification in young adult rat brain. J Neurosci 18, 8751-8757.

Lehre, K.P. and Rusakov, D.A. (2002) Asymmetry of glia near central synapses favors presynaptically directed glutamate escape. Biophys J 83 , 125-134.

Lehre, K.P., Levy, L.M., Ottersen, O.P., Storm-Mathisen, J., and Danbolt, N.C. (1995) Differential expression of two glial glutamate transporters in the rat brain: quantitative and immunocytochemical observations. J Neurosci 15, 1835-1853.

Levy, L.M., Lehre, K.P., Rolstad, B., and Danbolt, N.C. (1993) A monoclonal antibody raised against an [Na^+ - K^+]coupled L- glutamate transporter purified from rat brain confirms glial cell localization. FEBS Lett 317, 79-84.

Levy, L.M., Warr, O., and Attwell, D. (1998) Stoichiometry of the glial glutamate transporter GLT-1 expressed inducibly in a chinese hamster ovary cell line selected for low endogenous Na^+-dependent glutamate uptake. J Neurosci 18, 9620-9628.

Lin, C.I., Orlov, I., Ruggiero, A.M., Dykes-Hoberg, M., Lee, A., Jackson, M., and Rothstein, J.D. (2001) Modulation of the neuronal glutamate transporter EAAC1 by the interacting protein GTRAP3-18. Nature 410, 84-88.

MacAulay, N., Gether, U., Klaerke, D.A., and Zeuthen, T. (2001) Water transport by the human Na^+-coupled glutamate cotransporter expressed in Xenopus oocytes. Journal of Physiology - London 530, 367-378.

Maragakis, N.J. and Rothstein, J.D. (2001) Glutamate transporters in neurologic disease. Arch Neurol 58, 365-370.

Marie, H., Billups, D., Bedford, F.K., Dumoulin, A., Goyal, R.K., Longmore, G.D., Moss, S.J., and Attwell, D. (2002) The amino terminus of the glial glutamate transporter GLT-1 interacts with the LIM protein Ajuba. Mol Cell Neurosci 19, 152-164.

Martin, L.J., Brambrink, A.M., Lehmann, C., Portera-Cailliau, C., Koehler, R., Rothstein, J., and Traystman, R.J. (1997) Hypoxia-ischemia causes abnormalities in glutamate transporters and death of astroglia and neurons in newborn striatum. Ann Neurol 42, 335-348.

Massie, A., Vandesande, F., and Arckens, L. (2001) Expression of the high-affinity glutamate transporter EAAT4 in mammalian cerebral cortex. Neuroreport 12, 393-397.

McCullumsmith, R.E. and Meador-Woodruff, J.H. (2002) Striatal excitatory amino acid transporter transcript expression in schizophrenia, bipolar disorder, and major depressive disorder. Neuropsychopharmacology 26, 368-375.

Melone, M., Vitellaro-Zuccarello, L., Vallejo-Illarramendi, A., Pé rez-Samartin, A., Matute, C., Cozzi, A., Pellegrini-Giampietro, D.E., Rothstein, J.D., and Conti, F. (2001) The expression of glutamate transporter GLT-1 in the rat cerebral cortex is down-regulated by the antipsychotic drug clozapine. Mol Psychiatry 6, 380-386.

Milton, I.D., Banner, S.J., Ince, P.G., Piggott, N.H., Fray, A.E., Thatcher, N., Horne, C.H.W., and Shaw, P.J. (1997) Expression of the glial glutamate transporter EAAT2 in the human CNS:

an immunohistochemical study. Molecular Brain Research 52, 17-31.

Mitchell, S.J. and Silver, R.A. (2000) Glutamate spillover suppresses inhibition by activating presynaptic mGluRs. Nature 404, 498-502.

Munir, M., Correale, D.M., and Robinson, M.B. (2000) Substrate-induced up-regulation of Na^{+}-dependent glutamate transport activity. Neurochem Int 37, 147-162.

Nicholls, D. and Attwell, D. (1990) The release and uptake of excitatory amino acids. Trends Pharmacol Sci 11, 462-468.

Northington, F.J., Traystman, R.J., Koehler, R.C., and Martin, L.J. (1999) GLT1, glial glutamate transporter, is transiently expressed in neurons and develops astrocyte specificity only after midgestation in the ovine fetal brain. J Neurobiol 39, 515-526.

Oliet, S.H., Piet, R., and Poulain, D.A. (2001) Control of glutamate clearance and synaptic efficacy by glial coverage of neurons. Science 292, 923-926.

Peghini, P., Janzen, J., and Stoffel, W. (1997) Glutamate transporter EAAC-1-deficient mice develop dicarboxylic aminoaciduria and behavioral abnormalities but no neurodegeneration. EMBO J 16, 3822-3832.

Pines, G. and Kanner, B.I. (1990) Counterflow of L-glutamate in plasma membrane vesicles and reconstituted preparations from rat brain. Biochemistry US 29, 11209-11214.

Plachez, C., Danbolt, N.C., and Recasens, M. (2000) Transient expression of the glial glutamate transporters GLAST and GLT in hippocampal neurons in primary culture. J Neurosci Res 59, 587-593.

Ralph-Williams, R.J., Paulus, M.P., Zhuang, X.X., Hen, R., and Geyer, M.A. (2003) Valproate attenuates hyperactive and perseverative behaviors in mutant mice with a dysregulated dopamine system. Biol Psychiatry 53, 352-359.

Rauen, T., Fischer, F., and Wiessner, M. (1999) Glia-neuron interaction by high-affinity glutamate transporters in neurotransmission. Adv Exp Med Biol 468, 81-95.

Roberts, P.J., Storm-Mathisen, J., Johnston, G.A.R., Roberts, P.J., Storm-Mathisen, J., and Johnston, G.A.R. (1981>) Glutamate: Transmitter in the Central Nervous System. John Wiley & Sons.

Robinson, M.B. (2002) Regulated trafficking of neurotransmitter transporters: common notes but different melodies. J Neurochem 80, 1-11.

Rossi, D.J., Oshima, T., and Attwell, D. (2000) Glutamate release in severe brain ischaemia is mainly by reversed uptake. Nature 403, 316-321.

Rothstein, J.D., Martin, L., Levey, A.I., Dykes-Hoberg, M., Jin, L., Wu, D., Nash, N., and Kuncl, R.W. (1994) Localization of neuronal and glial glutamate transporters. Neuron 13, 713-725.

Rothstein, J.D., Van Kammen, M., Levey, A.I., Martin, L.J., and Kuncl, R.W. (1995) Selective loss of glial glutamate transporter GLT-1 in amyotrophic lateral sclerosis. Ann Neurol 38, 73-84.

Rusakov, D.A., Kullmann, D.M., and Stewart, M.G. (1999) Hippocampal synapses: do they talk to their neighbours? Trends Neurosci 22, 382-388.

Sakagawa, T., Ogawa, S., Kawashima, N., Hori, S., Takimoto, M., Wada, K., and Tanaka, K. (1998) Motor discoordination and increased susceptibility to cerebellar injury in GLAST mutant mice. Eur J Neurosci 10, 976-988.

Sato, H., Tamba, M., Ishii, T., and Bannai, S. (1999) Cloning and expression of a plasma membrane cystine/glutamate exchange transporter composed of two distinct proteins. J Biol Chem 274, 11455-11458.

Schlag, B.D., Vondrasek, J.R., Munir, M., Kalandadze, A., Zelenaia, O.A., Rothstein, J.D., and Robinson, M.B. (1998) Regulation of the glial Na^+ -dependent glutamate transporters by cyclic AMP analogs and neurons. Mol Pharmacol 53, 355-369.

Schmitt, A., Asan, E., Lesch, K.P., and Kugler, P. (2002) A splice variant of glutamate transporter GLT1/EAAT2 expressed in neurons: cloning and localization in rat nervous system. Neuroscience 109 , 45-61.

Schmitt, A., Asan, E., Puschel, B., and Kugler, P. (1997) Cellular and regional distribution of the glutamate transporter GLAST in the CNS of rats: nonradioactive in situ hybridization and comparative immunocytochemistry. J Neurosci 17, 1-10.

Schmitt, A., Asan, E., Puschel, B., Jons, T., and Kugler, P. (1996) Expression of the glutamate transporter GLT1 in neural cells of the rat central nervous system: non-radioactive in situ hybridization and comparative immunocytochemistry. Neuroscience 71, 989-1004.

Schneider, J.S., Wade, T., and Lidsky, T.I. (1998) Chronic neuroleptic treatment alters expression of glial glutamate transporter GLT-1 mRNA in the striatum. Neuroreport 9, 133-136.

Seal, R.P. and Amara, S.G. (1999) Excitatory amino acid transporters: a family in flux. Annu Rev Pharmacol Toxicol 39, 431-456.

Seal, R.P., Leighton, B.H., and Amara, S.G. (2000) A model for the topology of excitatory amino acid transporters determined by the extracellular accessibility of substituted cysteines. Neuron 25, 695-706.

Semyanov, A. and Kullmann, D.M. (2000) Modulation of GABAergic signaling among interneurons by metabotropic glutamate receptors. Neuron 25, 663-672.

Shimamoto, K., Lebrun, B., Yasudakamatani, Y., Sakaitani, M., Shigeri, Y., Yumoto, N., and Nakajima, T. (1998) DL-threo-beta-benzyloxyaspartate, a potent blocker of excitatory amino acid transporters. Mol Pharmacol 53, 195-201.

Sims, K.D. and Robinson, M.B. (1999) Expression patterns and regulation of glutamate transporters in the developing and adult nervous system. Crit Rev Neurobiol 13, 169-197

Sims, K.D., Straff, D.J., and Robinson, M.B. (2000) Platelet-derived growth factor rapidly increases activity and cell surface expression of the EAAC1 subtype of glutamate transporter through activation of phosphatidylinositol 3-kinase. J Biol Chem 275, 5228-5237.

Stern, J.R., Eggleston, L.V., Hems, R., and Krebs, H.A. (1949) Accumulation of glutamic acid in isolated brain tissue. Biochem J 44, 410-418.

Suchak, S.K., Baloyianni, N.V., Perkinton, M.S., Williams, R.J., Meldrum, B.S., and Rattray, M. (2003) The 'glial' glutamate transporter, EAAT2 (Glt-1) accounts for high affinity glutamate uptake into adult rodent nerve endings. J Neurochem 84, 522-532.

Sutherland, M.L., Delaney, T.A., and Noebels, J.L. (1996) Glutamate transporter mRNA expression in proliferative zones of the developing and adult murine CNS. J Neurosci 16, 2191-2207.

Takamori, S., Malherbe, P., Broger, C., and Jahn, R. (2002) Molecular cloning and functional characterization of human vesicular glutamate transporter 3. Embo Reports 3, 798-803.

Takamori, S., Rhee, J.S., Rosenmund, C., and Jahn, R. (2000) Identification of a vesicular glutamate transporter that defines a glutamatergic phenotype in neurons. Nature 407, 189-194.

Takamori, S., Rhee, J.S., Rosenmund, C., and Jahn, R. (2001) Identification of differentiation-associated brain-specific phosphate transporter as a second vesicular glutamate transporter (VGLUT2). J Neurosci 21, U7-U12.

Takumi, Y., Matsubara, A., Danbolt, N.C., Laake, J.H., Storm-Mathisen, J., Usami, S., Shinkawa, H., and Ottersen, O.P. (1997) Discrete cellular and subcellular localization of glutamine synthetase and the glutamate transporter GLAST in the rat vestibular end organ. Neuroscience 79, 1137-1144.

Tanaka, K., Watase, K., Manabe, T., Yamada, K., Watanabe, M., Takahashi, K., Iwama, H., Nishikawa, T., Ichihara, N., Hori, S., Takimoto, M., and Wada, K. (1997) Epilepsy and exacerbation of brain injury in mice lacking the glutamate transporter GLT-1. Science 276, 1699-1702.

Torp, R., Danbolt, N.C., Babaie, E., Bjørås, M., Seeberg, E., Storm-Mathisen, J., and Ottersen, O.P. (1994) Differential expression of two glial glutamate transporters in the rat brain: an in situ hybridization study. Eur J Neurosci 6, 936-942.

Torp, R., Hoover, F., Danbolt, N.C., Storm-Mathisen, J., and Ottersen, O.P. (1997) Differential distribution of the glutamate transporters GLT1 and rEAAC1 in rat cerebral cortex and thalamus: an in situ hybridization analysis. Anat Embryol (Berl) 195, 317-326.

Trotti, D., Rizzini, B.L., Rossi, D., Haugeto, Ø., Racagni, G., Danbolt, N.C., and Volterra, A. (1997) Neuronal and glial glutamate transporters possess an SH-based redox regulatory mechanism. Eur J Neurosci 9, 1236-1243.

Trotti, D., Volterra, A., Lehre, K.P., Rossi, D., Gjesdal, O., Racagni, G., and Danbolt, N.C. (1995) Arachidonic acid inhibits a purified and reconstituted glutamate transporter directly from the water phase and not via the phospholipid membrane. J Biol Chem 270, 9890-9895.

Ullensvang, K., Lehre, K.P., Storm-Mathisen, J., and Danbolt, N.C. (1997) Differential developmental expression of the two rat brain glutamate transporter proteins GLAST and GLT. Eur J Neurosci 9, 1646-1655.

Volterra, A., Bezzi, P., Rizzini, B.L., Trotti, D., Ullensvang, K., Danbolt, N.C., and Racagni, G. (1996) The competitive transport inhibitor L-trans-pyrrolidine-2,4- dicarboxylate triggers excitotoxicity in rat cortical neuron- astrocyte co-cultures via glutamate release rather than uptake inhibition. Eur J Neurosci 8, 2019-2028.

Volterra, A., Trotti, D., Tromba, C., Floridi, S., and Racagni, G. (1994) Glutamate uptake inhibition by oxygen free radicals in rat cortical astrocytes. J Neurosci 14, 2924-2932.

Volterra, A., Magistretti, P.J., and Haydon, P.G. (2002) The Tripartite Synapse - Glia in Synaptic Transmission. Oxford University Press: New York.

Voutsinosporche, B., Bonvento, G., Tanaka, K., Steiner, P., Welker, E., Chatton, J.Y., Magistretti, P.J., and Pellerin, L. (2003) Glial glutamate transporters mediate a functional metabolic crosstalk between neurons and astrocytes in the mouse developing cortex. Neuron 37, 275-286.

Wadiche, J.I. and Kavanaugh, M.P. (1998) Macroscopic and microscopic properties of a cloned glutamate transporter chloride channel. J Neurosci 18, 7650-7661.

Wadiche, J.I., Amara, S.G., and Kavanaugh, M.P. (1995) Ion fluxes associated with excitatory amino acid transport. Neuron 15, 721-728.

Watase, K., Hashimoto, K., Kano, M., Yamada, K., Watanabe, M., Inoue, Y., Okuyama, S.,

Ye, Z.C. and Sontheimer, H. (1999) Glioma cells release excitotoxic concentrations of glutamate. Cancer Res 59, 4383-4391.

Zerangue, N. and Kavanaugh, M.P. (1996) Flux coupling in a neuronal glutamate transporter. Nature 383, 634-637.

Zerangue, N., Arriza, J.L., Amara, S.G., and Kavanaugh, M.P. (1995) Differential modulation of human glutamate transporter subtypes by arachidonic acid. J Biol Chem 270, 6433-6435.

Chapter 3
NMDA RECEPTORS

Fabrizio Gardoni, Flaminio Cattabeni and Monica Di Luca
Center of Excellence on Neurodegenerative Diseases and Department of Pharmacological Sciences, University of Milano, via Balzaretti 9, 20133 Milano, Italy.

Abstract: The N-methyl-D-aspartate (NMDA) type glutamate receptors are abundant, ubiquitously distributed throughout the CNS, fundamental to excitatory glutamatergic transmission and critical for normal brain function. Activation of NMDA receptor is complex; both glutamate and glycine are required to open the ion channel and allow calcium entry. NMDA receptor is voltage dependently blocked by magnesium and depolarization is needed to remove this block. In the CNS synapses, NMDA receptors are localized in a highly organized subcellular fraction, the Post Synaptic Density (PSD), an electron dense and highly differentiated structure associated with the postsynaptic membrane, where clustering of ionotropic glutamate receptors to scaffolding proteins and signalling elements is dynamically regulated. To date, although molecular basis of glutamate toxicity remain uncertain, there is general agreement that NMDA subtype of ionotropic glutamate receptors plays a key role in mediating at least some aspects of glutamate excitotoxicity. NMDA receptor antagonists are effective in many experimental animal models of disease, and some of these compounds have moved into clinical trials. However, the initial enthusiasm for this approach has waned, because the therapeutic ratio for most NMDA antagonists is poor, with significant adverse effects at clinically effective doses, thus limiting their utility. Understanding the biochemical properties of the multitude of NMDA receptor subtypes offers the possibility of developing more effective and clinically useful drugs. In fact, NMDA receptor is part of a postsynaptic signaling complex including kinases, and other signaling and scaffolding proteins. This finding may improve the identification of specific molecular targets whose pharmacological or genetic manipulation might lead to innovative therapies for both acute and neurodegenerative disorders.

Key words: stsynaptic density, hippocampus, ischemia, PSD-95, CaMKII.

INDEX

1. Structure, subunit composition and functional properties of NMDA receptors.
2. Organization and regulation of NMDA receptors at the synapse.
3. Trafficking of NMDA receptors.

4. NMDA receptor role in excitotoxicity.
5. NMDA receptor mediated mechanisms in different disease states.
6. NMDA receptor pathways as drug target.
7. Conclusions.

1. Structure, subunit composition and functional properties of NMDA receptors.

NMDA receptors are oligomeric complexes formed by the coassembly of members of 3 receptor subunit families: NR1, NR2 subfamily and NR3 (Figure 1). One of these, the NR1 subunit, is a ubiquitous and necessary component of functional NMDA receptor channels. Each subunit possesses a large extracellular N-terminal domain, four membrane (M) regions and an intramembrane loop between the first and the second complete transmembrane domains. M-2 region contains a membrane-reentrant hairpin structure that contributes to the receptor channel pore. The N-terminal domain, which is large, glycosylated, and extracellular, contributes to the agonist binding site. NMDA receptors bind two agonist ligands: glutamate and the coagonist, glycine. The NR1 subunit contains the binding site for glycine and the NR2 subunit, the binding site for glutamate. The carboxyl-terminal region is intracellular and may control regulation of the receptor by second messenger systems, perhaps through multiple tyrosine and serine phosphorylation sites. This intracellular domain anchors NMDA receptor to specific subcellular compartments, i.e. the postsynaptic density fraction, and binds the receptor to different signal transduction systems.

Molecular diverse forms of the NR1 subunit are present and generated by alternative RNA splicing; differential splicing of three exons generate at least eight NR1 splice variants; the spliced exons encode a 21 amino acid sequence in the N-terminus domain (N1) and adjacent sequences of 37 and 38 amino acids in the C-terminus (C1 and C2 respectively). Splicing out the exon segment that encodes for the C2 cassette removes the first stop codon resulting in a new open reading frame that encodes an unrelated sequence of 22 aminoacids (C2') before a second stop codon is reached. These alternative splice processes cause alteration of the structural, physiological and pharmacological properties of NR1. Additional diversity is given by the multiplicity of NR2 subunits composing the receptor. There are four species of the second subunit type, NR2A-D, each encoded by a different gene. The different NR2 subunits are differentially expressed during different stages of development and in different tissues. NR2 subunits alone have no electrophysiologic activity but combine with NR1 to produce functional receptors in either *Xenopus* oocytes or mammalian cells.

NMDA receptors respond to agonists slowly than AMPA receptors, and require greater than 2 ms to open. However, they have a higher affinity for glutamate, and their currents persist longer than AMPA receptor currents. The NMDA receptor admits both Na+ and Ca2+ ions. At the resting potential of the cell, a Mg2+ ion blocks the NMDA receptor pore, but the Mg2+ ion is released from the pore upon cell depolarization. Therefore, opening of the channel requires binding of ligand and simultaneous depolarization of the cell. The channel thus operates as a coincidence detector that admits current only when agonist binding and cell depolarization take place simultaneously.

NMDA receptors display activity-dependent current decreases of several types, two of which are ligand dependent. One of these, glycine-dependent desensitization, occurs following receptor stimulation by glutamate when glycine concentrations are subsaturating, in the nanomolar range. The second, glycine-independent desensitization takes place in the presence of saturating glycine, at concentrations in the micromolar range. When glutamate binds to the receptor in the presence of low concentrations of glycine, the receptor rapidly enters a desensitized (low-conductance) state. This desensitizing transition is blocked by glycine. NMDA receptor desensitization may limit receptor currents during persistent stimulation by glutamate. Domains of NR2 that influence receptor desensitization characteristics have been defined.

A different activity-dependent NMDA receptor current decrease is Ca2+-dependent inactivation. Ca2+-dependent inactivation may be induced by increases in intracellular [Ca2+] that follow activity-dependent fluxes of Ca2+ through the receptor. The increase in intracellular [Ca2+] triggers biochemical modifications of the receptor that decrease receptor mean opening time. It has recently been shown that Ca2+-dependent inactivation results from binding of Ca2+-calmodulin to the membrane proximal region of the C-terminal domain of the NR1 subunit.

Coexpression of NR3A or NR3B with NR1 and NR2 subunits modulates NMDAR activity. NR3B is expressed predominantly in motor neurons, whereas NR3A is more widely distributed. Remarkably, when co-expressed in *Xenopus* oocytes, NR3A or NR3B co-assembles with NR1 to form excitatory glycine receptors that are unaffected by glutamate or NMDA, and inhibited by D-serine, a co-activator of conventional NMDA receptors. Moreover, NR1/NR3A or NR1/NR3B receptors form relatively Ca2+-impermeable cation channels that are resistant to Mg2+, MK-801, memantine and competitive antagonists. In cerebrocortical neurons containing NR3 family members, glycine triggers a burst of firing, and membrane patches manifest glycine-responsive single channels that are suppressible by D-serine. By itself, glycine is normally thought of as an inhibitory neurotransmitter. In contrast, these NR1/NR3A or NR1/NR3B NMDA receptors constitute a type of excitatory glycine receptor.

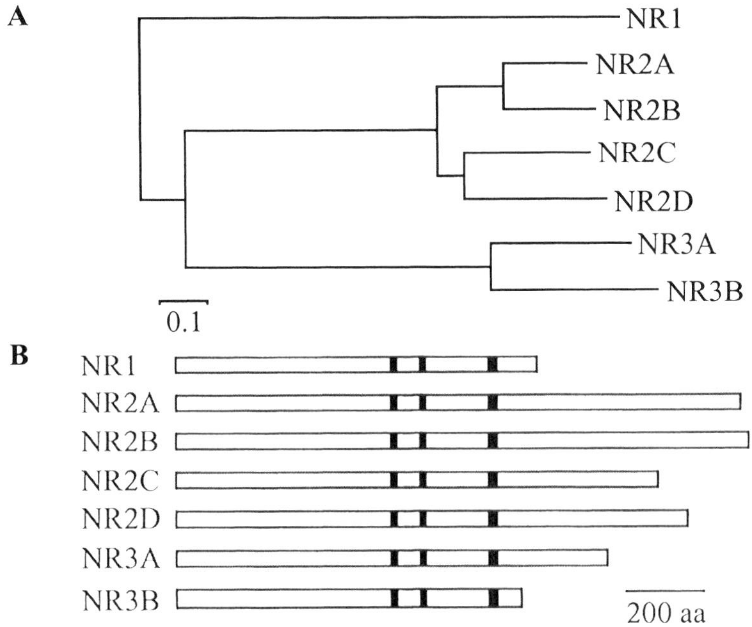

Figure 1 A: dendrogram of complete amino-acid sequences for NMDA receptor subunits, B: representation of NMDA receptor subunits polypeptides. Black boxes indicate transmembrane domains and grey boxes show the TM2 re-entrant loop.

2. Organization and regulation of NMDA receptors at the synapse.

NMDA receptors are embedded in the postsynaptic density (PSD), an electron dense and highly differentiated structure associated with the postsynaptic membrane (Figure 2). The PSD consists of numerous cytoskeletal and signaling proteins, some of which are in close contact with the cytoplasmic domain of glutamate ionotropic receptors in the postsynaptic membrane (Sheng and Kim, 2002). This accumulation of NMDA receptors at the postsynaptic sites ensures a rapid response to neurotransmitter release and provides a molecular mechanism for linking the transmembrane ion flux to the signaling machinery responsible for specific second messenger pathways.

In the PSD, the COOH-terminal cytoplasmic tails of NR2 subunits directly interact with PDZ domains of the membrane-associated guanylate kinases (MAGUK) family of scaffolding proteins such as PSD-95, Chapsyn-110, SAP102 and SAP97. In particular, the last three amino acids of the carboxyl termini of NR2A and NR2B subunits have a C-terminal consensus motif threonine/serine-X-valine (T/SXV, where X is any amino acid) that has been demonstrated to be responsible for efficient binding to PDZ domains of MAGUK protein family. MAGUK proteins, in turn, interact with a host of cytoplasmic signaling molecules, including neuronal nitric oxide sinthase (nNOS) and SynGAP, thereby connecting NMDA receptors to different signal transduction pathways (Kennedy, 2000).

nNOS, a Ca2+-activated form of NOS, can bind to PSD-95 through a PDZ domain interaction in which its own amino-terminal PDZ domain binds to a PDZ domain of PSD-95. nNOS is not abundant in the PSD fraction and is not expressed at high levels in pyramidal neurons. However, it is highly expressed in certain [gamma]-aminobutyric acid-containing neurons, which also express members of the PSD-95 family. Therefore, PSD-95 may concentrate nNOS near the NMDA receptor at postsynaptic sites in these neurons (Christopherson et al., 1999).

The cytosolic tails of the NR2A and NR2B subunits of the NMDA receptor bind directly also to signaling proteins including alpha subunit of Calcium-calmodulin-dependent protein kinase II (CaMKII). In addition, CaMKII phosphorylates both the NR2A and NR2B subunits (Gardoni et al., 2001). CaMKII-NMDA receptor interaction has received much attention in the last decade because CaMKII persistent activation after NMDA stimulation is essential for NMDA-dependent long-term potentiation processes. In fact, docking of CaMKII to the tail of the NMDA receptor would position its catalytic domains near the receptor mouth, ideally located for activation by Ca2+ flowing through the channel.

In the PSD fraction, phosphotyrosine-mediated pathways interact both physically and biochemically with NMDA receptor. The NMDA receptor NR2B subunit is the major synaptic phosphotyrosine protein of the excitatory synapse, although the precise effect of tyrosine phosphorylation on NR2B function at the synapse is not yet known. The Src family of nonreceptor tyrosine kinases stimulates NMDA receptor activity by direct phosphorylation (Cheung and Gurd, 2001).

NR1, in turn, provides a direct link to cytoskeletal proteins through the interaction with alpha-actinin-2 and other filamentous proteins. In fact, the rundown of NMDA channel activity can be prevented by the presence of the microfilament-stabilizing drug, phalloidin.

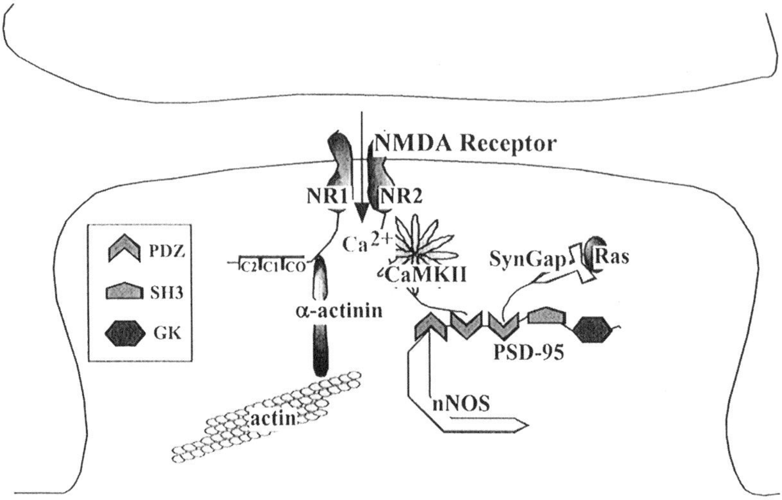

Figure 2. NMDA receptor complex in the postsynaptic density

In addition, NMDA currents are increased in hippocampal neurons lacking gelsolin, an actin-severing protein. Inside the PSD, also brain spectrin, and fodrin, are able to interact directly with NMDA receptor complex. Actin filaments bind directly to the PSD cytoplasmic face and associate with PSD components. The integrity of microfilament web secures the synaptic localization of ionotropic glutamate receptors; dissolution of microfilaments within the dendritic spines shifts a portion of NMDA clusters from a synaptic to a non-synaptic localization over a 24h period.

In this way, NMDA receptors are integrated in the complex protein network of the PSD, the components of which likely mediate main aspects of postsynaptic signaling by NMDA receptors. In addition, PSD is a dynamic structure, in which protein interactions may be transient allowing trafficking of ionotropic glutamate receptors through the secretory pathway as well as the clustering of receptors at synaptic sites.

3. Trafficking of NMDA receptors.

NMDA receptors are not static resident of the synapse, but the number of synaptic NMDA receptors is modulated by several factors (Wenthold et al., 2003). In fact, interaction of NMDA receptors with MAGUK proteins, generally thought to occur at the synapse, appears to occur early in the secretory pathway; this interaction may play a role in the assembly of the receptor complex and its exit from the endoplasmic reticulum (ER). NR2 subunits do not form functional receptors when expressed alone in heterologous cells, and those that have been studied (NR2A and NR2B) do not reach the cell surface. NR1 subunits also do not form functional receptors alone, but their cell surface expression depends on the splice variant. The ER retention of NR1 is dependent on an RXR motif in one of the alternatively expressed cassettes, C1. In addition, it has been suggested that another signal, present in C2', can negate the RXR retention signal in C1. This signal was shown to be the PDZ-binding domain at the distal end of C2' (STVV). This could result from masking of the RXR motif by the PDZ proteins that interact with the C2' cassette or the PDZ protein could be acting as an export protein that simply overrides the ER retention. The last six amino acids of C2' are sufficient to suppress ER retention, as shown by deleting the rest of C2'. However, the functional significance of the multiple C-terminal configurations of NR1 and their various behaviors in the ER remain unclear. Interestingly, NR1 can assemble with NR3 to produce functional glycine receptors; this assembly is also sufficient to relieve the ER retention of NR1.

The NR2 subunit is retained in the ER, and like NR1, its retention appears to depend on the C-terminal domain. The mechanism by which receptor assembly overcomes ER retention of both subunits is not clear. Recent data show that the entire NR1 C-terminal domain truncated allow the formation of functional heteromeric receptors on the cell surface. Similar results are obtained when NR2 C-terminal domains are truncated. For example, mice lacking the NR2A C-terminal domain still have functional receptors containing this subunit, although they are not correctly clustered at the synapse.

After they are released from the ER, NMDA receptor are processed in the Golgi apparatus and then distributed to the endosomes. Because functional ER and Golgi complexes are present in many dendrites and even dendritic spines, it is possible that some of the processing related to these compartments actually occurs near the synapse. This would provide a mechanism for a rapid and local response. Most receptors, however, are likely processed in the cell body and then transported to the synapse. Several findings suggest that the NMDA receptors interact with MAGUK proteins before it reaches the synapses and that these interactions may be involved in the synaptic delivery. Very recent findings showed that one of the PDZ partners of the NR2 subunit, SAP102, interacts through its PDZ domains with sec8, a protein of the exocyst or sec6/8 complex, known to be involved in targeting of secretory vesicles to the plasma membrane. The fact that SAP102 interacts with sec8 suggests that also NMDA receptors can indirectly associate with the exocyst complex through SAP102. The presence of this NMDA receptor - SAP102 - sec8 complex in the brain suggests that formation of this complex is involved in the synaptic delivery of NMDA receptors. Receptors without the PDZ interacting domain bypass the exocyst interaction and reach the cell surface. This raises the possibility that there are two mechanisms for delivery of NMDA receptors to the cell surface, one which involves the PDZ interacting domain and one that does not.

4. NMDA receptor role in excitotoxicity.

Excitotoxicity is thought to be a major mechanism in many human disease states such as central nervous system ischemia, trauma, epilepsy and chronic neurodegenerative disorders. Briefly, synaptic overactivity leads to the excessive release of glutamate that activates a number of postsynaptic cell membrane receptors, which upon activation open their associated ion channel pore to produce ion influx. To date, although molecular basis of glutamate toxicity remain uncertain, there is general agreement that NMDA subtype of ionotropic glutamate receptors plays a key role in mediating at least some aspects of glutamate neurotoxicity, possibly owing to their high calcium permeability (Sattler and Tymianski, 2001).

The original "calcium load hypothesis" suggests that neurodegeneration is simply a function of the quantity of calcium which enters the cell. This hypothesis has been best corroborated by studies demonstrating that in neurons exposed to glutamate or anoxia calcium measurements are correlated precisely with cell death. Different studies demonstrate a very strong linear correlation between calcium accumulation and cell death when neurons are exposed to differing concentrations of glutamate, or to 500 μM glutamate for various exposure duration. Furthermore, agents known to attenuate acute glutamate neurotoxicity such as the NMDA antagonist D-APV reduced neuronal calcium accumulation in a manner proportional to their protective effect. In paradigms using glutamate or NMDA to produce toxicity, calcium accumulation measurements appear presently to have resolved methodological difficulties in correlating calcium quantity with neurotoxicity. However, certain inconsistencies remain. For example, if a linear 1:1 relationship between calcium content and cell death is assumed, calcium influx evoked by pathways other than through NMDA receptor channels should evoke cell death exactly in proportion to the degree of calcium accumulation.

The correlation between neuronal cell death and total calcium loading as determined by calcium accumulation measurements for pharmacologically isolated calcium influx pathways revealed that distinct influx pathways, and not the total calcium loading, determine neuronal vulnerability to calcium-dependent neurotoxicity. In addition to these data, it is known that early glutamate neurotoxicity is triggered by [Ca2+]i increase in subcellular microdomains near NMDA receptors, as only fast calcium chelators are able to protect against NMDA receptors mediated neurotoxicity. Research has therefore focused for calcium-dependent neurotoxic mechanisms on calcium-activated processes that are preferentially associated with NMDA receptors. In particular, "the source-specificity hypothesis" proposed that specific targets, consisting of rate-limiting enzymes or substrates, are physically linked or colocalized with NMDA receptors.

Evidence for the existence of signaling molecules with the potential to act as neurotoxic triggers has arisen from recent knowledge of the molecular composition of the PSD fraction at the postsynaptic side of the synapse. This molecular framework was thought to enable neurons to target ion fluxes in very distinct patterns to achieve specific physiological effects such as triggering of one particular enzyme.

It has becoming increasingly evident that the PSD structure in general, and NMDA receptor complex in PSD in particular, are intimately involved not only in the regulation of synaptic plasticity events. In fact, the macromolecular protein complex of the PSD can play a role in the dynamic response to different excitotoxyc insults in the brain. Because many of the initial events (i.e. activation of glutamate receptors, Calcium influx, NOS activation) occur at/or in the vicinity of the PSD this appears to be a likely site for the pathogenic biochemical cascade that leads to neuronal cell death in numerous neurological disorders.

5. NMDA receptor mediated mechanisms in different disease states.

NMDA receptors have been implicated as a mediator of neuronal injury associated with many neurological disorders including ischemia, epilepsy, brain trauma, dementia, and neurodegenerative disorders. Neuronal injury caused by cerebral ischemia is believed to be mediated by excessive activation of glutamate receptors. In the ischemic brain, extracellular glutamate is elevated rapidly after the onset of ischemia and declines following reperfusion. In recent years, it has been shown that loss of calcium homeostasis may be an important mechanism of ischemic brain damage. Most of the studies focused on ischemia-induced changes in NMDA receptor complex in the PSD structure because of its temporal and spatial proximity to initial events, which occur after an ischemic challenge. Changes in the composition and morphology of forebrain PSD have been reported to occur after an ischemic challenge. In addition, molecular interactions involving PSD-95 are modified by an ischemic challenge, in the PSD of the most vulnerable CA1 region of the hippocampus. Ischemia also resulted in a decrease in the size of protein complexes containing PSD-95, but had only a small effect on the size distribution of complexes containing the NMDA receptor, indicating that molecular interactions involving PSD-95 and the NMDA receptor are modified by an ischemic challenge (Takagi et al., 1999).

On the other hand, PSD-95 reduced expression in hippocampal cultures is responsible for neuronal vulnerability mediated by direct activation of CaMKII transduction pathway in the postsynaptic compartment, suggesting a fine tuning of CaMKII/PSD-95 pathways in the modulation of NMDA receptor functions both in physiological and pathological conditions (Gardoni et al., 2002).

Cerebral ischemia differentially affects also the association of different tyrosine kinase with the PSD: the level of PSD-associated PYK2 and trkB increased during the ischemic episode whether FAK levels decreased. Src and Fyn levels appear increased with a short delay after reperfusion. In addition, transient cerebral ischemia increases tyrosine phosphorylation of NMDA receptor subunits NR2A and NR2B, but has no effect on the amounts of individual NMDA receptor subunits in the PSD. The ischemia-induced increase in the interaction of NR2A and NR2B with the SH2 domains of Src and Fyn suggests a possible mechanism for the recruitment of signaling proteins to the PSD and may contribute to altered signal transduction in the postischemic hippocampus.

NMDA receptors have been shown to play a critical role in hyperexcitability associated with several animal models of human epilepsy. Chronic changes in NMDA receptors are involved in human temporal lobe epilepsy and in the kindling model of temporal lobe epilepsy. In addition, acute models of spiking show sensitivity to NMDA receptor antagonists. The NMDA receptor antagonist D-2-amino-5-phosphonopentanoic acid (APV) has been reported to completely block hyperexcitability in slices from freeze-lesioned animals. In human cortical dysplasia, evidence exists for an increased expression of NR2 subunits within the region of the dysplasia, but not in surrounding or "nonspiking" cortex. The absence of NR2 immunoreactivity in nonspiking cortex compared with the abundance of NR2 immunoreactivity in actively epileptic cortex has led to the suggestion that increases in NR2 expression alone could explain hyperexcitability in human cortical dysplasia. Recent data obtained in human patients with periventricular nodular heterotopia (PNH), a brain dysgenesis, characterized by focal drug-resistant epilepsy, revealed a decreased expression of NR2A and NR2B subunits as well as alphaCaMKII, suggesting the existence of an alteration of the NMDA-receptor complex within the heterotopic and epileptogenic brain tissue. In addition, NR2A/B subunits not only were reduced but also were less efficiently phosphorylated, thus indicating a functional alteration of the NMDA-receptor complex.

6. NMDA receptor pathways as drug target.

In theory, a high number of central nervous system disease states such as ischemia, trauma, epilepsy and chronic neurodegenerative disorders in which neuronal cell death is associated to glutamate induced excitotoxicity could be treated by blocking NMDA receptors (Kemp et al, 2002). Nonselective NMDA receptor antagonists are effective in many experimental animal models of disease, and some of these compounds have moved into clinical trials. However, the initial enthusiasm for this approach has waned, because the therapeutic ratio for most NMDA antagonists is poor, with significant adverse effects at clinically effective doses, thus limiting their utility.

Subunit specific compounds, such as the NR2B selective antagonist ifenprodil, are neuroprotective in animal models without the usual side-effects seen with nonselective NMDA antagonists. However, enthusiasm for the use of new selective NR2B specific antagonisms has been recently tempered by the failure of clinical trials of traumatic brain injury. On the other hand, the improved side effect profile of NR2B-selective compounds in humans open the possibility to test them for other human CNS disorders. NMDA receptor antagonists have been indicated to have beneficial effects in different animal models of Parkinson's disease (PD). In particular, the low affinity NMDA receptor blocker Amantadine is used to treat PD in humans and it's known to improve levodopa-induced dyskinesias. Another low-affinity channel blocker, Memantine, has been indicated for the treatment of Alzheimer's disease.

Recent studies suggested treating stroke transducing neurons with peptides able to disrupt the interaction of NMDA receptors with the postsynaptic density protein PSD-95 (Aarts et al., 2002). This procedure dissociated NMDA receptors from downstream neurotoxic signaling without blocking synaptic activity or calcium influx and protected cultured neurons from excitotoxicity, reducing focal ischemic brain damage in rats, and improved their neurological function. This approach circumvents the negative consequences associated with blocking NMDA receptors and may constitute a practical stroke therapy.

Recently, an adeno-associated virus (AAV) vaccine was used to generate autoantibodies that targeted a specific brain protein, the NR1 subunit of the NMDA receptor (During et al., 2000). After peroral administration of the AAV vaccine, single-dose vaccine was associated with strong anti-epileptic and neuroprotective activity in rats for both a kainate-induced seizure model and also a middle cerebral artery occlusion stroke model at 1 to 5 months following vaccination. Thus, a vaccination strategy targeting NMDA receptor subunits is feasible and may have therapeutic potential for neurological disorders.

7. Conclusions.

The recent development in genomics and proteomics techniques in the last few years has allowed to identify in the structure of the excitatory synapse an high number of proteins that might modify glutamate neurotransmission downstream NMDA receptor activation. In fact, NMDA receptor is part of a postsynaptic signaling complex composed of MAGUK protein family and many other interacting proteins, including kinases, and other signaling and scaffolding proteins. This finding may improve the identification of specific molecular targets whose pharmacological or genetic manipulation might lead to innovative therapies for both acute and neurodegenerative disorders.

REFERENCES

Aarts M., Liu Y., Liu L., Besshoh S., Arundine M., Gurd J.W., Wang Y.T., Salter M.W., Tymianski M. Treatment of ischemic brain damage by perturbing NMDA receptor- PSD-95 protein interactions. Science. 2002, 298: 846-50.

Cheung H.H., Gurd J.W. Tyrosine phosphorylation of the N-methyl-D-aspartate receptor by exogenous and postsynaptic density-associated Src-family kinases. J Neurochem. 2001 78:524-34.

Christopherson K.S., Hillier B.J., Lim W.A., Bredt D.S. PSD-95 assembles a ternary complex with the N-methyl-D-aspartic acid receptor and a bivalent neuronal NO synthase PDZ domain. J.Biol.Chem. 1999, 274:27467-73.

During M.J., Symes C.W., Lawlor P.A., Lin J., Dunning J., Fitzsimons H.L., Poulsen D., Leone P., Xu R., Dicker B.L., Lipski J., Young D. An oral vaccine against NMDAR1 with efficacy in experimental stroke and epilepsy. Science. 2000, 287: 1453-60.

Gardoni F., Schrama L.H., Kamal A., Gispen W.H., Cattabeni F. and Di Luca M. Hippocampal synaptic plasticity involves competition between alphaCaMKII and PSD-95 for binding to the NR2A subunit of the NMDA receptor. J. Neurosci. 2001, 21: 1501-1509.

Gardoni F., Bellone C., Viviani B., Marinovich M., Meli E., Pellegrini-Giampietro D.E., Cattabeni F., Di Luca M. Lack of PSD-95 drives hippocampal neuronal cell death through activation of an alphaCaMKII transduction pathway. Eur J Neurosci. 2002, 16:777-86.

Kemp J.A., McKernan R.M. NMDA receptor pathways as drug targets. Nat Neurosci. 2002, 5:1039-42.

Kennedy M.B. Signal-processing machines at the postsynaptic density. Science 2000, 290:750-4

Sattler R., and Tymianski M. Molecular mechanisms of glutamate receptor-mediated excitotoxic neuronal cell death. Mol Neurobiol. 2001, 24:107-29.

Sheng M. and Kim M.J. Postsynaptic signaling and plasticity mechanisms. Science. 2002 Oct 25;298(5594):776-80.

Takagi N., Logan R., Teves L., Wallace M.C., Gurd J.W. Altered interaction between PSD-95 and the NMDA receptor following transient global ischemia. J Neurochem. 2000, 74:169-78.

Wenthold R.J., Prybylowski K., Standley S., Sans N., Petralia R. Trafficking of NMDA receptors. Annu Rev Pharmacol Toxicol. 2003, 43: 335-58.

Chapter 4
AMPA RECEPTOR AND SYNAPTIC PLASTICITY

Maria Passafaro[1] and Carlo Sala[2]
(1) DTI-CNR Institute of Neuroscience, Cellular and Molecular Pharmacology. (2) CNR Institute of Neuroscience, Cellular and Molecular Pharmacology, Department of Pharmacology, University of Milan, 20129,Italy.

Abstract: AMPA receptors mediate the majority of excitatory synaptic transmission in the brain. Thus, the mechanisms that control the developmental and activity-dependent changes in the functional synaptic expression of AMPA receptors are of fundamental importance. Recent advances have shown a critical role for AMPA receptor trafficking in long-term potentiation (LTP) and in long-term depression (LTD). Here we focus on the mechanisms that underlie the modulation of AMPA receptors during the synaptic plasticity.

Keywords: AMPA receptor, LTP, LTD, excitatory synapses, memory exocytosis, endocytosis.

INDEX

1. Introduction.
2. AMPA receptor structure.
3. AMPA receptor interactions.
   3.1 Interactions with PDZ proteins.
   3.2 Interactions with NSF, 4.1N, Stargazin and NARP.
4. Role of AMPA receptor phosphorylation on synaptic plasticity.
5. LTD and AMPA receptor endocytosis.
   5.1 Intracellular signalling in AMPA receptor endocytosis.
6. LTP and AMPA receptor exocytosis
7. Conclusions

1. Introduction.

α-Amino-3-hydroxy-5-methylisoxazolepropionate (AMPA) receptors are ionotropic glutamate receptors, typically concentrated at post-synaptic sites of excitatory synapses. They mediate the majority of the fast excitatory synaptic trasmission in the vertebrate central nervous system (CNS) (Hollmann and Heinemann 1994; Dingledine et al 1999) and play an important role in many different functions in the brain, such as learning and memory. Recent studies indicate that synaptic AMPA receptors content is much more heterogeneous than that of NMDA receptors. Many excitatory synapses contain NMDA receptors but not AMPA receptors (silent synapses) and the content of AMPA receptors in AMPA receptor-positive synapses is variable (Malinow et al.2000). In addition, they are not statically localized at synapses, but move in and out of the post-synaptic membrane on a rapid time-scale. Moreover, LTD and a variety of factors that induce depression, also promote rapid AMPA receptor endocytosis in cultured hippocampal neurons. LTP, on the other hand, is correlated with a recruitment of AMPA receptor to the synapse, involving regulated exocytosis. This chapter aims to describe the recent progress made in the study of AMPA receptor regulation during synaptic plasticity.

2. AMPA receptor structure.

AMPA receptors (AMPARs) are composed of heteromeric combinations of GluR1 to GluR4 (also known as GluRA-D) (Hollman and Heinemann 1994; Wisden and Seeburg 1993), whose membrane topology is similar to that of NMDA receptor subunits. Each subunit occurs in one of two molecular forms, termed Flip and Flop, with respect to a 35 amino acid segment preceding M3 (third transmembrane) (Sommer et al. 1990). The Flip version of all subunits are expressed prominently before birth and remain largely invariant in their expression during postnatal development and in the adult. The expression of Flop variants increases throughout the developing brain, and reaches adult levels by postnatal day 14 (Monyer et al. 1991). These splice forms affect the desensitization kinetics of AMPARs (Sommer et al. 1990). Each receptor complex is thought to contain four subunits, which combine in different stoichiometries to form distinct receptor subtypes (Dingledine et al. 1999; Rosenmund et al. 1998). In the adult hippocampus, the majority of AMPARs appear to consist of GluR1-GluR2 and GluR2-GluR3 (Wenthold et al. 1996). The immature hippocampus, as well as other mature brain regions, express GluR4, which also complexes with GluR2 to form a receptor (Zhu et al. 2000). The typical GluR subunit consists of an amino-terminal domain (NTD or ATD), a ligand-binding domain (S1 and S2), three membrane-spanning domains (M1, M2 and M3), one cytoplasmic re-entrant loop (P), and a carboxy-terminal intracellular region (Figure 1).

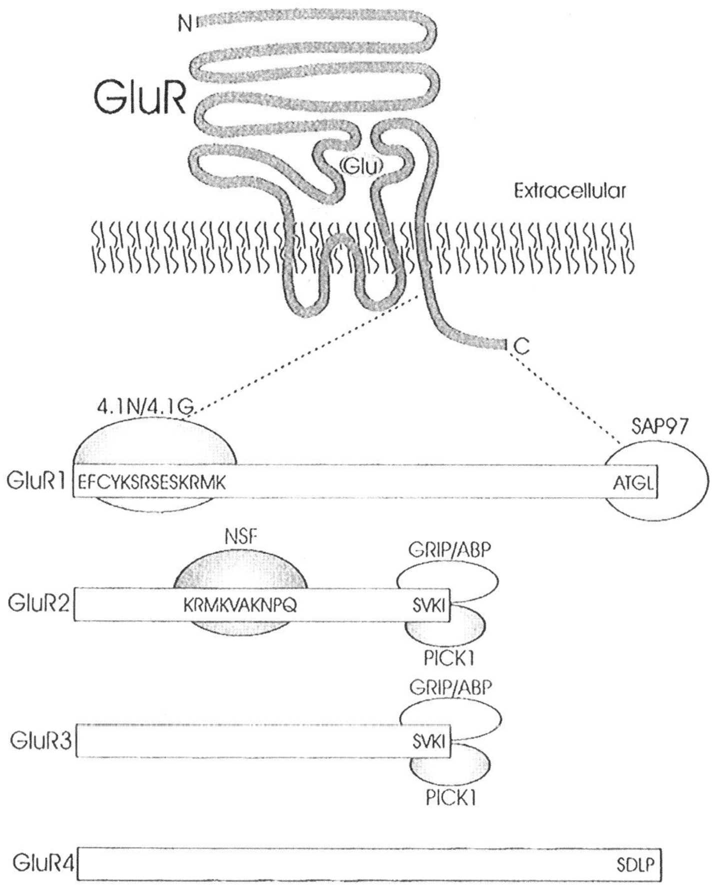

Fig.1 Schematic representation of AMPA receptor subunit structure and the proteins that interact with the cytoplasmic tails of GluR1, GluR2 and GluR3. S1 and S2 domains constitute the glutamate-binding site. NTD, N-terminal domain.

The NTD shows sequence homology to the leucine, isoleucine, valine-binding protein LIVBP, one of the bacterial periplasmic binding proteins (PBPs) (O'Hara et al. 1993). The only function ascribed to the

NTD of AMPA receptors is in subunit dimerization and subtype-specific assembly (i.e. preventing heteromerization between AMPA receptor and kainate receptor subunits) (Stern-Bach et al. 2001). The S1/S2 ligand-binding domain is formed by two sequence and structural homology domains with the glutamine-binding protein QBP (Armstrong et al. 1998; Nakanishi et al.1990).

The re-entrant pore loop is homologous to the P-loop sequences of the other channels, and is located between the first and second transmembrane domains, forming the pore domain. In the channel-forming segment M2, GluR2 carries a positively charged residue (arginine) in a position (Q/R site), whereas the other AMPA receptor subunits harbor a glutamine. This position is subject to RNA editing and is the main site controlling divalent ion permeation of recombinant AMPA receptors (Dingledine et al. 1999). In addition, GluRs contain a third transmembrane domain and a cytoplasmic carboxy-terminal domain. The C-terminal intracellular domains of the subunit are distinct. GluR1, GluR4 and a minor splice variant of GluR2 (GluR2L) have long cytoplasmic tails and are homologous, whereas the predominant splice form of GluR2, GluR3, and an alternative splice form of GluR4 (GluR4c) have short C-terminal domains (Fig.1).The C-termini of these receptor subunits interact with PDZ-domain proteins of the post-synaptic density and are crucial for the regulation of AMPA receptor function.

3. AMPA receptor interactions.

3.1 Interactions with PDZ proteins.

Use of yeast two–hybrid methods to identify AMPA-receptor-interacting proteins was crucial for the progress in this field and helped to identify a large complex of such proteins (Fig.1). Most of these identified AMPA-interacting proteins have single or multiple PDZ domains, which are protein-protein interaction motifs that bind the C-tail of their target proteins (Sheng and Sala 2001). The C-terminus of GluR1 binds via a class I PDZ domain interaction, to SAP97 (synapse-associated protein 97 kDa) a member of the membrane associated guanylate kinase (MAGUK) protein family (Leonard et al. 1998) and RIL (reversion-induced LIM gene (Schulz et al. 2001). GluR2 and GluR3 share a C-terminal sequence (-SVKI) that interacts with GRIP1 (glutamate receptor-interacting protein), a protein containing seven PDZ domain (Dong et al. 1997) and with ABP (AMPA receptor binding protein) (Dong et al.1999; Srivastava et al. 1998), a protein with six PDZ domains. ABP appears to be a splice variant of a

GRIP-related protein (also called GRIP2) that contains seven PDZs; ABP lacks the N-terminus and PDZ7 of GRIP2 (Dong et al.1999). ABP and GRIP are found at the synaptic membrane and in intracellular punctate structures resembling endosomes. The function of the GluR2/3-GRIP interaction seems important for the synaptic accumulation of the receptor, presumably by anchoring it to the cytoskeleton through other PDZ domains or promoting its insertion into the synaptic membrane. In addition, GRIP binds to Liprins, proteins that bind to the LAR family of receptor tyrosine phoshatase (Serra-Pages et al. 1998), also bind to GRIP1 (Wyszynski et al. 2001). Recently, has reported that GRIP binds to KIF5 (a microtubule-based motor protein that is important for vescicular transport along axons and dendrites), selectively directing it to dendrites (Setou et al. 2002). The KIF5-GRIP1 complex interacts with GluR2 and appears to be involved in the transport of GluR2 to dendrites. GluR3 and GluR4c also bind PICK1 (protein interacting with C-kinase)(Dev et al. 1999; Xia et al. 1999) and rDLG6 (Inagaki et al. 1999). For GluR4 and GluR2L no binding partners have yet been reported. These PDZ-based interactions have been implicated in the synaptic targeting and/or surface stability of AMPA receptors.

3.2 Interactions with NSF, 4.1N, Stargazin and NARP.

GluR1 subunit also forms a complex with 4.1 N, a neuronal specific form of the red blood cell actin cytoskeleton-associated protein, 4.1R. Interaction of GluR1 with 4.1N appears to stabilize AMPA receptors at the cell surface. However the C-terminus of GluR2 also binds to NSF (NEM-sensitive-factor) (Nishimune et al. 1998; Osten et al.1998; Song et al. 1998), an ATPase known to play an important role in membrane fusion processes. NSF binding is mediated by a membrane proximal segment of GluR2's cytoplasmic tail, distinct from the C-terminus that binds to GRIP or PICK1. Surface expression of AMPA receptors is inhibited by peptides that block the GluR-NSF interaction, suggesting that NSF is involved in the insertion of AMPA receptors into the post-synaptic membrane (Noel et al. 1999). Inhibition of NSF activity prevents LTP (Lledo et al. 1998), and the abundance of NSF in the PSD (post-synaptic density) appears to be dynamically regulated. It is possible that the NSF-GluR2 interaction is relevant to synaptic plasticity by regulating the vesicle trafficking or protein unfolding of AMPA receptors. Interestingly, the product of Stargazin, a gene that is mutated in the ataxic and epileptic mutant mouse stargazer, has been reporteted to interact with all four AMPA-receptor subunits (Chen et al. 2000). This discovery resulted from the observation that the stargazer mice have defective AMPA-receptor responses in cerebellar granule cells, apparently because of a lack of proper synaptic targeting of the receptors. The structural domains of the stargazin protein and the AMPA-receptor subunits that are important for this interaction are not clear, but in addition to binding to the AMPA-receptor subunits, Stargazin also has a sequence in its C-terminus that binds PDZ domain-containing proteins, such as the postynaptic-density protein PSD95 (Fig.1). The interaction between stargazin and AMPA-receptor subunits appears to be important for the surface delivery of AMPA receptors, whereas the interaction between stargazin and PDZ-containing proteins is important for the synaptic targeting of AMPA receptors.

Another protein that interacts with all four AMPA-receptor subunits is the neuronal immediate-early gene NARP (neuronal-activity-regulated pentraxin). NARP is a secreted neuronal protein that is localized at synapses and induces AMPA receptor clustering in heterologous cells and in neurons. In addition, overxpression of the gene encoding NARP in neurons increases the number of synaptic AMPA-receptor clusters (Malinow et al. 2002). These results suggest that NARP could be important to stabilize synaptic AMPA-receptor clusters at excitatory synapses.

4. Role of AMPA receptor phosphorylation on synaptic plasticity.

The role of protein phosphorylation in the regulation of AMPA-receptor function was initially studied by manipulating the activity of protein kinases in neurons and analyzing the resulting changes in AMPA-receptor function or in excitatory synaptic transmission. Application of peptides that inhibit PKA (cAMP-dependent protein kinase) was shown to inhibit AMPA-receptor-mediated currents in neurons, and intracellular perfusion of constitutively active CaMKII (Ca2+-calmodulin-dependent protein kinaseII) or PKM (a constitutively active form of protein kinase C) was shown to increase AMPA-receptor mediated synaptic responses. There are ten identified phosphorylation sites on the intracellular carboxy-terminal domains of GluR1-GluR4 (Carvalho et al 1999; Chung et al 2000). In particular, the functional role of serine 831 and serine 845, two sites localized in C-terminal of GluR1 subunit (Roche et al.1996) has been extensively characterized. Ser845 is phosphorylated by PKA, whereas PKC and CaMKII phosphorylate Ser831 (Roche et al. 1996). In addition, mutant mice with double phosphomutation (Ser831 and Ser845) essentially lack NMDA receptor-dependent LTD and show reduced LTP in the CA1 region of the hippocampus (Lee et al. 2003). Furthermore, Ser863 and Ser880 on the C-terminal of the GluR2 subunit were shown to be phosphorylated by

PKC.The GluR4 subunit is also phosphorylated in its C-terminal domain at two sites: Ser830 and Ser842 (Carvalho et al. 1999). Ser830 is phosphorylated by PKC and Ser842 is phosphorylated by PKA (Carvalho et al. 1999).

5. LTD and AMPA receptor endocytosis

Several lines of evidence indicate that the AMPA receptor internalization occurs through clathrin-coated pits and requires the activity of dynamin (Beattie et al. 2000; Ehlers et al. 2000; Lin et al. 2000). This was established by experiments showing that the regulated internalization of AMPARs was associated with an increased colocalization and interaction with the clathrin adaptor AP2 (Carroll et al. 2001). Furthermore, regulated AMPAR internalization was completely blocked when clathrin-mediated endocytosis was inhibited, either by a high concentration of sucrose, expression of a dominant-negative form of dynamin, or the disruption of the dynamin-amphiphysin complex that is essential to this form of internalization. The ability of AMPARs to undergo rapid endocytosis, in response to glutamate receptor activation, is consistent with a contribution of this mechanism to the reduction in synaptic strength observed with LTD.

Now there is direct evidence that LTD is associated with dephosphorylation of the GluR1 subunit of the AMPAR, the consequence of which is known to be depression of glutamate-evoked currents. Several lines of evidence suggest that AMPAR internalization is an expression mechanism for LTD. First, in hippocampal cell cultures, field stimulation at 5 Hz causes an NMDAR-dependent depression of spontaneous miniature EPSC amplitudes and causes a removal of surface AMPARs from the post-synaptic membrane while having no effect on the distribuition of synaptic NMDARs (Carroll et al. 1999). In the hippocampus in vivo, LTD causes an NMDAR-dependent depression and the loss in the number of AMPAs in synaptoneurosomes, providing further evidence for the role of AMPA endocytosis. The involvement of AMPAR endocytosis in several forms of LTD suggests that similar intracellular signalling pathways should mediate both AMPAR transport and plasticity.

5.1 Intracellular signaling in AMPA receptor endocytosis.

The hypothesis for the triggering of NMDAR-dependent LTD proposes that LTD requires an NMDAR-dependent rise in post-synaptic calcium, which preferentially activates a protein-phosphatase cascade that includes calcineurin (PP2B) a Ca^{2+}/calmodulin-dependent phosphatase1(PP1). However, the mechanism by which calcineurin affects AMPAR endocytosis is unknown. One attractive hypothesis, based on work on the mechanisms mediating the activity-dependent increase in the endocytosis of presynaptic vesicles, is that calcineurin facilitates endocytosis via its association with dynamin-amphiphysin and the consequent dephosphorylation of components of the endocytic machinery (Beattie et al. 2000). The role of PP1 in AMPAR endocytosis is less clear. Pharmacological inhibition of PP1 has been reported to block the endocytosis of AMPARs triggered by NMDA application (Elhers 2000), but also has been found to increase the AMPAR endocytosis (Lin et al. 2000; Beattie et al. 2000). Further complexity to the signalling pathways triggering AMPAR endocytosis have been provided by studies of LTD in cerebellar purkinje cells. In Purkinje cells, LTD requires activation of PKC (Linden et al. 1994), which seems to drive the internalization of AMPARs in these cells (Matsuda et al. 1999; Xia et al. 2000). Considering all these data together, it appears that the regulation of AMPAR endocytosis may be cell-type specific. In addition to the regulated endocytosis of AMPARs that are important in LTD, AMPARs also undergo two other forms of internalization: a constitutive endocytosis that contributes to the basal cycling of AMPARs (Luscher et al. 1999; Elhers 2000) and an activity-dependent endocytosis that does not require calcium influx or calcineurin (Elhers et al.2000; Lin et al. 2000).The mechanisms of the above AMPAR internalization mechanisms are distinguishable from regulated endocytosis.

6. LTP and AMPA receptor exocytosis.

Several studies over the past fews years have tested the notion that silent synapses lack AMPA receptors and that AMPARs, are rapidly inserted in to synapses during LTP. Recent works have revealed two general mechanisms of synaptic insertion, both of which depend strongly on subunit composition. The first is de novo insertion, manifest by GluR1-containing receptors and dependent upon tetanic stimulation. The second consists of continuous, constitutive receptor cycling in and out of the synapse, and is manifest by GluR2 homomers and GluR2/3 heteromers. GluR2/3 receptors appear to serve as a highly mobile species that can enter and then leave sites initially occupied by GluR1-containing receptors, independently of activity-dependent signals.

One recent elegant work tagged AMPAR subunits such that the channel rectified at positive potentials, hence, AMPAR subunit synaptic expression could be monitored (Shi et al. 2001).When the recipient synapse was silent and lacked pre-existing AMPARs, GluR1 homomers were inserted into the synapse, in a manner dependent upon constitutively active Ca^{2+}/calmodulin-dependent protein kinase II (CaMKII). On the other hand, GluR2 homomers incorporates into the synapse independently of activity, but causes no change in synaptic strength (Shi et al. 1999), suggesting that GluR2 subunits exchange constitutively with existing synaptic AMPA receptors. The synaptic expression of homomeric GluR2 receptors depends on its C-terminal PDZ binding motif (Shi et al. 2001), suggesting involvement of a PDZ (such as GRIP or PICK-1). It is not clear, however, whether the PDZ protein is required for the delivery of GluR2 to synapses, or for its retention in synapses after insertion. The GluR1 appears to govern the trafficking behavior of heteromeric GluR1/GluR2 receptors, preventing constitutive exchange and conferring inducible delivery of the heteromer (Shi et al. 2001). Similar findings were made through the use of a thrombin cleavage assay to visualize receptor exocytosis (Passafaro et al. 2001). This study found that GluR2 insertion is constitutively rapid, whereas GluR1 insertion is slow but inducible; the GluR1 phenotype is dominant in heteromers. Activation of NMDARs enhanced the rate of exocytosis of GluR1 homomers but not GluR2 homomers. GluR2 accumulated more in synapses, whereas GluR1 accumulated in nonsynaptic locations.

Thus, both the subunit composition of the incoming receptor and the prior synaptic AMPA content appear to govern the requirements for synaptic receptor insertion (Figure 2). However the molecular mechanisms responsible for the delivery of AMPARs to the synaptic membrane are not clear.

7. Conclusions.

Changes in synaptic strength that occur following LTP or LTD induction can be related to changes in the exocytosis and endocytosis of AMPARs. The assignment of specific functions to transport and anchorage factors, and their regulation by phosphorylation, will be crucial to understanding the underlying mechanisms of synaptic plasticity.

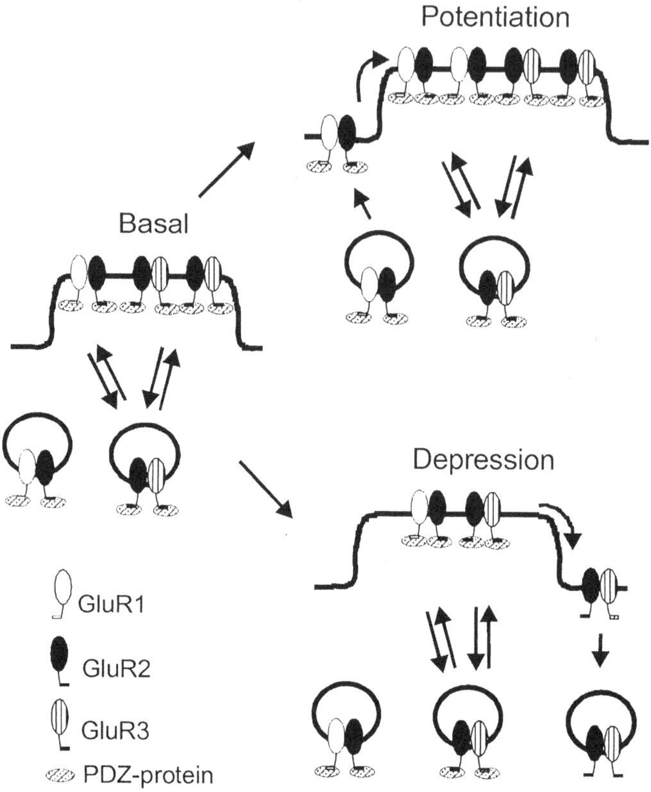

Figure 2. A model for post-synaptic AMPA receptor trafficking. AMPA receptors are represented as heteromers of GluR1/GluR2 or GluR2/GluR3. GluR1 controls the regulated exocytosis of AMPA receptors (acting dominantly in GluR1/GluR2 heteromers). GluR2 mediates the constitutive recycling and regulated endocytosis of AMPA receptors

REFERENCES

Ayalon G, Stern-Bach Y. Functional assembly of AMPA and kainate receptors is mediated by several discrete protein-protein interactions. Neuron. 2001 Jul 19;31(1):103-13.

Beattie EC, Carroll RC, Yu X, Morishita W, Yasuda H, von Zastrow M, Malenka RC. Regulation of AMPA receptor endocytosis by a signaling mechanism shared with LTD. Nat Neurosci. 2000 Dec;3(12):1291-300.

Carroll RC, Beattie EC, von Zastrow M, Malenka RC. Role of AMPA receptor endocytosis in synaptic plasticity. Nat Rev Neurosci. 2001 May;2(5):315-24.

Carvalho AL, Kameyama K, Huganir RL. Characterization of phosphorylation sites on the glutamate receptor 4 subunit of the AMPA receptors. J Neurosci. 1999 Jun 15;19(12):4748-54.

Chen L, Chetkovich DM, Petralia RS, Sweeney NT, Kawasaki Y, Wenthold RJ, Bredt DS, Nicoll RA. Stargazin regulates synaptic targeting of AMPA receptors by two distinct mechanisms. Nature. 2000 Dec 21-28;408(6815):936-43.

Chung HJ, Xia J, Scannevin RH, Zhang X, Huganir RL.Phosphorylation of the AMPA receptor subunit GluR2 differentially regulates its interaction with PDZ domain-containing proteins. J Neurosci. 2000 Oct 1;20(19):7258-67.

Dev KK, Nishimune A, Henley JM, Nakanishi S.The protein kinase C alpha binding protein PICK1 interacts with short but not long form alternative splice variants of AMPA receptor subunits. Neuropharmacology. 1999 May;38(5):635-44.

Dingledine R, Borges K, Bowie D, Traynelis SF. The glutamate receptor ion channels. Pharmacol Rev. 1999 Mar;51(1):7-61.

Dong H, O'Brien RJ, Fung ET, Lanahan AA, Worley PF, Huganir RL. GRIP: a synaptic PDZ domain-containing protein that interacts with AMPA receptors Nature. 1997 Mar 20;386(6622):279-84.

Dong H, Zhang P, Song I, Petralia RS, Liao D, Huganir RL. Characterization of the glutamate receptor-interacting proteins GRIP1 and GRIP2. J Neurosci. 1999 Aug 15;19(16):6930-41.

Ehlers MD. Reinsertion or degradation of AMPA receptors determined by activity-dependent endocytic sorting. Neuron. 2000 Nov;28(2):511-25.

Hollman,M and Heinemann,S. Cloned glutamate receptors. Ann.Rev.Neurosci. (1994).17,31-108.

Inagaki H, Maeda S, Lin KH, Shimizu N, Saito T. rDLG6: a novel homolog of Drosophila DLG expressed in rat brain. Biochem Biophys Res Commun. 1999 Nov 19;265(2):462-8.

J Biol Chem. 1998 Jun 19;273(25):15611-20.

Lee HK, Takamiya K, Han JS, Man H, Kim CH, Rumbaugh G, Yu S, Ding L, He C, Petralia RS, Wenthold RJ, Gallagher M, Huganir RL. Phosphorylation of the AMPA receptor GluR1 subunit is required for synaptic plasticity and retention of spatial memory. Cell. 2003 Mar 7;112(5):631-43.

Leonard AS, Davare MA, Horne MC, Garner CC, Hell JW. SAP97 is associated with the alpha-amino-3-hydroxy-5-methylisoxazole-4-propionic acid receptor GluR1 subunit. J Biol Chem. 1998 Jul 31;273(31):19518-24.

Lin JW, Ju W, Foster K, Lee SH, Ahmadian G, Wyszynski M, Wang YT, Sheng M. Distinct molecular mechanisms and divergent endocytotic pathways of AMPA receptor internalization. Nat Neurosci. 2000 Dec;3(12):1282-90.

Linden DJ.Long-term synaptic depression in the mammalian brain. Neuron. 1994 Mar;12(3):457-72.

Lledo PM, Zhang X, Sudhof TC, Malenka RC, Nicoll RA. Postsynaptic membrane fusion and long-term potentiation. Science. 1998 Jan 16;279(5349):399-403.

Luscher C, Xia H, Beattie EC, Carroll RC, von Zastrow M, Malenka RC, Nicoll RA. Role of AMPA receptor cycling in synaptic transmission and plasticity. Neuron. 1999 Nov;24(3):649-58.

Malinow R, Mainen ZF, Hayashi Y. LTP mechanisms: from silence to four-lane traffic. Curr Opin Neurobiol. 2000 Jun;10(3):352-7.

Malinow R, Malenka RC. AMPA receptor trafficking and synaptic plasticity. Annu Rev Neurosci. 2002;25:103-26. Epub 2002 Mar 04.

Monyer H, Seeburg PH, Wisden W. Glutamate-operated channels: developmentally early and mature forms arise by alternative splicing. Neuron. 1991 May;6(5):799-810.

Nishimune A, Isaac JT, Molnar E, Noel J, Nash SR, Tagaya M, Collingridge GL, Nakanishi S, Henley JM NSF binding to GluR2 regulates synaptic transmission. Neuron. 1998 Jul;21(1):87-97.

Novel interaction of the GluR-A AMPA receptor subunit with the PDZ-LIM domain protein RIL. 2001 Presented at 27th Soc. Neurosci.Mtg., San Diego.

O'Hara PJ, Sheppard PO, Thogersen H, Venezia D, Haldeman BA, McGrane V, Houamed KM, Thomsen C, Gilbert TL, Mulvihill ER. The ligand-binding domain in metabotropic glutamate receptors is related to bacterial periplasmic binding proteins. Neuron. 1993 Jul;11(1):41-52.

Osten P, Srivastava S, Inman GJ, Vilim FS, Khatri L, Lee LM, States BA, Einheber S, Milner TA, Hanson PI, Ziff EB. The AMPA receptor GluR2 C terminus can mediate a reversible, ATP-dependent interaction with NSF and alpha- and beta-SNAPs. Neuron. 1998 Jul;21(1):99-110.

Passafaro M, Piech V, Sheng M.Subunit-specific temporal and spatial patterns of AMPA receptor exocytosis in hippocampal neurons. Nat Neurosci. 2001 Sep;4(9):917-26.

Roche KW, O'Brien RJ, Mammen AL, Bernhardt J, Huganir RL. Characterization of multiple phosphorylation sites on the AMPA receptor GluR1 subunit. Neuron. 1996 Jun;16(6):1179-88.

Rosenmund C, Stern-Bach Y, Stevens CF. The tetrameric structure of a glutamate receptor channel. Science. 1998 Jun 5;280(5369):1596-9.

Schulz W, Nakagawa T, Kim J-H, Sheng M, Seeburg PH, Osten P.

Serra-Pages C, Medley QG, Tang M, Hart A, Streuli M. Liprins, a family of LAR transmembrane protein-tyrosine phosphatase-interacting proteins

Setou M, Seog DH, Tanaka Y, Kanai Y, Takei Y, Kawagishi M, Hirokawa N. Glutamate-receptor-interacting protein GRIP1 directly steers kinesin to dendrites. Nature. 2002 May 2;417(6884):83-7.

Sheng M, Sala C. PDZ domains and the organization of supramolecular complexes Annu Rev Neurosci. 2001;24:1-29.

Shi S, Hayashi Y, Esteban JA, Malinow R. Subunit-specific rules governing AMPA receptor trafficking to synapses in hippocampal pyramidal neurons. Cell. 2001 May 4;105(3):331-43.

Sommer B, Keinanen K, Verdoorn TA, Wisden W, Burnashev N, Herb A, Kohler M, Takagi T, Sakmann B, Seeburg PH. Flip and flop: a cell-specific functional switch in glutamate-operated channels of the CNS. Science. 1990 Sep 28;249(4976):1580-5.

Song I, Kamboj S, Xia J, Dong H, Liao D, Huganir RL. Interaction of the N-ethylmaleimide-sensitive factor with AMPA receptors. Neuron. 1998 Aug;21(2):393-400.

Srivastava S, Osten P, Vilim FS, Khatri L, Inman G, States B, Daly C, DeSouza S, Abagyan R, Valtschanoff JG, Weinberg RJ, Ziff EB. Novel anchorage of GluR2/3 to the postsynaptic density by the AMPA receptor-binding protein ABP. Neuron. 1998 Sep;21(3):581-91.

Wenthold RJ, Petralia RS, Blahos J II, Niedzielski AS. Evidence for multiple AMPA receptor complexes in hippocampal CA1/CA2 neurons. J Neurosci. 1996 Mar 15;16(6):1982-9.

Wisden W, Seeburg PH. Mammalian ionotropic glutamate receptors. Curr Opin Neurobiol. 1993 Jun;3(3):291-8. Review.

Wyszynski M, Kim E, Dunah AW, Passafaro M, Valtschanoff JG, Serra-Pages C, Streuli M, Weinberg RJ, Sheng M. Interaction between GRIP and liprin-alpha/SYD2 is required for AMPA receptor targeting. Neuron. 2002 Mar 28;34(1):39-52.

Xia J, Zhang X, Staudinger J, Huganir RL. Clustering of AMPA receptors by the synaptic PDZ domain-containing protein PICK1 Neuron. 1999 Jan;22(1):179-87

Zhu JJ, Esteban JA, Hayashi Y, Malinow R. Postnatal synaptic potentiation: delivery of GluR4-containing AMPA receptors by spontaneous activity. Nat Neurosci. 2000 Nov;3(11):1098-106.

Chapter 5
METABOTROPIC GLUTAMATE RECEPTORS AND NEURODEGENERATION

Valeria Bruno[1,2], Giuseppe Battaglia[1] and Ferdinando Nicoletti[1,2]
[1]I.N.M. Neuromed, Pozzilli, Italy; [2]Department of Human Physiology and Pharmacology, University of Rome "La Sapienza", Rome, Italy.

Abstract: Metabotropic glutamate (mGlu) receptors, which exert a modulatory effect on excitatory synaptic transmission, are considered as potential targets for neuroprotective drugs and the advent of potent and centrally available subtype-selective ligands has lead to an extensive investigation of the role of individual mGlu receptor subtypes in neurodegeneration. Pharmacological blockade of mGlu1 or -5 receptors or pharmacological activation of mGlu2/3 or mGlu4/7/8 receptors produces neuroprotection in a variety of *in vitro* or *in vivo* models. MGlu1 receptor antagonists are promising drugs for the treatment of brain ischemia or for the prophylaxis of neuronal damage induced by synaptic hyperactivity. MGlu5 receptor antagonists may limit neuronal damage induced by a hyperactivity of NMDA receptors, because mGlu5 and NMDA receptors are physically and functionally connected in neuronal membranes, suggesting a potential application of mGlu5 receptor antagonists in chronic neurodegenerative disorders, such as Amyotrophic Lateral Sclerosis (ALS) and Alzheimer's disease (AD). MGlu2/3 receptor agonists inhibit glutamate release, and also promote the synthesis and release of neurotrophic factors in astrocytes. These drugs may therefore have a broad application as neuroprotective agents in a variety of CNS disorders. Finally, mGlu4/7/8 receptor agonists potently inhibit glutamate release and have a potential application in seizure disorders.

Keywords: mGlu receptors; neuroprotection; subtype-selective ligands.

INDEX

1. Introduction.
2. Role of group-I mGlu receptors in neuroprotection.
   2.1 mGlu1 receptor antagonists are neuroprotective
   2.2 mGlu5 receptor antagonists are neuroprotective
3. Role of group-II mGlu receptors in neuroprotection

3.1 mGlu2/3 receptor agonists are neuroprotective
4. Role of group-III mGlu receptors in neuroprotection
4.1 mGlu4/7/8 receptor agonists are neuroprotective

1. Introduction.

The neurotoxicity of excitatory amino acids is well established in the central nervous systems (CNS) (Lipton and Rosenberg, 1994). Glutamate, the main excitatory neurotransmitter in the mammalian brain, activates ionotropic and metabotropic glutamate (iGlu and mGlu) receptors, which mediate its physiological and pathological effects. Activation of iGlu receptors results in fast excitatory synaptic transmission via cation channel gating, whereas mGlu receptors play a modulatory role by controlling membrane enzymes and second messengers. This class of receptor is recruited at the glutamatergic synapse only when high levels of extracellular glutamate are present, i.e. during hyperactivity or in pathological conditions. The G protein-coupled mGlu receptors form a family of currently eight subtypes (mGlu1 to -8), subdivided into three groups (I-III) on the basis of their sequence homology, pharmacological profiles and signal transduction mechanisms (Nakanishi, 1994; Conn and Pin, 1997). Group-I mGlu receptors (mGlu1 and mGlu5) are positively coupled to the polyphosphoinositide/Ca^{2+} cascade. Group-II mGlu receptors (mGlu2 and mGlu3) and group-III mGlu receptors (mGlu4, mGlu6, mGlu7 and mGlu8) are both negatively coupled to adenylate cyclase in heterologous expression systems.

The three groups of mGlu receptors can be discriminated pharmacologically with the use of selective agonists and antagonists. 3,5-Dihydroxyphenylglycine (3,5-DHPG) selectively activates group-I mGlu receptors, but cannot discriminate between mGlu1 and mGlu5, whereas 2-chlorohydroxyphenylglycine (CHPG) selectively activates mGlu5 receptors, but is less potent than DHPG. 2-Methyl-4-carboxyphenylglycine (LY367385) and (R,S)-1-aminoindan-1,5-dicarboxylic acid (AIDA) behave as competitive mGlu1 antagonists, whereas 7-(hydroxyimino)cyclopropa[b]chromen-1a-carboxylate ethyl ester (CPCCOEt) and [(3aS,6aS)-6a-naphtalen-2-ylmethyl-5-methyliden-hexahydro-cyclopental[c]furan-1-on] (BAY36-7620) are non-competitive antagonists of mGlu1 receptors (Annoura et al., 1996; Litschig et al., 1999; Schoepp et al., 1999; Carroll et al., 2001). Interestingly, BAY36-7620 can cross the blood-brain barrier, and is therefore suitable for *in vivo* studies (Carroll et al., 2001). A series of phenylpyridine derivatives, including 2-methyl-6-(phenylethynyl)pyridine (MPEP), (E)-2-methyl-6-stryrylpyridine

(SIB-1893) and 6-methyl-2-(phenylazo)pyridin-3-ol (SIB-1757), behave as highly potent and selective non-competitive mGlu5 receptor antagonists, and at least MPEP is proven to be centrally available (Gasparini et al., 1999a; Spooren et al., 2001). Group-II mGlu receptors are selectively activated by 2*R*,4*R*-4-aminopyrrolidine-2,4-dicarboxylic acid (2R,4R-APDC), (1S,2S,5R,6S)-(+)-2-aminobicyclo[3.1.0]hexane-2,6-dicarboxylic acid (LY354740), (-)-2-oxa-4-aminocyclo[3.1.0]hexane-4,6-dicarboxylic acid (LY379268), the latter two being highly potent and systemically available. (2S,1'S,2'S)-2-(9-Xanthylmethyl)-2-(2'-carboxycyclopropyl)glycine (LY341495) shows a high potency in antagonizing mGlu2/3 receptors, but can recruit all other receptor subtypes at micromolar concentrations (Schoepp et al., 1999). Finally, all group-III mGlu receptors are activated by phosphonoamino acids, such as L-2-amino-4-phosphonobutanoate (L-AP4), L-serine-O-phosphate (L-SOP) and (+)-4-phosphonophenylglycine (PPG) (Schoepp et al., 1999; Gasparini et al., 2000). Subtype-selective antagonists of group-III mGlu receptors are lacking, although (RS)-α-methyl-4-phosphonophenylglycine (MPPG) and (RS)-α-methylserine-O-phosphate (MSOP) are generally used to inhibit this class of receptors (Jane et al., 1994; Thomas et al., 1996).

The individual mGlu receptor subtypes show a wide but distinct regional distribution throughout the mammalian nervous system. Subtypes of all three groups are highly expressed in neocortical layers, hippocampus, basal ganglia, thalamus/hypothalamus, cerebellum and spinal cord. Only mGlu6 receptors appear to be exclusively restricted to retinal ON bipolar cells, where they couple to a cGMP-phosphodiesterase and amplify visual transmission (Nakanishi, 1994). Group-I mGlu receptors are found postsynaptically with the highest receptor density at perisynaptic locations outside the postsynaptic membrane specializations (Baude et al., 1993; Shigemoto et al., 1997). MGlu2/3 receptors are located presynaptically in the preterminal region of the axon and mGlu3 are also expressed in glial cells (Lujan et al., 1997; Ohishi et al., 1993). Group-III mGlu receptors are localized presynaptically near the active zones of neurotransmitter release and are thought to mediate presynaptic depression of glutamatergic synaptic potentials, most likely via inhibition of voltage-gated Ca^{2+} channels and regulation of glutamate release (Shigemoto et al., 1997; Trombley and Westbrook, 1992).

Activation or inhibition of distinct groups or subtypes of metabotropic glutamate receptors results in neuroprotective effects. Synaptic localization and signal transduction of the individual receptor subtypes will be considered to explain possible mechanisms of neuroprotection and the potential use of mGlu receptor-selective compounds in the treatment of acute and chronic degenerative disorders of the CNS.

2. Role of group-I mGlu receptors in neuroprotection.

The role of mGlu1 and mGlu5 receptors in neuroprotection has been debated for several years because of the contrasting effects of agonists in different *in vitro* models of neuronal toxicity (Nicoletti et al., 1999). The evidence that subtype-selective mGlu1 or mGlu5 receptor antagonists are consistently neuroprotective, suggests that endogenous activation of group-I receptors is permissive to cell death, knowing that activation of both subtypes induces intracellular Ca^{2+} mobilization and regulates Ca^{2+} and K^{+} channels, thus enhancing neuronal excitability. Particular attention deserves the description of a positive modulation of NMDA-gated ion currents by mGlu5 receptors, which provides the most likely substrate for the permissive action of mGlu5 receptor agonists on neuronal toxicity (Pisani et al., 1997; Awad et al., 2000; Salt and Binns, 2000). Interestingly, NMDA receptor activation can also potentiate mGlu5 receptor responses by inducing a calcineurin-dependent dephosphorylation of a PKC phosphorylation site that participates in mGlu5 desensitization (Alagarsamy et al., 1999; De Blasi et al., 2001) The SHANK family of postsynaptic density proteins may cross-link Homer and PSD-95, thus allowing the functional coupling between mGlu5 and NMDA receptors in the same region of the dendritic spine (Tu et al., 1999). A question still debated is whether or not group-I mGlu receptor subtypes are presynaptically localized and modulate neurotransmitter release. This issue is fundamental because the extent of excitotoxic degeneration is strictly related to glutamate and GABA release in a number of CNS disorders, such as ischemia-induced neurodegeneration and temporal lobe epilepsy. Interestingly it has been reported that in the CA1 region of the hippocampus, mGlu1 receptors are preferentially localized on GABAergic interneurons (Baude et al., 1993), and their activation depresses inhibitory synaptic transmission (Gereau IV and Conn, 1995; Morishita et al., 1998). Activation of mGlu1 receptors can also depress transmission at GABAergic autapses (Bushell et al., 1999). Moreover, the evidence that mGlu5 receptor is found in the axon when co-transfected with Homer-1a raises the intriguing possibility that homer proteins regulate the dendritic or axonal targeting of mGlu5 receptors (Ango et al., 2000).

2.1 mGlu1 receptor antagonists are neuroprotective.

Endogenous activation of mGlu1 receptors contributes to the induction and/or progression of neuronal death. Both competitive and non-competitive mGlu1 receptor antagonists (such as AIDA, LY367385 and CPCCOEt) reduce excitotoxic neuronal death in mouse cortical cultures challenged with NMDA (Strasser et al., 1998; Bruno et al., 1999; Battaglia et al., 2001) and are also effective in protecting striatal neurons against NMDA or quinolinic acid toxicity (Bruno et al., 1999; Orlando et al., 2001; Battaglia et al., 2001). Neuroprotection by LY367385, AIDA and CBPG is also observed in models that are closer to human pathology, such as oxygen-glucose deprivation in mouse cortical cultures or in rat organotypic hippocampal cultures, and the 4-vessel occlusion model in gerbils (Pellegrini-Giampietro et al., 1999a, b; Bruno et al., 1999). As mGlu1 receptors are expressed in GABAergic neurons, we have hypothesized that mGlu1 receptor antagonists are neuroprotective by enhancing GABA release (Battaglia et al., 2001). This hypothesis has been supported by the observations that neuroprotection by LY367385 or CPCCOEt against NMDA toxicity in culture is obliterated by GABA or SKF89976A (an inhibitor of GABA transporter) and prevented by bicuculline and 2-hydroxysaclophen, which antagonize $GABA_A$ and $GABA_B$ receptors, respectively; the same cocktail of GABA receptor blockers also prevents the protective activity of CPCCOEt against excitotoxic death of striatal neurons in *in vivo* experiments. Moreover, toxic concentrations of NMDA can stimulate GABA release in the caudate nucleus of freely moving rats only if LY367385 or CPCCOEt are present in the perfusate and in cortico-striatal slices, and DHPG depresses GABAergic IPSCs through the activation of mGlu1 receptors (Battaglia et al., 2001). This contrasts with the protective action of mGlu5 receptor antagonists, which does not involve a GABAergic component (Battaglia et al., 2001). Based on these findings, one expects that mGlu1 receptor antagonists are particularly helpful in the treatment of CNS disorders that are primarily caused by an impairment of inhibitory synaptic transmission, such as temporal lobe epilepsy associated with Ammon's horn sclerosis and mossy fiber sprouting. A potential drawback in the use of mGlu1 receptor antagonists is the induction of ataxia, therefore *in vivo* studies with centrally available mGlu1 receptor antagonists are necessary to establish whether pharmacological inhibition of mGlu1 receptors impairs motor coordination in experimental animals.

2.2 mGlu5 receptor antagonists are neuroprotective.

The non-competitive mGlu5 receptor antagonists, MPEP, SIB-1757 and SIB-1893 are highly potent (substantial activity in the nanomolar range) in reducing NMDA toxicity in mixed cultures of cortical cells, and relatively low doses of MPEP (5-20 nmol) are protective against NMDA or quinolinic acid toxicity when locally injected into the rat caudate nucleus (Bruno et al., 2000a). Neuroprotection by MPEP *in vitro* and *in vivo* is not affected by GABAergic drugs (Battaglia et al., 2001) and can be ascribed to the ability of the drug to disrupt the functional coupling between mGlu5 and NMDA receptors. A growing body of evidence indicates that mGlu5 receptors positively modulate NMDA receptors (Awad et al., 2000; Ugolini et al., 1999; Salt and Binns, 2000; Doherty et al., 2000; Jia et al., 1998; Pisani et al., 1997), and that the two receptors are co-expressed in most of the neurons. A chain of anchoring proteins may physically link NMDA and mGlu5 receptors. The NR2 subunits of NMDA receptors interact with a series of "membrane-associated guanlyl kinase" (MAGUK) proteins (such as PSD-95, PSD-93 and SAP-102), which, via their PDZ domains, may assemble with the SHANK proteins. These proteins can interact with the long isoforms of Homer proteins, thus linking NMDA receptors with mGlu5 receptors (Tu et al., 1999). One mechanism by which mGlu5 receptors positively modulate NMDA receptors is mediated by the formation of DAG with ensuing activation of PKC. PKC can phosphorylate NMDA receptors, thus relieving the Mg^{2+} blockade of the NMDA-gated ion channel (Chuang et al., 2000). This modulation mediated by PKC has some structural requirements. In recombinant cells, only NMDA receptors containing the NR2A or -2B subunits are positively modulated by PKC, whereas receptors containing the NR2C or NR2D subunits are not (Kutsuwada et al., 1992; Mori et al., 1993). The Ca^{2+} released from intracellular stores through the InsP3-sensitive channels (that also bind to Homer proteins) can induce changes in NMDA receptor activity after binding to calmodulin. Ca^{2+}-calmodulin can compete with α-actinin-2 for a common binding site on the NR1 subunit, thus affecting the anchorage of NMDA receptors to the cytoskeleton (Niethammer et al., 1996; Wyszynski et al., 1997; Dunah et al., 2000). The interaction between NMDA and mGlu5 receptors is reciprocal. Activation of NMDA enhances mGlu5 receptor responses by preventing the homologous desensitization of mGlu5 receptors (De Blasi et al., 2001). This effect is mediated by calcineurin, which is activated by Ca^{2+} and dephosphorylates mGlu5 receptors at a PKC phosphorylation site (Alagarsamy et al., 1999). Thus, a vicious cycle exists in which NMDA and mGlu5 receptors potentiate each other. This particular form of receptor crosstalk has important implications in synaptic plasticity and also in neurotoxicity. We believe that mGlu5 receptor antagonists may be beneficial under those conditions in which a toxic insult, mediated by an excitotoxic component, homogeneously impairs the survival of a specific neuronal population. This is typified by toxicological Parkinsonism induced by 1-methyl-4-phenyl-1,2,3,6-tetrahydropyridine (MPTP) or by some amphetamine derivatives (such as methamphetamine), which selectively destroy the nigro-striatal dopaminergic pathway. We have recently found that systemically administered MPEP or SIB-1893 (but not CPCCOEt and LY379268) are protective against degeneration of striatal dopaminergic terminals produced by three consecutive injections of methamphetamine in mice. MPEP can also reduce the formation of reactive oxygen species induced by methamphetamine in the striatum of freely moving mice without interfering with the uptake of methamphetamine into dopaminergic terminals (Battaglia et al., 2002). As mGlu5 receptor antagonists do not reduce methamphetamine-stimulated dopamine release, neuroprotection cannot be reconducted to changes in the amount of extravesicular dopamine available as a substrate for oxidative processes. A possible scenario is that glutamate released in response to methamphetamine (Nash and Yamamoto, 1992; Abekawa et al., 1994; Stephans and Yamamoto 1994) activates mGlu5 receptors, thereby triggering a chain of intracellular reactions that contribute to damage dopaminergic terminals. As both NMDA (Sonsalla et al., 1989; 1998) and mGlu5 receptor antagonists are protective against methamphetamine toxicity, we conclude that endogenous activation of both receptors is required for the induction of damage. Activation of mGlu5 receptors generates oscillatory increases in intracellular Ca^{2+}, as a result of inositoltrisphosphate formation and PKC activation (Kawabata et al., 1996). A sustained activity of Ca^{2+}-dependent enzymes is classically implicated in the development of neuronal damage through a variety of mechanisms, which include the formation of reactive oxygen species (Choi, 1994). This chain of events may readily contribute to the overall production of reactive oxygen species induced by methamphetamine, and, therefore, to the oxidative damage of striatal dopaminergic terminals (De Vito and Wagner, 1989). MPEP and SIB-1757 are more effective in reducing the formation of hydroxyl radicals in the striatum of freely moving mice injected with methamphetamine than in decreasing the depletion in striatal dopamine. This suggests that oxidative damage contributes to, but is not the only mechanism responsible for methamphetamine neurotoxicity. Methamphetamine may damage striatal dopaminergic terminals by increasing the formation of reactive oxygen species *via* two distinct mechanisms: autooxidation of extravesicular dopamine; and activation of mGlu5 and NMDA receptors by endogenous glutamate. We speculate that activation of Ca^{2+}-dependent enzymes with ensuing formation of radical oxygen species that follows the stimulation of NMDA and mGlu5 receptors combines with autoxidation of dopamine to reach the threshold of neuronal death. This would explain why mGlu5 receptor antagonists are protective without reducing methamphetamine-stimulated dopamine release (Figure 1).

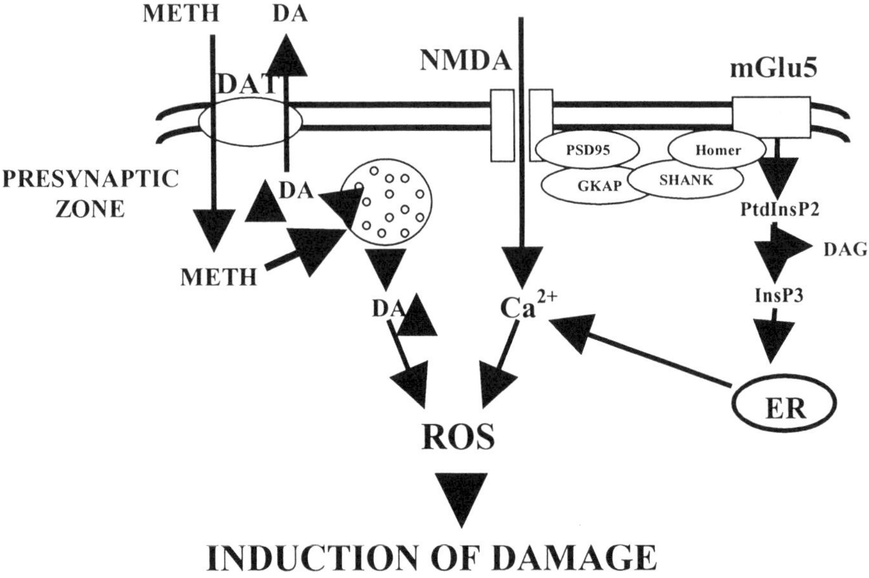

Figure 1. Speculative model on the interaction between NMDA and mGlu5 receptors in the induction of nigro-striatal damage induced by methamphetamine. Combined activation of NMDA and mGlu5 receptors by methamphetamine-induced glutamate release may induce reactive oxygen species formation, which, in concert with those produced by autooxidation of extravesicular dopamine, reaches the threshold for the induction of neuronal death. DA: dopamine; DAG: diacylglycerol; DAT: high affinity dopamine transporter; ER: endoplasmic reticulum; InsP3: inositoltrisphosphate; METH: methamphetamine; ROS: reactive oxygen species.

A series of studies suggests that mGlu5 receptors are present on striatal dopaminergic terminals (Yu et al., 2001; Page et al., 2001), thus providing a possible substrate for this hypothesis. Therefore, mGlu5 receptors may represent a novel target for drugs aimed at preventing the development of methamphetamine neurotoxicity in drug abusers. MGlu5 receptor antagonists do not limit the enhanced release of dopamine induced by methamphetamine, and should not therefore interfere with the rewarding properties of methamphetamine. This should make mGlu5 receptor antagonists more "acceptable" by methamphetamine abusers.

MGlu5 receptor antagonists can also act as symptomatic drugs in experimental models of nigro-striatal degeneration. Activation of mGlu5 receptors induces direct excitation, and selectively potentiates NMDA currents in neurons of the subthalamic nucleus (Awad et al., 2000). As an increased activity of the subthalamic nucleus is implicated in the pathophysiology of bradykinesia, one can predict that mGlu5 receptor antagonists can improve motor symptoms resulting from the loss of nigro-striatal dopaminergic fibers. It will be interesting to study the efficacy of mGlu5 antagonists in the MPTP model of Parkinsonism in mice or primates.

Finally, studies carried out in cortical cultures show that MPEP, SIB-1893 and SIB-1757 are highly potent against neurotoxicity induced by the active fragment of β-amyloid peptide ($\beta AP_{(25-35)}$) (Bruno et al., 2000a). These experiments are carried out in the presence of a cocktail of NMDA and AMPA receptor antagonists (MK-801 and DNQX), thus providing a model in which NMDA receptors do not contribute to neuronal death. Interestingly, inhibition of mGlu1 receptors by AIDA does not reduce but rather exacerbates β-amyloid toxicity in cortical cells (Allen et al., 2000). Although the mechanism(s) underlying the neuroprotective effect of mGlu5 receptor antagonists is unknown, this evidence encourages the use of MPEP or its congeners in experimental animal models of AD, such as mice carrying a double mutation in the amyloid precursor protein (APP) and presenilin-1 genes.

3. Role of group-II mGlu receptors in neuroprotection.

Group-II mGlu receptors may be considered as a potential target for drugs aimed at reducing the progression of neuronal degeneration (Bruno et al., 2001). A safe neuroprotective drug with a favorable therapeutic window is particularly needed in the experimental therapy of acute CNS disorders, such as brain ischemia. Group-II mGlu receptor agonists might meet these criteria because: (i) they are neuroprotective against *in vitro* degeneration induced by excitotoxins and oxygen-glucose deprivation and in *in vivo* models of global ischemia (Buisson and Choi, 1995 Henrich-Noach et al., 1998; Bond et al., 1998, 1999; Lam et al., 1998) (ii) they do not interfere with the normal excitatory synaptic transmission, in contrast to NMDA or AMPA receptor antagonists; (iii) they protect cultured neurons when applied after a toxic pulse with NMDA, at times at which NMDA antagonists are no longer protective, and they are even active *in vivo* when applied after the induction of an ischemic damage (Bruno et al., 1997; Kingston et al., 1999); (iiii) by locally increasing, through the activation of the mGlu3 subtype, the glial production of neurotrophic factors such as TGF-β, they may provide a broad spectrum mechanism of protection, as TGF-β is known to exert neuroprotective activities against neuronal degeneration induced by excitotoxins, oxygen-glucose deprivation, β-amyloid peptide and the HIV capside protein gp120 (Bruno et al., 1998; D'Onofrio et al., 2001; Buisson et al., 1998; Chao et al., 1994; Prehn et al., 1996; Ren et al., 1997; Scorziello et al 1997; and (iiiii) mGlu3 receptors are localized on the vascular side of astrocytes, in proximity of endothelial cells, so they can be easily reached by drugs present in the blood stream and able to cross the blood-brain-barrier. However, LY354740 and LY379268 do not reduce the infarct size in rat models of focal ischemia, i.e. electrocoagulation, (Lam et al., 1998) and endothelin-1-induced (Bond et al., 1999) middle cerebral artery occlusion. This casts doubt on the possible efficacy of mGlu2/3 receptor agonists in the experimental therapy of stroke.

3.1 mGlu2/3 receptor agonists are neuroprotective.

Group-II mGlu receptors have been recently proposed as targets for the therapy of Parkinson's disease (PD). Overactive glutamatergic afferents from the subthalamic nucleus could cause both an excitotoxic loss of dopaminergic neurons in the substantia nigra, and hyperactivation of GABAergic neurons in the globus pallidus, which leads to a reduction of motor activity (Albin et al., 1989). Drugs that can reduce glutamate release by activating presynaptic mGlu2/3 receptors might at one time delay the degeneration of substantia nigra neurons and improve motor activity (Konieczny et al., 1999; Wolfarth et al., 2000). We have recently found that neurodegeneration induced by MPTP in mice is affected by mGlu2/3 receptor ligands (Battaglia et al., 2003). Neuroprotection by LY379268 is dependent on the dose and regimen of MPTP administration. The protective activity of LY379268 could be disclosed under conditions in which MPTP produces a massive degeneration of nigral dopaminergic neuronal cell bodies and striatal dopaminergic terminals, whereas a milder form of toxicity is insensitive to LY379268, but is worsened by the antagonist, LY341495. This finding is somehow paradoxical because one expects a stronger neuroprotection by the agonist under conditions in which the toxin produces a moderate extent of degeneration.

We speculate that activation of mGlu2/3 receptors is neuroprotective against MPTP toxicity through at least two mechanisms. Receptors located on cortico-striatal glutamatergic terminals might limit the excitotoxic component of MPTP toxicity (Turski et al., 1991; Lange et al., 1997; Sonsalla et al., 1998) by inhibiting glutamate release (Battaglia et al., 1997; Cozzi et al., 1997). In addition, activation of mGlu3 receptors present in reactive astrocytes or microglia might reduce the extent of nigro-striatal damage by stimulating the production of neurotrophic factors, such as BDNF (Matarredona et al., 2001), transforming-growth factor-β (TGF- β) (Bruno et al., 1997; 1998; D'Onofrio et al., 2001), or nerve growth factor (Ciccarelli et al., 1999) (Figure 2).

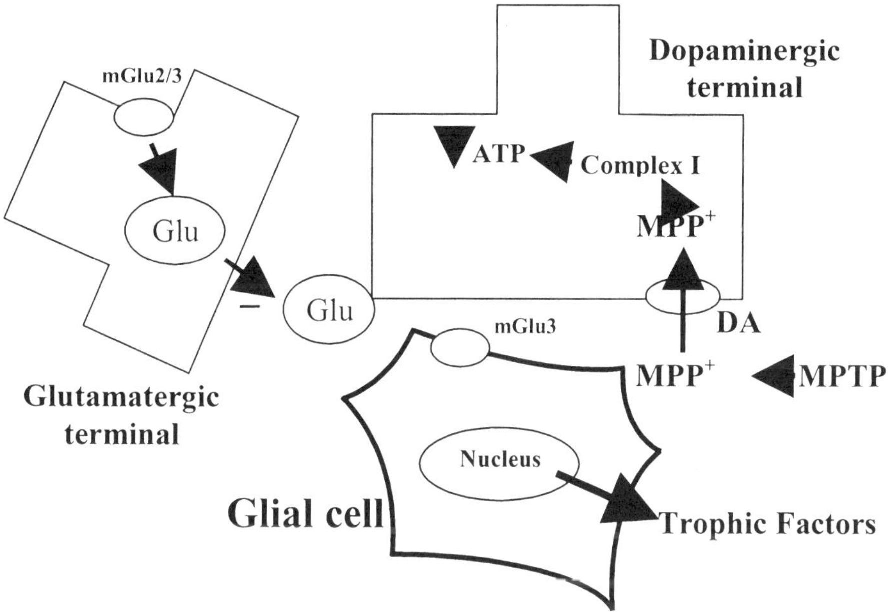

Figure 2. Possible mechanisms by which activation of mGlu2/3 receptors induces neuroprotection in the MPTP model of Parkinsonism. Activation of presynaptic mGlu2/3 receptors decreases MPTP-induced glutamate release, thus reducing the excitotoxic component of neuronal striatal damage. Activation of glial mGlu3 receptors induces the release of neuroprotective trophic factors, which further decrease striatal neurodegeneration. In this model, activation of mGlu2/3 receptors does not interfere with the primary mechanism of action of MPTP. DA: dopamine; Glu: glutamate; MPP^+: 1-methyl-4-phenylpyridinium ion; MPTP: 1-methyl-4-phenyl-1,2,3,6-tetrahydropyridine.

With doses of MPTP that produce a mild lesion of nigro-striatal terminals, activation of neuronal mGlu2/3 receptors by endogenous glutamate might be sufficient to limit the extent of degeneration, and this could explain why the lesion is aggravated by the antagonist LY341495, but is not affected by the agonist, LY379268. With a stronger insult, the rescuing components mediated by trophic factors of glial origin may become relevant, and glial mGlu2/3 receptors may not be easily saturated by the endogenous glutamate because they are not located in the synaptic side of the glial surface (Shigemoto et al., 1999). This might explain why LY379268 shows neuroprotective activity under these conditions. This activity, however, is only partial and is not improved by increasing the i.p. dose of the drug or by continuously delivering the drug through an osmotic minipump. This suggests that activation of mGlu2/3 receptors controls some of the components that are permissive to the pathophysiology of nigro-striatal damage, but does not limit the primary mechanism of MPTP toxicity.

Other than playing a protective role in the MPTP model of parkinsonism, mGlu2/3 receptors are effective in preventing neurodegeneration also in the 6-OHDA model of experimental Parkinsonism in rats (Murray et al., 2002), but this may not be extended to all experimental models of Parkinsonism because nigro-striatal degeneration induced by methamphetamine in mice is attenuated by mGlu5 receptor antagonists, but not by systemic injection of LY379268 (Battaglia et al., 2002). Finally, we have recently found that nigro-striatal denervation enhances the expression and activity of mGlu2/3 receptors in the striatum (Picconi et al., 2002). Thus, we expect that mGlu2/3 receptors become progressively more sensitive to agonist activation under conditions in which nigro-striatal fibers slowly degenerate, as occurs in PD.

The effect of the activation of mGlu2/3 receptors has not been investigated in others animal models of extrapiramidal disorders such as Huntington's disease (HD). However, the observation of an early reduction of mGlu2 receptors in the striatum of transgenic mice carrying expanded CAG repeats in the huntingtin gene (Cha et al., 1998) raises the intriguing possibility that the loss of these receptors contributes to the pathophysiology of neuronal damage in HD. The evidence that mGlu2/3 receptor agonists protect cultured spinal motoneurons against kainate toxicity (Pizzi et al., 2000) encourages the use of these drugs in animal models of ALS, such as transgenic mice carrying mutations of superoxide dismutase. Group-II mGlu receptor agonists are also protective against seizures in experimental animal models of epilepsy (Tang et al., 1997; Monn et al., 1997; Klodzinska et al., 1999, 2000; Suzuki et al., 1999). It will be interesting to examine whether activation of mGlu2/3 receptors protects against neuronal damage that is secondary to neuronal hyperactivity, as observed in models of temporal lobe epilepsy.

Furthermore, it's interesting to add that group-II mGlu receptor agonists could interfere with the pathophysiology of neuropsychiatric disorders such as anxiety and schizophrenia, and they may also exert anti-addictive effects (Monn et al., 1997; Klodzinska et al., 1999; Helton et al., 1997, 1998; Moghaddam et al., 1998; Shekhar and Keim, 2000). LY354740, a selective mGlu2/3 receptor agonist, will be soon available in the market as anxiolitic drug.

4. Role of group-III mGlu receptors in neuroprotection.

Among the three groups of mGlu receptors, the pharmacology of group-III mGlu receptors (mGlu4, -6, -7 and -8) is the least developed because of the requirement for an ω-phosphonic group in molecules acting on these receptors, such as the agonists, L-AP4 and PPG. In addition, the strong acidic moiety limits the penetration of these drugs across the blood-brain barrier, and no centrally available agonists or antagonists have been developed so far.

4.1 mGlu4/7/8 receptor agonists are neuroprotective.

Most of the available information derives from *in vitro* studies or from the effect of intracerebral injections of group-III mGlu receptor agonists. Group-III mGlu receptor agonists protect cultured cortical cells and cultured cerebellar granule cells against excitotoxic death (Bruno et al., 1995; 1996; 2000b; Gasparini et al., 1999b; Lafon-Cazal et al., 1999), and apoptosis induced by $\beta AP_{(25\text{-}35)}$ (Copani et al., 1995). In addition, group-III mGlu receptor agonists are protective against "mechanical injury" in neuronal cultures (Faden et al., 1997). In mixed cortical cultures challenged with NMDA, neuroprotection by L-AP4 or PPG is mediated by mGlu4 receptors, because it is no longer observed in cultures prepared from mGlu4 knockout mice (Bruno et al., 2000b). Interestingly, cultures from mGlu4 knockout mice are more sensitive to NMDA toxicity and release greater amounts of glutamate under basal conditions and in response to NMDA. High concentrations of PPG, which can recruit mGlu7 receptors, reduce NMDA-stimulated glutamate release in cultures lacking mGlu4 receptors, but never below the levels observed in wild-type cultures challenged with NMDA (Bruno et al., 2000b). These data suggest that mGlu4 receptors are critically involved in the homeostatic regulation of glutamate release, whereas mGlu7 receptors are recruited only in response to high levels of extracellular glutamate. *In vivo* studies support a role for mGlu4 receptors in the control of glutamate release and excitotoxic neuronal death. MGlu4 receptor knockout mice are refractory to the protective activity of low doses of PPG co-injected with NMDA into the caudate nucleus. In addition, these animals show a much higher stimulation of striatal glutamate release by NMDA in microdialysis studies (Bruno et al., 2000b). Thus, mGlu4 receptor agonists should effectively prevent the component of excitotoxic neuronal death mediated by an excessive release of glutamate.

Although mGlu4 receptors emerge as promising drug targets, the pharmacological use of agonists is limited by a number of factors, including competition with endogenous glutamate and induction of adaptation mechanisms at receptor level. Electron microscopy studies have shown that mGlu4 and -7 receptors are located in the central portion of presynaptic terminals, in close vicinity to the active zone of neurotransmitter release (Shigemoto et al., 1997). Hence, it is likely that mGlu4 receptors are easily saturated by the endogenous glutamate, particularly under conditions that are associated with an enhanced glutamate release, such as epilepsy, brain ischemia or head trauma (Bruno et al., 2001). In addition, agonist activation may result in mGlu4 receptor desensitization (De Blasi et al., 2001; Iacovelli et al., personal communication), a process possibly leading to development of pharmacodynamic tolerance. The identification of positive allosteric modulators for mGlu4 receptors may offer a strategy to overcome these limitations. We have recently identified the first positive allosteric modulator selective for mGlu4 receptors, (-)-PHCCC, which acts on the transmembrane regions of the receptor. To assess the effect of (-)-PHCCC at native mGlu4 receptors, we have examined its action against neurotoxicity in mixed cultures of mouse cortical cells challenged with NMDA or $\beta AP_{(25\text{-}35)}$. (-)-PHCCC applied alone protects cultured cortical neurons against NMDA toxicity, and shows a synergistic effect when applied in combination with L-AP4 (Maj et al., 2003), thus representing a novel tool to better study the role of group-III mGlu receptors in neurodegeneration, as it can be used even when saturating concentrations of extracellular glutamate are present, i.e. under pathological conditions.

Other possible mechanisms underlying neuroprotection induced by activation of group-III mGlu receptors include the inhibition of NMDA responses by postsynaptic group-III mGlu receptors (Martin et al., 1997) and the modulation of nitric oxide formation (Maiese et al., 1995; 2000). The agonist (R,S)-PPG improves the functional recovery of CA1 pyramidal cells of hippocampal slices subjected to a severe hypoxic/hypoglycemic insult (Henrich-Noak et al., 2000; Sabelhaus et al., 2000), suggesting a possible protective activity of group-III mGlu receptor activation against ischemic brain damage.

Interestingly, activation of group-III mGlu receptors supports the survival of cultured cerebellar granule cells undergoing apoptosis by trophic deprivation (Graham and Burgoyne, 1994; Iacovelli et al., 2002), and the positive allosteric modulator, (-)-PHCCC, is also neuroprotective against β-amyloid toxicity under conditions in which the endogenous excitotoxic component is minimized by the presence of MK-801 and DNQX (Maj et al., 2003). These effects cannot be explained by any of the mechanisms outlined above, and suggest that activation of group-III mGlu receptors supports neuronal survival via some additional intracellular mechanisms. We have recently found that activation of native group-III mGlu receptors stimulates the PI-3-kinase and the MAP-kinase pathways in cultured cerebellar granule cells (Iacovelli et al., 2002), which have an established role in supporting cell survival and are activated in response to neurotrophic agents (such as insulin-like growth factor-I) that prevent apoptosis of cultured cerebellar granule cells (D'Mello et al., 1997). Interestingly, the activation of both pathways is required to confer neuroprotection to cultured granule cells against apoptosis induced by trophic deprivation (Iacovelli et al., 2002).

Recently, it has been shown that mGlu4, -6, -8, but not mGlu7 receptors are expressed in microglial cells, where their activation reduces microglial reactivity induced by lipopolysaccharide, chromogranin A or β-amyloid. This leads to a reduction of microglial neurotoxicity probably due to a reduced production of neurotoxins fom microglia (Taylor et al., 2003). These results, together with the inhibition of β-amyloid toxicity induced by neuronal group-III mGlu receptor activation, suggests that modulation of this class of receptors could be beneficial in neuroinflammatory diseases, such as AD. Moreover, we have recently shown that activation of astrocytic mGlu4 receptors leads to a reduced production of the β-chemokine RANTES, which is involved in the amplification of the inflammatory process, and whose expression correlates with the intensity of neuroinflammation, in experimental allergic encephalomyelitis, an animal model of multiple sclerosis (Glabinski et al., 1998; Ransohoff, 1999). In this model, activation of group-III mGlu receptors with L-AP4 reduces RANTES mRNA levels, as well as immunostaining for MHC class II and CD4, all markers of neuroinflammation, in cerebellum and spinal cord and induces a faster recovery from the disease (Besong et al., 2002).

Similar to group-I antagonists or group-II agonists, group-III mGlu receptor agonists exert anticonvulsant activity (Abdul-Ghani et al., 1997; Tang et al., 1997; Chapman et al., 1999; Gasparini et al., 1999b; Claudio et al., 2000). Although these effects are common to many drugs, a growing body of evidence suggests that mGlu4 receptors are directly involved in the pathophysiology of epilepsy (Lie et al., 2000). mGlu7 receptors have also been implicated in the pathophysiology of epilepsy, as mGlu7 knockout mice show increased seizure-susceptibility (H. van der Putten, personal communication).

Finally, recent evidences strongly suggest that the group-III receptor subtypes, mGlu4 and mGlu7, modulate basal ganglia function at multiple sites (Bradley et al., 1999; Kosinski et al., 1999), not only by controlling extracellular glutamate levels but also by modulating GABA release at GABAergic synapses (Valenti et al., 2002). Therefore, activation of mGlu4 and/or -7 receptors may increase viability of selective cell populations, such as GABAergic projecting neurons in the striatum, which are selectively lost in HD. Alternatively, modulation of group-III mGlu receptors activity may balance basal ganglia circuits under pathological conditions, for instance in PD, where subthalamic nuclei are overactive. Thus, reduction of glutamatergic transmission or enhancement of GABA ergic transmission by group-III mGlu receptor activation may well be a novel mechanism for experimental therapy of HD and PD.

REFERENCES

Abdul-Ghani A.S., Attwell P.J., Singh-Kent N., Bradford H.F., Croucher M.J., Jane D.E. Anti-epileptogenic and anticonvulsant activity of L-2-amino-4-phosphonobutyrate, a presynaptic glutamate receptor agonist. Brain Res 1997; 755:202-212

Abekawa T., Ohmori T., Koyama T. Effects of repeated administration of a high dose of methamphetamine on dopamine and glutamate release in rat striatum and nucleus accumbens. Brain Res 1994; 643:276-281

Alagarsamy S., Marino M.J., Rouse S.T., Gereau IV R.W., Heinemann S.F., Conn P.J. Activation of NMDA receptors reverses desensitization of mGluR5 in native and recombinant systems. Nature Neurosci 1999; 2:234-240

Albin R.L., Young A.B., Penney J.B. The functional anatomy of basal ganglia disorders. Trends Neurosci 1989; 12:366-375

Allen J.W., Knoblach S.M., Faden A.I. Activation of group I metabotropic glutamate receptors reduces neuronal apoptosis but increases necrotic cell death in vitro. Cell Death Diff 2000; 7:470-476

Ango F., Pin J.P., Tu J.C., Xiao B., Worley P.F., Bockaert J., Fagni L. Dendritic and axonal targeting of type 5 metabotropic glutamate receptor is regulated by homer 1 protein and neuronal excitation. J Neurosci 2000; 20:8710-8716

Annoura H., Fukunaga A., Uesugi M., Tatsuoka T., Horikawa Y. A novel class of antagonists for metabotropic glutamate receptors, 7-(hydroxyimino)cyclopropa[b]chromen-1a-carboxylates. Bioorg Med Chem Lett 1996; 6:7763-7766

Awad H., Hubert G.W., Smith Y., Levey A.I., Conn P.J. Activation of metabotropic glutamate receptor 5 has direct excitatory effects and potentiates NMDA receptor currents in neurons of the subthalamic nucleus. J Neurosci 2000; 20:7871-7879

Battaglia G., Bruno V., Pisani A., Centonze D., Catania M.V., Calabresi P., Nicoletti F. Selective blockade of type-I metabotropic glutamate receptors induces neuroprotection by enhancing GABAergic transmission. Mol Cell Neurosci 2001; 17:1071-1083

Battaglia G., Busceti C.L., Pontarelli F., Biagioni F., Fornai F., Paparelli A., Bruno V., Ruggieri S., Nicoletti F. Protective role of group-II metabotropic glutamate receptors against nigro-striatal degeneration induced by 1-methyl-4-phenyl-1,2,3,6-tetrahydropyridine in mice. Neuropharmacology 2003; 45:155-166

Battaglia G., Fornai F., Busceti C.L., Aloisi G., Cerrito F., De Blasi A., Melchiorri D., Nicoletti F. Selective blockade of mGlu5 metabotropic glutamate receptors is protective against methamphetamine neurotoxicity. J Neurosci 2002; 22:2135-2141

Battaglia G., Monn J.A., Schoepp D.D. In vivo inhibition of veratridine-evoked release of striatal excitatory amino acids by the group II metabotropic glutamate receptor agonist LY354740 in rats. Neurosci Lett 1997; 229:161-164

Baude A., Nusser Z., Roberts J.D., Mulvihill E., McIlhinney R.A., Somogyi P. The metabotropic glutamate receptor (mGluR1 alpha) is concentrated at perisynaptic membrane of neuronal subpopulations as detected by immunogold reaction. Neuron 1993; 11:771-787

Besong G., Battaglia G., D'Onofrio M., Di Marco R., Ngomba R.T., Storto M., Castiglione M., Mangano K., Busceti C.L., Nicoletti F.R., Bacon K., Tusche M., Valenti O., Conn P.J., Bruno V., Nicoletti F. Activation of group III metabotropic glutamate receptors inhibits the production of RANTES in glial cell cultures. J Neurosci 2002; 22:5403-5411

Bond A., O'Neill M.J., Hicks C.A., Monn J.A., Lodge D. Neuroprotective effects of a systemically active group II metabotropic glutamate receptor agonist LY354740 in a gerbil model of global ischaemia. Neuroreport 1998; 9:1191-1193

Bond A., Ragumoorthy N., Monn J.A., Hicks C.A., Ward M.A., Lodge D., O'Neill M.J. LY379268, a potent and selective group II metabotropic glutamate receptor agonist, is neuroprotective in gerbil global, but not focal, cerebral ischaemia. Neurosci Lett 1999; 273:191-194

Bradley S.R., Standaert D.G., Rhodes K.J., Rees H.D., Testa C.M., Levey A.I., Conn P.J. Immunohistochemical localization of subtype 4a metabotropic glutamate receptors in the rat and mouse basal ganglia. J Comp Neurol 1999; 407:33-46

Bruno V., Battaglia G., Copani A., Giffard R.G., Raciti G., Raffaele R., Shinozaki H., Nicoletti F. Activation of class II or class III metabotropic glutamate receptors protects cultured cortical neurons against excitotoxic degeneration. Eur J Neurosci 1995; 7:1906-1913

Bruno V., Battaglia G., Kingston A., O'Neill M.J., Catania M.V., Di Grezia R., Nicoletti F. Neuroprotective activity of the potent and selective mGlu1a metabotropic glutamate receptor antagonist, (+)-2-methyl-4-carboxyphenylglycine (LY367385): comparison with LY367366, a broader spectrum antagonist with equal affinity for mGlu1a and mGlu5 receptors. Neuropharmacology 1999; 38:199-207

Bruno V., Battaglia G., Ksiazek I., van der Putten H., Catania M.V., Giuffrida R., Lukic S., Leonhardt T., Inderbitzin W., Gasparini F., Kuhn R., Hampson D.R., Nicoletti F., Flor P.J. Selective activation of mGlu4 metabotropic glutamate receptors is protective against excitotoxic neuronal death. J Neurosci 2000b; 20:6413-6420

Bruno V., Copani A., Bonanno L., Knöpfel T., Kuhn R., Roberts P.J., Nicoletti F. Activation of group III metabotropic glutamate receptors is neuroprotective in cortical cultures. Eur J Pharmacol 1996; 310:61-66

Bruno V., Ksiazek I., Battaglia G., Lukic S., Leonhardt T., Sauer D., Gasparini F., Kuhn R., Nicoletti F., Flor P.J. Selective blockade of metabotropic glutamate receptor subtype 5 is neuroprotective. Neuropharmacology 2000a; 39:2223-2230

Bruno, V., Battaglia, G., Casabona, G., Copani, A., Caciagli, F., Nicoletti, F. Neuroprotection by glial metabotropic glutamate receptors is mediated by transforming growth factor-β. Journal of Neuroscience 1998; 18: 9594-9600

Bruno, V., Battaglia, G., Copani, A., D'Onofrio, M., Di Iorio, P., De Blasi, A., Melchiorri, D., Flor, P.J., Nicoletti, F. Metabotropic glutamate receptor subtypes as targets for neuroprotective drugs. J Cereb Blood Flow Metab 2001; 21:1013-1033

Bruno, V., Sureda, F.X., Storto, M., Casabona, G., Caruso, A., Knopfel, T., Kuhn, R., Nicoletti, F. The neuroprotective activity of group-II metabotropic glutamate receptors requires new protein synthesis and involves a glial-neuronal signaling. J Neurosci 1997; 17:1891-1897

Buisson A., Choi D.W. The inibitory mGluR agonist, S-4-carboxy-3-hydroxy-phenylglycine selectively attenuates NMDA neurotoxicity and oxygen-glucose deprivation-induced neuronal death. Neuropharmacology 1995; 34:1081-1087

Buisson A., Nicole O., Docagne F., Sartelet H., MacKenzie E.T., Vivien D. Up-regulation of a serine protease inhibitor in astrocytes mediates the neuroprotective activity of transforming growth factor-β1. FASEB J 1998; 12:1683-1691

Bushell T.J., Lee C.C., Shigemoto R., Miller R.J. Modulation of synaptic transmission and differential localisation of mGluRs in cultured hippocampal autapses. Neuropharmacology 1999; 38:1553-1567.

Carroll F.Y., Stolle A., Beart P.M., Voerste A., Brabet I., Mauler F., Joly C., Antonicek H., Bockaert J., Miller T., Pin J.P., Prezeau L. BAY36-7620: a potent non-competitive mGlu1 receptor antagonist with inverse agonist activity. Mol Pharmacol 2001; 59:965-973

Cha J.H., Kosinski C.M., Kerner J.A., Alsdorf S.A., Mangiarini L., Davies S.W., Penney J.B., Bates G.P., Young A.B. Altered brain neurotransmitter receptors in transgenic mice expressing a portion of an abnormal human huntington disease gene. Proc Natl Acad Sci USA 1998; 95:6480-6485

Chao C.C., Hu S., Kravitz F.H., Tsang M., Anderson W.R., Peterson P.K. Transforming growth factor-beta protects human neurons against beta-amyloid-induced injury. Mol Chem Neuropathol 1994; 23:159-178

Chapman A.G., Nanan K., Yip P., Meldrum B.S. Anticonvulsant activity of a metabotropic glutamate receptor 8 preferential agonist, (R,S)-4-phosphonophenylglycine. Eur J Pharmacol 1999; 383:23-27

Choi D.W. Calcium and excitotoxic neuronal injury. Ann N Y Acad Sci 1994; 747:162-171

Chuang S.C., Bianchi R., Wong R.K. Group-I mGluR activation turns on a voltage-gated inward current in hippocampal pyramidal cells. J Neurophysiol 2000; 83:2844-2853

Ciccarelli R., Di Iorio P., Bruno V., Battaglia G., D'Alimonte I., D'Onofrio M., Nicoletti F., Caciagli F. Activation of A(1) adenosine or mGlu3 metabotropic glutamate receptors enhances the release of nerve growth factor and S-100beta protein from cultured astrocytes. Glia 1999; 27:275-281

Claudio O.I., Ferchmin P., Velisek L., Sperber E.F., Moshe S.L., Ortiz J.G. Plasticity of excitatory amino acid transporters in experimental epilepsy. Epilepsia 2000; 41:S104-S110

Conn P.J., Pin J.-P. Pharmacology and functions of metabotropic glutamate receptors. Annu Rev Pharmacol Toxicol 1997; 37:205-237

Copani A., Bruno V., Battaglia G., Leanza G., Pellitteri R., Russo A., Stanzani S., Nicoletti F. Activation of metabotropic glutamate receptors protects cultured neurons against apoptosis induced by beta-amyloid peptide. Mol Pharmacol 1995; 47:890-897

Cozzi, A., Attucci, S., Peruginelli, F., Marinozzi, M., Luneia, R., Pellicciari, R., Moroni, F. Type 2 metabotropic glutamate (mGlu) receptors tonically inhibit transmitter release in rat caudate nucleus: in vivo studies with (2S,1'S,2'S,3'R)-2-(2'-carboxy-3'-phenylcyclopropyl)glycine, a new potent and selective antagonist. Eur J Neurosci 1997; 9:1350-1355

D'Onofrio M., Cuomo L., Battaglia G., Ngomba T.R., Storto M., Kingston A.E., Orzi F., De Blasi A., Nicoletti F., Bruno V. Neuroprotection mediated by group-II metabotropic glutamate receptors involves the activation of the MAP kinase pathway and requires the induction of TGF-β1. J Neurochem 2001; 78:435-445

De Blasi A., Conn P.J., Pin J.P., Nicoletti F. Molecular determinant of metabotropic glutamate receptor signaling. Trends Pharmacol Sci 2001; 22:114-120

De Vito M.J., Wagner G.C. Methamphetamine-induced neuronal damage: a possible role for free radicals. Neuropharmacology 1989; 28:1145-1150

D'Mello S.R., Borodezt K., Soltoff S.P. Insulin-like growth factor and potassium depolarization maintain neuronal survival by distinct pathways: possible involvement of PI 3-kinase in IGF-1 signaling. J Neurosci 1997; 17:1548-1560

Doherty A.J., Palmer M.J., Bortolotto Z.A., Hargreaves A., Kingston A.E., Ornstein P.L., Schoepp D.D., Lodge D., Collingridge G.L. A novel, competitive mGlu5 receptor antagonist (LY344545) blocks DHPG-induced potentiation of NMDA responses but not the induction of LTP in rat hippocampal slices. Br J Pharmacol 2000; 131:239-244

Dunah A.W., Wyszynski M., Martin D.M., Sheng M., Standaert D.G. Alpha-actinin-2 in rat striatum: localization and interaction with NMDA glutamate receptor subunits. Brain Res Mol Brain Res 2000; 79:77-87

Faden A.I., Ivanova S.A., Yakovlev A.G., Mukhin A.G. Neuroprotective effects of group III mGluR in traumatic neuronal injury. J Neurotrauma 1997; 14:885-895

Gasparini F., Bruno V., Battaglia G., Lukic S., Leonhardt T., Inderbitzin W., Laurie D., Sommer B., Varney M.A., Hess S.D., Johnson E.C., Kuhn R., Urwyler S., Sauer D., Portet C., Schmutz M., Nicoletti F., Flor P.J. (R,S)-4-phosphonophenylglycine, a potent and selective group III metabotropic glutamate receptor agonist is anticonvulsive and neuroprotective in vivo. J Pharmacol Exp Ther 1999b; 290:1678-1687

Gasparini F., Inderbitzin W., Francotte E., Lecis G., Richert P., Dragic Z., Kuhn R., Flor P.J. (+)-4-Phosphonophenylglycine (PPG) a new group III selective metabotropic glutamate receptor agonist. Bioorg Med Chem Lett 2000; 10:1241-1244

Gasparini F., Lingenhohl K., Stoehr N., Flor P.J., Heinrich M., Vranesic I., Biollaz M., Allgeier H., Heckendorn R., Urwyler S., Varney M.A., Johnson E.C., Hess S.D., Rao S.P., Sacaan A.I., Santori E.M., Velicelebi G., Kuhn R. 2-Methyl-6-(phenylethynyl)-pyridine (MPEP), a potent, selective and systemically active mGlu5 receptor antagonist. Neuropharmacology 1999a; 38: 1493-1503

Gereau IV R.W., Conn P.J. Multiple presynaptic metabotropic glutamate receptors modulate excitatory and inhibitory synaptic transmission in hippocampal area CA1. J Neurosci 1995; 15:6879-6889

Glabinski A.R., Tuohy V.K., Ransohoff R.M. Expression of chemokines RANTES, MIP-1 alpha and GRO-alpha correlates with inflammation in acute experimental autoimmune encephalomyelitis. Neuroimmunomodulation 1998; 5:166-171

Graham M.E., Burgoyne R.D. Activation of metabotropic glutamate receptors by L-AP4 stimulates survival of rat cerebellar granule cells in culture. Eur J Pharmacol 1994; 288:115-123

Helton D.R., Tizzano J.P., Monn J.A., Schoepp D.D., Kallman M.J. LY354740: a metabotropic glutamate receptor agonist which ameliorates symptoms of nicotine withdrawal in rats. Neuropharmacology 1997; 36:1511-1516

Helton D.R., Tizzano J.P., Monn J.A., Schoepp D.D., Kallman M.J. Anxiolytic and side-effect profile of LY354740: a potent, highly selective, orally active agonist for group II metabotropic glutamate receptors. J Pharmacol Exp Ther 1998; 284:651-660

Henrich-Noack P., Flor P.J., Sabelhaus C.F., Prass K., Dirnagl U., Gasparini F., Sauter A., Rudin M., Reymann K.G. Distinct influence of the group III metabotropic glutamate receptor agonist (R,S)-4-phosphonophenylglycine [(R,S)-PPG] on different forms of neuronal damage. Neuropharmacology 2000; 39:911-917

Henrich-Noack P., Hatton C.D., Reymann K.G. The mGlu receptor ligand (S)-4C3HPG protects neurons after global ischaemia in gerbils. Neuroreport 1998; 9:985-988

Iacovelli L., Bruno V., Salvatore L., Melchiorri D., Gradini R., Caricasole A., Barletta E., De Blasi A., Nicoletti F. Native group-III metabotropic glutamate receptors are coupled to the mitogen-activated protein kinase/phosphatidylinositol-3-kinase pathways. J Neurochem 2002; 82:216-223

Jane D.E., Jones P.L., Pook P.C., Tse H.W., Watkins J.C. Actions of two new antagonists showing selectivity for different sub-types of metabotropic glutamate receptor in the neonatal rat spinal cord. Br J Pharmacol 1994; 112:809-816

Jia Z., Lu Y., Henderson J., Taverna F., Romano C., Abramow-Newerly W., Wojtowicz J.M., Roder J. Selective abolition of the NMDA component of long-term potentiation in mice lacking mGluR5. Learn Mem 1998; 5:331-343

Kawabata S., Tsutsumi R., Kohara A., Yamaguchi T., Naknishi S., Okada M. Control of calcium oscillations by phosphorylation of metabotropic glutamate receptors. Nature 1996; 383:89-92

Kingston A.E., O'Neill M.J., Lam A., Bales K.R., Monn J.A., Schoepp D.D. Neuroprotection by metabotropic glutamate receptor agonists: LY354740, LY379268 and LY389795. Eur J Pharmacol 1999; 377:155-165

Klodzinska A., Bijak M., Chojnacka-Wojcik E., Kroczka B., Swiader M., Czuczwar S.J., Pilc A. Roles of group II metabotropic glutamate receptors in modulation of seizure activity. Naunyn Schmiedebergs Arch Pharmacol 2000; 361:283-288

Klodzinska A., Chojnacka-Wojcik E., Pilc A. Selective group II glutamate metabotropic receptor agonist LY354740 attenuates pentetrazole- and picrotoxin-induced seizures. Pol J Pharmacol 1999; 51:543-545

Konieczny J., Ossowska K., Wolfarth J., Pilc A. LY354740, a group II metabotropic glutamate receptor agonist with potential antiparkinsonian properties in rats. Naunyn-Schmiedebergs Arch Pharmacol 1999; 358:500-502

Kosinski C.M., Risso Bradley S., Conn P.J., Levey A.I., Landwehrmeyer G.B., Penney J.B. Jr, Young A.B., Standaert D.G. Localization of metabotropic glutamate receptor 7 mRNA and mGluR7a protein in the rat basal ganglia. J Comp Neurol 1999; 415:266-284

Kutsuwada T., Kashiwabuchi N., Mori H., Sakimura K., Kushiya E., Araki K., Meguro H., Masaki H., Kumanishi T., Arakawa M., et al. Molecular diversity of the NMDA receptor channel. Nature 1992; 358:36-41

Lafon-Cazal M., Fagni L., Guiraud M.J., Mary S., Lerner-Natoli M., Pin J.P., Shigemoto R., Bockaert J. mGluR7-like metabotropic glutamate receptors inhibit NMDA-mediated excitotoxicity in cultured mouse cerebellar granule neurons. Eur J Neurosci 1999; 11:663-672

Lam A.G., Soriano M.A., Monn J.A., Schoepp D.D., Lodge D., McCulloch J. Effects of the selective metabotropic glutamate agonist LY354740 in a rat model of permanent ischaemia. Neurosci Lett 1998; 254:121-123

Lange K.W., Kornhuber J., Riederer P. Dopamine/glutamate interactions in Parkinson's disease. Neuroscience Biobehaviour Review 1997; 21:393-400

Lie A.A., Becker A., Behle K., Beck H., Malitschek B., Conn P.J., Kuhn R., Nitsch R., Plaschke M., Schramm J., Elger C.E., Wiestler O.D., Blumcke I. Up-regulation of the metabotropic glutamate receptor mGluR4 in hippocampal neurons with reduced seizure vulnerability. Ann Neurol 2000; 47:26-35

Lipton S.A., Rosenberg P.A. Excitatory amino acids as a final common pathway for neurologic disorders. N Engl J Med 1994; 330:613-622

Litschig S., Gasparini F., Rüegg D., Munier N., Flor P.J., Vranesic I.-T., Prezeau L., Pin J.-P., Thomsen C., Kuhn R. CPCCOEt, a noncompetitive metabotropic glutamate receptor 1 antagonist, inhibits receptor signaling without affecting glutamate binding. Mol Pharmacol 1999; 55:453-461

Lujan R., Roberts J.D., Shigemoto R., Ohishi H., Somogyi P. Differential plasma membrane distribution of metabotropic glutamate receptors mGluR1 alpha, mGluR2 and mGluR5, relative to neurotransmitter release sites. J Chem Neuroanat 1997; 13:219-241

Maiese K., Greenberg R., Boccone L., Swiriduk M. Activation of the metabotropic glutamate receptor is neuroprotective during nitric oxide toxicity in primary hippocampal neurons of rats. Neurosci Lett 1995; 194:173-176

Maiese K., Vincent A., Lin S.H., Shaw T. Group I and group III metabotropic glutamate receptor subtypes provide enhanced neuroprotection. J Neurosci Res 2000; 62:257-272

Maj M., Bruno V., Dragic Z., Yamamoto R., Battaglia G., Inderbitzin W., Stoehr N., Stein T., Gasparini F., Vranesic I., Kuhn R., Nicoletti F., Flor J.P. (-)-PHCCC, a positive allosteric modulator of mGluR4: Characterization, mechanism of action, and neuroprotection. Neuropharmacology 2003; in press

Martin G., Nie Z., Siggins G.R. Metabotropic glutamate receptors regulate N-methyl-D-aspartate-mediated synaptic transmission in nucleus accumbens. J Neurophysiol 1997; 78:3028-3038

Matarredona E.R., Santiago M., Venero J.L., Cano J., Machado A. Group II metabotropic glutamate receptor activation protects striatal dopaminergic nerve terminals against MPP^+-induced neurotoxicity along with brain-derived neurotrophic factor induction. Journal of Neurochemistry 2001; 76:351-360

Moghaddam B., Adams B.W. Reversal of phencyclidine effects by a group II metabotropic glutamate receptor agonist in rats. Science 1998; 281:1349-1352

Monn J.A., Valli M.J., Massey S.M., Wright R.A., Salhoff C.R., Johnson B.G., Howe T., Alt C.A., Rhodes G.A., Robey R.L., Griffey K.R., Tizzano J.P., Kallman M.J., Helton D.R., Schoepp D.D. Design, synthesis, and pharmacological characterization of (+)-2-aminobicyclo[3.1.0]hexane-2,6-dicarboxylic acid (LY354740): a potent, selective, and orally active group 2 metabotropic glutamate receptor agonist possessing anticonvulsant and anxiolytic properties. J Med Chem 1997; 40:528-537

Mori H., Yamakura T., Masaki H., Mishina M. Involvement of the carboxy-terminal region in modulation by TPA of the NMDA receptor channel. Neuroreport 1993; 4:519-522

Morishita W., Kirov S.A., Alger B.E. Evidence for metabotropic glutamate receptor activation in the induction of depolarization-induced suppression of inhibition in hippocampal CA1. J Neurosci 1998; 18:4870-4882

Murray T.K., Messenger M.J., Ward M.A., Woodhouse S., Osborne D.J., Duty S., O'Neill M.J. Evaluation of the mGluR2/3 agonist LY379268 in rodent models of Parkinson's disease. Pharmacol Biochem Behav 2002; 73:455-466

Nakanishi S. Metabotropic glutamate receptors: synaptic transmission, modulation, and plasticity. Neuron 1994; 13:1031-1037

Nash J.F., Yamamoto B.K. Methamphetamine neurotoxicity and striatal glutamate release: comparison to 3,4-methylenedioxymethamphetamine. Brain Res 1992; 581:237-243

Nicoletti F., Bruno V., Catania M.V., Battaglia G., Copani A., Barbagallo G., Ceña V., Sanchez-Prieto J., Spano P.F., Pizzi M. Group-I metabotropic glutamate receptors: hypotheses to explain their dual role in neurotoxicity and neuroprotection. Neuropharmacology 1999; 38:1477-1484

Niethammer M., Kim E., Sheng M. Interaction between the C terminus of NMDA receptor subunits and multiple members of the PSD-95 family of membrane-associated guanylate kinases. J Neurosci 1996; 16:2157-2163

Ohishi H., Shigemoto R., Nakanishi S., Mizuno N. Distribution of the mRNA for a metabotropic glutamate receptor (mGluR3) in the rat brain: an in situ hybridization study. J Comp Neurol 1993; 335:252-266

Orlando L.R., Alsdorf S.A., Penney J.B. Jr, Young A.B. The role of group I and group II metabotropic glutamate receptors in modulation of striatal NMDA and quinolinic acid toxicity. Exp Neurol 2001; 167:196-204

Page G., Peters M., Najimi G.M., Maloteaux J.M., Hermans E. Modulation of the neuronal dopamine transporter activity metabotropic glutamate receptors mGuR5 in rat striatal synaptosomes through phosphorylation mediated processes. J Neurochem 2001; 76:1282-1290

Pellegrini-Giampietro D.E., Cozzi A., Peruginelli F., Leonardi P., Meli E., Pellicciari R., Moroni F. 1-Aminoindan-1,5-dicarboxylic acid and (S)-(+)-2-(3'-carboxyicyclo(1.1.1.)pentyl)-glycine, two mGlu1 receptor-preferring antagonists, reduce neuronal death in in vitro and in vivo models of cerebral ischemia. Eur J Neurosci 1999a; 11:3637-3647

Pellegrini-Giampietro D.E., Peruginelli F., Meli E., Cozzi A., Albani-Torregrossa S., Pellicciari R., Moroni F. Protection with metabotropic glutamate 1 receptor antagonists in models of ischemic neuronal death: time-course and mechanisms. Neuropharmacology 1999b; 38:1607-1619

Picconi B., Pisani A., Centonze D., Battaglia G., Storto M., Nicoletti F., Bernardi G., Calabresi P. Striatal metabotropic glutamate receptor function following experimental parkinsonism and chronic levodopa treatment. Brain 2002; 125:2635-2645

Pisani A., Calabresi P., Centonze D., Bernardi G. Enhancement of NMDA responses by group-I metabotropic glutamate receptor activation in striatal neurones. Br J Pharmacol 1997; 120:1007-1014

Pizzi M., Benarese M., Boroni F., Goffi F., Valerio A., Spano P.F. Neuroprotection by metabotropic glutamate receptor agonists on kainate-induced degeneration of motor neurons in spinal cord slices from adult rat. Neuropharmacology 2000; 39:903-910

Prehn J.H.M., Bindokas V.P., Jordan J., Galindo M.F., Ghadge G.D., Roos R.P., Boise L.H., Thomson C.B., Krajewski S.W., Reed J.C., Miller R.J. Protective effect of transforming growth factor-β1 on β-amyloid neurotoxicity in rat hippocampal neurons. Mol Pharmacol 1996; 49:319-328

Ransohoff R.M. Mechanism of inflammation in MS tissue: adhesion molecules and chemokines. J Neuroimmunol 1999; 98:57-68

Ren R.F., Hawver D.B., Kim R.S., Flanders K.C. Transforming growth factor-*β* protects human hNT cells from degeneration induced by β-amyloid peptide: involvement of the TGF-β type II receptor. Brain Res Mol Brain Res 1997; 48:315-322

Sabelhaus C.F., Schroder U.H., Breder J., Henrich-Noack P., Reymann K.G. Neuroprotection against hypoxic/hypoglycaemic injury after the insult by the group III metabotropic glutamate receptor agonist (R, S)-4-phosphonophenylglycine. Br J Pharmacol 2000; 131:655-658

Salt T.E., Binns K.E. Contributions of mGlu1 and mGlu5 receptors to interactions with N-methyl-D-aspartate receptor-mediated responses and nociceptive sensory responses of rat thalamic neurons. Neuroscience 2000; 100:375-380

Schoepp D.D., Jane D.E., Monn A.J. Pharmacological agents acting at subtypes of metabotropic glutamate receptors. Neuropharmacology 1999; 38:1431-1476

Scorziello A., Florio T., Bajetto A., Thellung S., Schettini G. TGF-beta1 prevents gp120-induced impairment of Ca^{2+} homeostasis and rescues cortical neurons from apoptotic death. J Neurosci Res 1997; 49:600-607

Shekhar A., Keim S.R. LY354740, a potent group II metabotropic glutamate receptor agonist prevents lactate-induced panic-like response in panic-prone rats. Neuropharmacology 2000; 39:1139-1146

Shigemoto R., Kinoshita A., Wada E., Nomura S., Ohishi H., Takada M., Flor P.J., Neki A., Abe T., Nakanishi S., Mizuno N. Differential presynaptic localization of metabotropic glutamate receptor subtypes in the rat hippocampus. J Neurosci 1997; 17:7503-7522

Shigemoto R., Masugi M., Fujimoto K. Assembly-disassembly of metabotropic glutamate receptor 3 and water channel aquaporin 4 in astrocyte cell membrane. Neuropharmacology 1999; 38:A42

Sonsalla P.K., Albers D.S., Zeevalk G.D. Role of glutamate in neurodegeneration in dopamine neurons in several animal models of parkinsonism. Amino Acids 1998; 14:69-74

Sonsalla P.K., Nicklas W.J., Heikkila R.E. Role for excitatory amino acids in methamphetamine-induced nigrostriatal dopaminergic toxicity. Science 1989; 243:398-400

Spooren W.P., Gasparini F., Salt T.E., Kuhn R. Novel allosteric antagonists shed light on mGlu(5) receptors and CNS disorders. Trends Pharmacol Sci 2001; 22:331-337

Stephans S.E., Yamamoto B.K. Methamphetamine-induced neurotoxicity: roles for glutamate and dopamine efflux. Synapse 1994; 17:203-209

Strasser U., Lobner D., Berhens M.M., Canzoniero L.M., Choi D.W. Antagonists of group-I mGluRs attenuate exicitotoxic neuronal death in cortical cultures. Eur J Neurosci 1998; 10:2848-2855

Suzuki T., Shimizu N., Tsuda M., Soma M., Misawa M. Role of metabotropic glutamate receptors in the hypersusceptibility to pentylentetrazole-induced seizures during diazepam withdrawal. Eur J Pharmacol 1999; 369:163-168

Tang E., Yip P.K., Chapman A.G., Jane D.E., Meldrum B.S. Prolonged anticonvulsant action of glutamate metabotropic receptor agonists in inferior colliculus of genetically epilepsy-prone rats. Eur J Pharmacol 1997; 327:109-115

Taylor D.L., Diemel L.T., Pocock J.M. Activation of microglial group III metabotropic glutamate receptors protects neurons against microglial neurotoxicity. J Neurosci 2003; 23:2150-2160

Thomas N.K., Jane D.E., Tse H.W., Watkins J.C. Alpha-methyl derivatives of serine-O-phosphate as novel, selective competitive metabotropic glutamate receptor antagonists. Neuropharmacology 1996; 35:637-642

Trombley P.Q., Westbrook G.L. L-AP4 inhibits calcium currents and synaptic transmission via a G-protein-coupled glutamate receptor. J Neurosci 1992; 12:2043-2050

Tu J.C., Xiao B., Naisbitt S., Yuan J.P., Petralia R.S., Brakman P., Doan A., Aakalu V.K., Lanahan A.A., Sheng M., Worley P.F. Coupling of mGluR/Homer and PSD-95 coplexes by the Shank family of postsynaptic density proteins. Neuron 1999; 23:583-592

Turski L., Bressler K., Rettig K.J., Loschmann P.A., Wachtel H. Protection of substantia nigra from MPP^+ neurotoxicity by N-methyl-D-aspartate antagonists. Nature 1991; 349:414-418

Ugolini A., Corsi M., Bordi F. Potentiation of NMDA and AMPA responses by the specific mGluR5 agonist CHPG in spinal cord motoneurons. Neuropharmacology 1999; 38:1569-1576

Valenti O., Wittmann M., Marino M.J., Conn P.J. Modulation of inhibitory transmission in the rat globus pallidus by group III metabotropic glutamate receptors. Neuropharmacology 2002; 43:311

Wolfarth S., Konieczny J., Lorenc-Koci E., Ossowaska K., Pilc A. The role of metabotropic glutamate receptor (mGluR) ligands in parkinsonian muscle rigidity. Amino Acids 2000; 19:95-101

Wyszynski M., Lin J., Rao A., Nigh E., Beggs A.H., Craig A.M., Sheng M. Competitive binding of alpha-actinin and calmodulin to the NMDA receptor. Nature 1997; 385:439-42

Yu T.S., Wang S.D., Liu J.C., Yin H.S. Changes in the gene expression of GABA(A) receptor alpha1 and alpha2 subunits and metabotropic glutamate receptor 5 in the basal ganglia of the rats with unilateral 6-hydroxydopamine lesion and embryonic mesencephalic grafts. Exp Neurol 2001; 168:231-241

Chapter 6
MECHANISMS OF EXCITOTOXICITY AND EXCITOPROTECTION

Mark P. Mattson
Laboratory of Neurosciences, National Institute on Aging Gerontology Research Center, 5600 Nathan Shock Drive, Baltimore, MD 21224.

Abstract: All neurons in the central nervous system of mammals express receptors for the excitatory amino acid glutamate. Although glutamatergic neurotransmission is therefore essential for the functioning of neuronal circuits in the brain and spinal cord, under certain conditions activation of glutamate receptors can trigger the death of neurons. Such excitotoxicity most often occurs when cells are coincidentally subjected to reduced levels of oxygen or glucose, increased levels of oxidative stress, trauma, or exposure to toxins or other pathogenic agents. Excitotoxicity is mediated by excessive calcium influx and release from internal organelles, oxyradical production and the activation of a form of programmed cell death called apoptosis. Proteins such as p53, Bax and Par-4 induce mitochondrial membrane permeability changes resulting in the release of cytochrome c and the activation of proteases such as caspase-3. Essentially all subcellular compartments, including the endoplasmic reticulum, mitochondria and nucleus are involved in the excitotoxic process. Excitotoxic cascades are initiated in postsynaptic dendrites where glutamate receptors are most highly concentrated, and may either cause local degeneration or plasticity of those synapses or may propagate the signals to the cell body resulting in cell death. The nervous system protects itself against excitotoxicity by deploying multiple anti-excitotoxic signaling pathways including neurotrophic signaling pathways, intrinsic stress-response pathways, and survival proteins such as protein chaperones, calcium-binding proteins and inhibitor of apoptosis proteins. A rapid accumulation of information on the molecular underpinnings of the excitotoxic process is leading to the development of novel therapeutic approaches for neurodegenerative disorders, as well as unexpected insight into mechanisms of synaptic plasticity.

Keywords: Alzheimer's disease, apoptosis, calcium, glutamate, mitochondria, neurotrophic factor, oxidative stress, Parkinson's disease, stroke.

INDEX

1. Introduction.
2. Excitotoxic mechanisms.
   2.1 Calcium.

2.2 Oxidative stress
2.3 Energy impairment
2.4 DNA damage
2.5 Apoptosis
3. Excitotoxic mechanisms in synaptic and dendritic plasticity.
4. Excitoprotective mechanisms.
4.1 Neurotrophic factors and cytokines
4.2 Preconditioning/stress response pathways
4.3 Cytoskeleton-mediated neuroprotection
5. Glial cells and excitotoxicity.
5.1 Astrocytes
5.2 Oligodendrocytes
5.3 Microglia
6. Implications for the prevention and treatment of neurological disorders.

1. Introduction.

The preceding chapters reviewed the cellular and molecular mechanisms underlying glutamatergic neurotransmission. The purpose of the present chapter, much of the content of has been modified from Mattson (2003), is to review the current understanding of how, in certain physiological and pathological settings, excessive activation of glutamate receptors causes the death of neurons. This chapter is not intended to be comprehensive and draws largely on mechanisms which my colleagues and I have focused our own work. The process of glutamate receptor-mediated neuronal death, now called excitotoxicity, was described more than 3 decades ago (Olney, 1969). Cell culture studies have shown that glutamate can kill many different types of mammalian CNS neurons in a concentration-dependent, receptor-mediated, manner. The identification of natural excitotoxins such as kainic acid and domoic acid, and the development of analogs of glutamate that act as either agonists (e.g., NMDA and AMPA) or antagonists (e.g., APV and CNQX) of NMDA and AMPA/kainate receptors provided tools that have been used to demonstrate roles for activation of glutamate receptors in various models of acute neurodegenerative conditions, such as stroke and traumatic brain and spinal cord injury (Hayes et al., 1992; Dirnagl et al., 1999), and in chronic neurodegenerative disorders, such as Alzheimer's, Parkinson's and Huntington's diseases and amyotrophic lateral sclerosis (Mattson, 1997; Sonsalla et al., 1998; Ludolph et al., 2000; Sieradzan and Mann, 2001). The cloning and molecular characterization of the various types of glutamate receptor proteins (Michaelis, 1998), combined with advances in technologies for electrophysiological evaluation of ion channel functions (Sommer et al., 1992) and measurements of intracellular calcium levels (Lipscombe et al., 1988; Mattson et al., 1989) and reactive oxygen species (ROS) (Cheng and Mattson, 1995; Dugan et al., 1995), has led to a major advances in our understanding of mechanisms of developmental (Mattson et al., 1988) and synaptic (Kullman et al., 2000) plasticity, as well as the pathogenesis of an array of neurological disorders (Doble, 1999).

2. Excitotoxic mechanisms.

At the subcellular and molecular levels, excitotoxicity is a complex process. The cascade of events triggered by glutamate receptor activation, that ultimately results in cell death, includes changes in different subcellular compartments including the cytosol, mitochondria, endoplasmic reticulum and nucleus. The following description of the excitotoxic process covers events that have clearly been shown to be pivotal in the cell death process, but is by no means comprehensive.

2.1 Calcium

Activation of glutamate receptors results in an increase in the concentration of cytoplasmic Ca^{2+} as the result of Ca^{2+} influx through AMPA and NMDA receptor channels and through voltage-dependent Ca^{2+} channels (Fig. 1). Such opening of ionotropic glutamate receptors, and the consequent influx of Na^{+} and Ca^{2+}, are the first step in the excitotoxic process. In addition, activation of metabotropic glutamate receptors stimulates the release of Ca^{2+} from the endoplasmic reticulum (ER). Depending upon the particular type of neuron, its developmental stage and various environmental factors, activation of either AMPA/kainate or NMDA receptors may mediate Ca^{2+} influx and excitotoxicity. Accordingly, a glutamate receptor antagonist that acts on only one subtype of receptor may or may not be excitoprotective. For example, NMDA receptor antagonists are very effective in protecting cultured rat hippocampal neurons against excitotoxicity (Wang et al., 1994; Mattson et al., 1995), but are less effective in protecting motor neurons in organotypic spinal cord cultures (Saroff et al., 2000). The ability of voltage-dependent Ca^{2+} channels (VDCC) antagonists to protect several different types of neurons against excitotoxicity, including hippocampal pyramidal and cortical neurons (Weiss et al., 1990; Blanc et al., 1998a), demonstrates the importance of VDCC in excitotoxicity. Release of Ca^{2+}

from the ER plays a pivotal role in some cases of excitotoxicity as demonstrated in studies showing that blockers of the two different types of ER Ca^{2+} channels, IP_3 receptors and ryanodine receptors, can protect neurons against excitotoxic injury (Frandsen and Schousboe, 1991; Mattson et al., 2000). Mitochondria play important roles in the regulation of neuronal calcium homeostasis (Werth and Thayer, 1994) and it has been shown that genetic and pharmacological manipulations that enhance mitochondrial calcium sequestration can protect neurons against excitotoxicity (Bruce-Keller et al., 1998; Nicholls et al., 1999; Duchen, 2000).

There are multiple and, often cross-ampligying, mechanisms whereby Ca^{2+} kill neurons (Fig. 1).

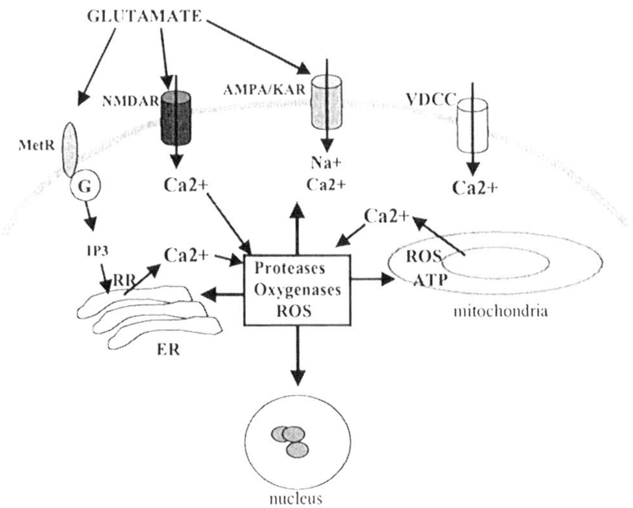

Figure 1. Mechanisms whereby glutamate increases cytoplasmic Ca^{2+} concentrations, and modulation by endoplasmic reticulum (ER) and mitochondria. Binding of glutamate to AMPA receptors and kainate receptors (KAR) opens the receptor channels resulting in Na^+ influx and consequent membrane depolarization and opening of voltage-dependent Ca^{2+} channels (VDCC). Some forms of AMPA receptor are also permeable to Ca^{2+}. Binding of glutamate to NMDA receptors under depolarizing conditions opens the NMDA receptor channel resulting in large amounts of Ca^{2+} influx. Activation of metabotropic glutamate receptors (MetR) induces IP3 production and activation of IP3 receptor and ryanodine receptor (RR) channels in the ER membrane resulting in release of Ca^{2+} from the ER into

the cytoplasm. The increases in cytoplasmic Ca^{2+} levels in response to glutamate receptor activation can induce Ca^{2+} uptake into the mitochondria which, if excessive, can induce the production of reactive oxygen species (ROS) and inhibit ATP production. By activating proteases and inducing oxidative stress Ca^{2+} is a key mediator of excitotoxic cell death.

Ca^{2+} activates cysteine proteases called calpains that degrade a variety of substrates including cytoskeletal proteins, membrane receptors and metabolic enzymes (Bi et al., 1996; Caba et al., 2002; Guttmann et al., 2002). Calpains may also play an important role in the triggering of apoptotic cascades by virtue of their ability to activate caspases (Leist et al., 1997; Volbracht et al., 2001). A second general mechanism involved in excitotoxicity is induction of oxidative stress by Ca^{2+}. Calcium induces oxidative stress by several different mechanisms including activation of oxygenases such as cyclooxygenases and lipoxygenases (Goodman et al., 1994), perturbation of mitochondrial calcium and energy metabolism (Sengpiel et al., 1998), and induction of membrane lipid peroxidation (Goodman et al., 1996). The ROS generated in response to glutamate-induced Ca^{2+} influx include superoxide anion radical, hydrogen peroxide, hydroxyl radical and peroxynitrite (Culcasi et al., 1994; Mattson et al., 1995; Dawson and Dawson, 1998; Sengpiel et al., 1998; Yu et al., 1998). Third, Ca^{2+} triggers apoptosis, a form of programmed cell death (Ankarcrona, 1998). This might occur by Ca^{2+}-mediated induction/activation of pro-apoptotic proteins such as Bax, Par-4 and p53 leading to mitochondrial membrane permeability changes, release of cytochrome c and caspase activation (Duan et al., 1999; Culmsee et al., 2001; Dargusch et al., 2001).

Oxidative Stress

Numerous studies have documented increased ROS production in neurons subjected to excitotoxic insults, in various in vitro and in vivo models. For example, cultured hippocampal, cortical and cerebellar neurons exhibit increased levels of superoxide and hydrogen peroxide when exposed to excitotoxic concentrations of glutamate or NMDA (Gunasekar et al., 1995; Mattson et al., 1995; Vergun et al., 2001). Hydrogen peroxide can be converted to hydroxyl radical in the Fenton reaction upon interaction with Fe^{2+} or Cu^{+}; hydroxyl radical is a highly reactive free radical and potent inducer of membrane lipid peroxidation. Microglial cells and some neurons produce nitric oxide in response to glutamate receptor activation; nitric oxide can react with superoxide to produce peroxynitrite which can damage membranes and proteins. ROS can also damage DNA and RNA, and may thereby disrupt gene expression and, if severe enough, may trigger apoptosis (Fig. 2).

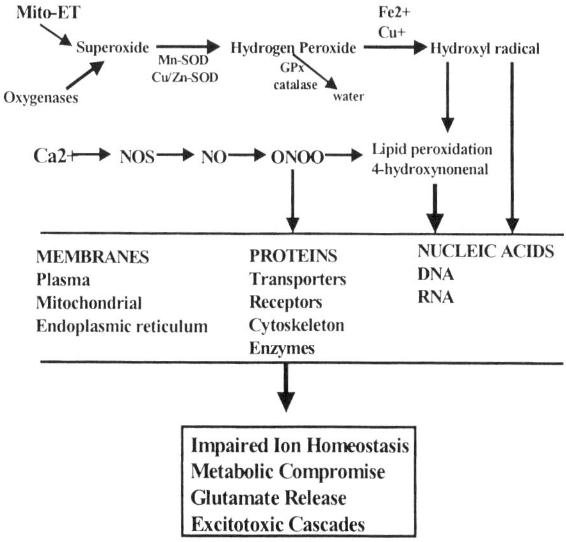

Figure 2. Roles of oxidative stress in excitotoxicity. A major source of reactive oxygen species include mitochondrial electron transport (Mito-ET) which generates superoxide. By activating certain oxygenases Ca^{2+} can induce superoxide production. Superoxide is converted to hydrogen peroxide by the action of Mn- and Cu/Zn-superoxide dismutases (SOD). In the presence of Fe^{2+} and Cu^{+}, hydrogen peroxide is converted to hydroxyl radical, a potent inducer of membrane lipid peroxidation. Ca^{2+} can also promote ROS production by activating nitric oxide synthase (NOS) resulting in the production of nitric oxide (NO); NO interacts with superoxide to generate peroxynitrite (ONOO), a potent inducer of lipid peroxidation. Lipid peroxidation generates the cytotoxic aldehyde 4-hydroxynonenal which covalently modifies and thereby impairs the function of membrane ion-motive ATPases, and glutamate and glucose transporters. ROS, particularly hydroxyl radical and peroxynitrite, can also directly damage proteins, lipids and nucleic acids resulting in impairment of the functions of those molecules, thus executing the excitotoxic process.

A pivotal role for oxidative stress in excitotoxicity has been demonstrated in various cell culture and in vivo models. Treatment of cultured neurons with antioxidants such as vitamin E, lipoic acid, and glutathione protects them against glutamate toxicity and ischemia-like conditions (Goodman et al., 1996). Manipulations of the expression of genes that encode antioxidant enzymes or proteins involved in the generation of ROS have provided evidence that oxidative stress is involved in excitotoxic injury in vivo. For example, neurons in the brains of mice lacking inducible nitric oxide synthase exhibit increased resistance to ischemic and excitotoxic injury (Dawson et al., 1996). In addition, overexpression of Cu/Zn (Chan et al., 1990) and manganese (Keller et al., 1998) superoxide dismutases in transgenic mice increases the resistance of hippocampal and cortical neurons to ischemic/excitotoxic death.

Oxidative stress not only contributes to the excitotoxic process downstream of glutamate receptor activation, but it can also act upstream, by causing dysregulation of ion homeostasis. Neurons are more readily killed by glutamate when they are under conditions of increased oxidative stress. Membrane lipid peroxidation is a particularly important consequence of oxidative stress that can impair the function of key proteins involved in the maintenance of cellular calcium homeostasis.

For example, membrane lipid peroxidation resulting from exposure to Fe^{2+} or amyloid β-peptide impairs the activities of the plasma membrane Na^{+}/K^{+}-ATPase, Ca^{2+}-ATPase and glucose transporter (Mark et al., 1995a, 1997a, 1997b). Impairment of such membrane transporters destabilizes Ca^{2+} homeostasis resulting in large and sustained elevation of intracellular Ca^{2+} concentrations following activation of glutamate receptors. Indeed, impairment of the Na^{+}/K^{+}-ATPase results in membrane depolarization and enhanced opening of NMDA receptors. The adverse effects of oxidative stress on ion homeostasis are particularly pronounced in synaptic terminals (Keller et al., 1997). In addition to disrupting membrane transporter functions in neurons, oxidative stress can impair glutamate transporter function in astrocytes (Blanc et al., 1998b), which promotes excitotoxicity as a result of increased concentrations of extracellular glutamate.

The mechanism whereby lipid peroxidation promotes excitotoxicity involves the non-enzymatic production of a lipid-derived aldehyde called 4-hydroxynonenal. By covalently modifying transporter proteins on cysteine, lysine and/or histidine residues, 4-hydroxynonenal may directly compromise protein function and can also cause protein cross-linking (Mark et al., 1997a, 1997b; Blanc et al., 1998b). 4-hydroxynonenal may also exacerbate Ca^{2+} influx through VDCC by an indirect mechanism involving inhibition of a protein phosphatase and a consequent increase in the phosphorylation of Ca^{2+} channel proteins (Lu et al., 2002a). Because most, if not all, neurodegenerative disorders are associated with increased oxidative stress due to age-related processes or tissue damage, it is likely that overactivation of glutamate receptors contributes to the deaths of many neurons in these disorders.

Although oxidative stress can increase the vulnerability of neurons to excitotoxicity by the mechanisms described above, certain actions of

ROS can be excitoprotective. Low levels of oxidative stress can stimulate the expression of genes encoding cytoprotective proteins including antioxidant enzymes, heat-shock proteins and proteins such as Bcl-2 and inhibitor of apoptosis proteins (IAPs) that antagonize apoptotic cascades (Mattson and Camandola, 2001). In addition, data suggest that NMDA receptors possess a redox modulatory site which, when oxidized, reduces channel activity (Aizenman et al., 1990; Lipton et al., 1993).

Energy Impairment

The ion-motive ATPases that maintain transmembrane ion gradients under resting conditions and restore the gradients following activation of glutamate receptors require large amounts of ATP. These ATPases are a major energy drain in neurons, and overactivation of glutamate receptors can therefore deplete energy reserves in neurons (Khodorov et al., 1996). In addition to the increased energy demands resulting from glutamate receptor activation, the excitotoxic process is characterized by impairment of mitochondrial function which manifests as membrane depolarization, calcium uptake and reduced ATP production (Mattson et al., 1995). Neurons are very sensitive to excitotoxicity when they are subjected to conditions of reduced energy availability such as hypoglycemia and ischemia (Novelli et al., 1988). Accordingly, glutamate receptor antagonists can prevent the death of neurons under such conditions of reduced energy availability (Nellgard and Wieloch, 1992). Certain pharmacological agents that promote maintenance of ATP levels can also protect neurons against excitotoxicity including creatine (Brustovetsky et al., 2001).

DNA Damage

DNA damage, such as occurs under conditions of oxidative stress or exposure to genotoxic agents, can render neurons vulnerable to excitotoxicity. DNA damage can stimulate the activation of poly (ADP-ribose) polymerase (PARP) which normally plays a role in the DNA repair process (Decker and Muller, 2002). However, PARP activity may lead to the depletion of nicotine adenine dinucleotide (NAD) and hence ATP depletion, thereby increasing the vulnerability of neuron to excitotoxicity. Overactivation of glutamate receptors may itself cause DNA damage, which likely results from calcium influx and oxyradical production (Didier et al., 1996). DNA damage can, in turn, induce the production and activation of a protein called p53 which can trigger apoptosis (Chen and Chuang, 1999; Kruman et al., 2000, 2002).

Apoptosis

Apoptosis is a form of programmed cell death that occurs normally during development of the nervous system, but may also occur in response to glutamate receptor activation (Mattson, 2000). Cells undergoing apoptosis exhibit characteristic morphological changes including cell body shrinkage, the formation of cell surface membrane blebs and nuclear chromatin condensation and fragmentation. During the process of apoptosis the plasma and organellar membranes retain their integrity; this contrasts with a form of death called necrosis in which the cells swell and membranes lyse. Studies in an array of organisms and cell types have revealed several key steps in apoptosis. In most cases of apoptosis the death process involves pivotal mitochondrial changes including increased membrane permeability and release of cytochrome c (Fig. 3).

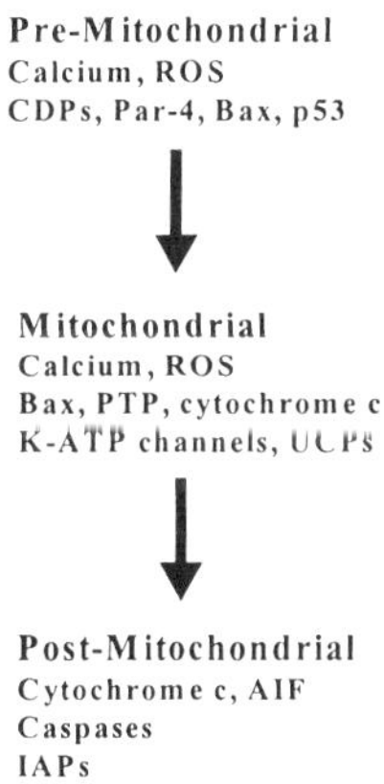

Figure 3. Three key phases in the process of excitotoxic apoptosis. Overactivation of glutamate receptors triggers the production and/or activation of pro-apoptotic proteins including prostate apoptosis response –4 (Par-4), p53, Bax and calcium-dependent proteases (CDPs). Such proteins promote mitochondrial membrane permeability changes resulting in the opening of permeability transition pores (PTP) and the release of

cytochrome c. Mitochondrial apoptotic changes can be facilitated or inhibited by calcium, ROS, ATP-sensitive K^+ channels (KATP) and uncoupling proteins (UCP). Cytochrome c is released from mitochondria and forms a complex with Apaf-1 and caspase-9 resulting in the activation of caspase-3. In addition, apoptosis-inducing factor (AIF) may be released from mitochondria and then translocates to the nucleus where it induces nuclear chromatin condensation and fragmentation. Postmitochondrial apoptotic events can be modulated by ROS which facilitate the effector phase of cell death, and by inhibitor of apoptosis proteins (IAPs) which directly inhibit caspases.

The mitochondrial changes are triggered by the actions of several different proteins such as Bax and Par-4 that can cause mitochondrial membrane permeability by directly interacting with the membrane or by indirect actions (Guo et al., 1998; Duan et al., 1999). Changes downstream of the mitochondrial changes that are believed execute the cell death process include the formation of a protein complex called the apoptosome which consists of cytochrome c, Apaf-1 and caspase-9. Caspase-9 is activated and, in turn, cleaves and activates caspase-3, the latter caspase being a key player in the excitotoxic cell death process in many cases (Du et al., 1997; Mattson et al., 1998).

3. Excitotoxic mechanisms in synaptic and dendritic plasticity

Glutamate receptors are concentrated in postsynaptic regions of dendrites, and are present at lower levels or absent from the cell body and axon. It is therefore not surprising that excitotoxic neuronal death is characterized by degeneration of dendrites with relative sparing of axons (Swann et al., 2000). This compartmentalization of the excitotoxic injury is manifest in neurons dying in several different neurodegenerative disorders including severe epileptic seizures, ischemic stroke and Alzheimer's disease (Scheibel, 1979). Because glutamatergic synapses play fundamental roles in the processes of learning and memory, there has been intense interest in understanding the cellular and molecular events that occur in synaptic terminals that are responsible for long-term changes in synaptic strength. Several of the same changes responsible for excitotoxic cell death are known to mediate synaptic plasticity. These include calcium influx and activation of protein kinases and proteases.

Our work has shown that apoptotic biochemical cascades can be activated in synaptic terminals and dendrites in response to glutamate receptor activation. Studies of synaptosome preparations and cultured hippocampal neurons have demonstrated that apoptotic insults, including glutamate receptor agonists can induce Par-4 production (Duan et al.,

1999), mitochondrial membrane depolarization and oxyradical production (Mattson et al., 1998), and activation of caspase-3 and the release from mitochondria of a factor(s) capable of inducing nuclear chromatin condensation and fragmentation (Mattson et al., 1998). Initiation of apoptotic pathways in postsynaptic dendritic spines may propagate to more proximal regions of dendrites and the cell body, resulting in cell death (Mattson et al., 1998). Interestingly, synaptic/dendritic apoptotic cascades need not lead to neuronal death; they can be activated reversibly and in synapses and dendrites without the neurons dying (Glazner et al., 2000). These and other findings have led to the realization that apoptotic pathways play important roles in structural remodeling and functional plasticity of synapses (Gilman and Mattson, 2002).

Specific mechanisms whereby apoptotic pathways can modulate synaptic plasticity are being identified. In searching for mechanisms whereby apoptotic cascades might regulate synaptic plasticity, we discovered that two subunits of the AMPA receptor are substrates of caspase-3 (Glazner et al., 2000; Lu et al., 2002b). Whole-cell patch clamp analyses and calcium imaging studies established that cleavage of the AMPA receptor subunits results in decreased ion current flow through the AMPA receptors and a decreased calcium response to glutamate. Reversible activation of caspases in dendritic spines, in response to stimulation of glutamate receptors, would be expected to transiently reduce postsynaptic responsivity at that synapse. With respect to a possible function of caspase-mediated AMPA receptor cleavage in excitotoxic cell death, our data suggest that this action of caspases can prevent excitotoxic necrosis and thereby ensure that the cell dies by apoptosis (Glazner et al., 2000; Lu et al., 2001).

4. Excitoprotective mechanisms

The cells of the nervous system possess many different mechanisms to prevent excitotoxic neuronal death. While neurons may undergo excitotoxic death in pathological conditions such as cerebral ischemia and epilepsy, they normally resist the calcium, oxidative and metabolic loads induced by glutamate receptor activation by deploying several different intercellular and intracellular signaling pathways.

4.1 Neurotrophic Factors and Cytokines

Neurotrophic factors, initially discovered based on their ability to prevent developmental neuronal death (Conover and Yancopoulos, 1997), can also protect neurons against pathological death (Mattson, 1997). The first neurotrophic factor shown to protect neurons against excitotoxicity was basic fibroblast growth factor (bFGF) (Mattson et al., 1989). Subsequent studies demonstrated the abilities of nerve growth factor (NGF), brain-derived neurotrophic factor (BDNF), neurotrophin-4/5, insulin-like growth factors and platelet-derived growth factors to protect neurons against metabolic and excitotoxic insults (Cheng and Mattson, 1991, 1992, 1994, 1995; Cheng et al., 1994a). Some cytokines can also protect neurons against excitotoxicity including tumor necrosis factor (TNF) and TGFβ (Cheng et al., 1994b; Bruce et al., 1996). In addition, novel excitoprotective factors have been discovered including secreted forms of β-amyloid precursor protein (sAPP) (Mattson et al., 1993a; Furukawa et al., 1996) and activity-dependent neurotrophic factor (Glazner and Mattson, 1999; Guo et al., 1999b). In several instances in vivo studies have confirmed excitoprotective actions of specific ligands. For example, administration of bFGF, TGFβ and sAPP reduced ischemic neuronal death in a rodent stroke models (Prehn et al., 1993; Koketsu et al., 1994; Smith-Swintosky et al., 1994). Traumatic spinal cord injury was reduced and functional outcome improved in rats administered with bFGF (Rabchevsky et al., 2000).

Excitoprotective neurotrophic factors and cytokines can stabilize cellular Ca^{2+} homeostasis such that glutamate-induced elevation of intracellular Ca^{2+} concentration resulted to be reduced (Mattson et al., 1989; Cheng and Mattson, 1991, 1992; Mattson et al., 1993a; Cheng et al., 1995). Excitoprotective factors may also suppress oxidative stress and preserve mitochondrial function (Mattson et al., 1993b). In several cases, the signal transduction pathways that mediate excitoprotection have been identified (Fig. 4).

Activation of bFGF receptors results in tyrosine phosphorylation of the receptors and activation of kinases such as mitogen-activated protein (MAP) kinases that phosphorylate transcription factors. Some of the genes known to be regulated by bFGF that are likely to contribute to its anti-excitotoxic action are NMDA and AMPA receptor subunits (Mattson et al., 1993c; Cheng et al., 1995), antioxidant enzymes (Mattson et al., 1995) and calcium-regulating proteins (Cheng et al., 1995; Pappas and Parnavelas, 1997; Furukawa and Mattson, 1998).

In the case of BDNF, it activates receptors coupled to the phosphatidylinositol-3-kinase (PI3K) – Akt kinase pathway which stimulates the expression of antioxidant enzymes and anti-apoptotic Bcl-2 family members (Allsopp et al., 1995; Mattson et al., 1995).

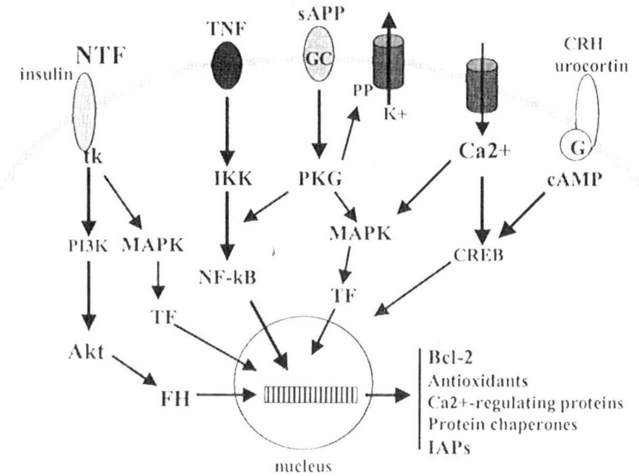

Figure 4. Examples of excitoprotective signaling pathways. Neurotrophic factors (NTF) and hormones such as insulin activate receptor tyrosine kinases (tk) resulting in the activation of the phosphatidylinositol-3 kinase (PI3K) – Akt kinase, and mitogen-activated protein kinase (MAPK), pathways. Tumor necrosis factor (TNF) stimulates receptors coupled to the activation of IkB kinases (IKK) which, in turn, activates the transcription factor NF-kB. Secreted forms of APP (sAPP) stimulate receptors coupled to the production of cyclic GMP and the activation of cyclic GMP-dependent protein kinase (PKG). PKG phosphorylates a protein phosphatase (PP) resulting in the dephosphorylation and activation of plasma membrane K^+ channels and a consequent membrane hyperpolarization. PKG can also induce the activation of NF-kB and MAPK. Corticotropin releasing hormone (CRH) and urocortin activate receptors coupled to adenylate cyclase and the production of cyclic AMP. Cyclic AMP can then activate the cyclic AMP-response element binding protein (CREB). Increases in the intracellular Ca^{2+} concentration can also activate, or enhance the activation of, several of these pathways including the MAPK, NF-kB and CREB pathways. Excitoprotective genes are induced by transcription factors (TF); examples include anti-apoptotic proteins of the Bcl-2 and IAP families, antioxidant enzymes such as glutathione peroxidase and superoxide dismutases, Ca^{2+}-regulating proteins, and protein chaperones such as heat-shock proteins and glucose-regulated proteins.

A similar transduction mechanism is employed by integrin receptors which are activated by binding to extracellular matrix and cell surface adhesion molecules (Gary et al., 2001, 2002). A different excitoprotective mechanism is employed by TNF, a cytokine that activates receptors coupled to transduction proteins that activate the transcription factor NF-κB which stimulates the transcription of genes encoding several different neuroprotective proteins including manganese superoxide dismutase, the calcium-binding protein calbindin and Bcl-2 (Mattson and Camandola, 2001). Studies of the excitoprotective mechanism of sAPP have revealed a novel pathway involving a presumptive membrane receptor guanylate cyclase which catalyzes the production of cyclic GMP. Cyclic GMP then activates kinases and phosphatases resulting in the activation of high-conductance K^+ channels and membrane hyperpolarization (Furukawa et al., 1996). In addition, activation of the cyclic GMP pathway by sAPP results in activation of NF-κB and induction of the same neuroprotective genes induced by TNF (Barger and Mattson, 1996).

4.2 Preconditioning/Stress Response Pathways

Glutamate induces cellular stress response pathways that can be excitoprotective. Pretreatment of cultured neurons with a subtoxic level of glutamate or NMDA protects the neurons against death when subsequently exposed to an otherwise neurotoxic concentration of NMDA (Marini and Paul, 1992). Several other types of mild stress can also protect neurons against excitotoxic death including heat stress (Lowenstein et al., 1991), metabolic stress (Gorgias et al., 1996) and even intermittent fasting, a diet-induced stress (Bruce-Keller et al., 1999). Several different, and presumably complementary, stress response pathways that mediate such excitoprotective "preconditioning" have been identified. NMDA-induced preconditioning in cerebellar cells has been shown to involve stimulation of BDNF production and activation of NF-κB (Marini et al., 1998; Lipsky et al., 2001). Several different stress-responsive protein chaperones have been shown to be upregulated in neurons subjected to mild neuroprotective stressors, including heat-shock protein 70 and glucose-regulated protein 78. Studies in which levels of these protein chaperones are manipulated suggest that they play key roles in excitoprotection (Yu et al., 1999). Interestingly, increased production of BDNF and protein chaperones also play important role in the neuroprotective effect of dietary restriction (Duan and Mattson, 1999; Yu and Mattson, 1999; Duan et al., 2001; Lee et al., 2002).

4.3 Cytoskeleton-Mediated Neuroprotection

Activation of glutamate receptors results in structural changes in the cytoskeleton of neurons. The changes include depolymerization of actin filaments and microtubules, and reorganization of dendritic spines. While such structural changes are thought to be important for adaptive modifications in the structure of neuronal circuits, they may also influence the vulnerability of neurons to excitotoxicity. Electrophysiological recordings of glutamate-induced currents and Ca^{2+} currents in cultured hippocampal neurons exposed to agents that effect the polymerization states of actin and tubulin have demonstrated roles for these cytoskeletal polymers in regulating activity of NMDA receptor channels and voltage-dependent Ca^{2+} channels. The actin-depolymerizing agent cytochalasin D reduces currents through NMDA receptor channels and voltage-dependent Ca^{2+} channels by enhancing channel rundown and, conversely, stabilization of actin filaments with jasplakinolide reduces channel rundown (Furukawa et al., 1995). Because Ca^{2+} influx induces actin depolymerization, the latter studies suggested that actin depolymerization serves as part of an excitoprotective feedback pathway. Gelsolin is a Ca^{2+}-activated actin-severing protein. Studies of gelsolin-deficient mice have demonstrated an excitoprotective role for gelsolin, and have solidified the evidence that actin filaments regulate NMDA receptor channels and voltage-dependent Ca^{2+} channels (Furukawa et al., 1997). Not only do actin filaments regulate Ca^{2+} influx through plasma membrane channels, they regulate Ca^{2+} release from endoplasmic reticulum stores. When actin filaments in cultured hippocampal neurons are depolymerized using cytochalasin D, Ca^{2+} release from IP_3-sensitive stores is reduced (Wang et al., 2002). The molecular interactions responsible for modulation of NMDA receptors, VDCC and IP_3 receptors by actin have not been established. Actin might directly interact with channel proteins, but it is more likely that there are actin-binding proteins that mediate the effects of actin filaments on the channels.

As for microtubules, they also modulate neuronal Ca^{2+} homeostasis and responsivity to glutamate. Calcium responses to glutamate are increased in cultured hippocampal neurons that are treated with the microtubule-depolymerizing agent colchicines (Furukawa and Mattson, 1995). Conversely, stabilization of microtubules with taxol suppresses glutamate-induced Ca^{2+} influx and protects neurons against excitotoxicity (Furukawa and Mattson, 1995). The microtubule-associated protein tau regulates microtubule polymerization. Abnormalities in tau, including hyperphosphorylation and aggregation, occur in Alzheimer's disease and other dementing disorders. Calcium imaging studies of cultured human neuroblastoma cells overexpressing either wild-type tau or a mutant form of tau that causes inherited fronto-temporal lobe dementia and Parkinsonism linked to chromosome 17 suggest roles for tau and microtuble polymerization state in modulating VDCC (Furukawa et al., 2003). The latter study demonstrated that mutant tau enhances Ca^{2+} influx through VDCC by a mechanism involving enhanced microtubule depolymerization.

5. Glial cells and excitotoxicity

Neurons are not the only cells in the nervous system that express glutamate receptors. Indeed, each of three major types of glial cells, astrocytes, oligodendrocytes and microglia express one or more types of glutamate receptor.

5.1 Astrocytes

Non-NMDA ionotropic (AMPA/kainate) receptors and metabotropic receptors are expressed in astrocytes, and mediate Ca^{2+} influx and release from ER, respectively (Gallo and Russell, 1995). Astrocytes are very resistant to glutamate-induced cell death (Mattson and Rychlik, 1990), but can be killed by activation of AMPA/kainate receptors under certain conditions (David et al., 1996). Interestingly, astrocytes can propagate intercellular calcium waves in response to glutamate receptor activation (Haydon, 2001). Astrocytes play important roles in modifying the excitotoxic process in neurons. Astrocytes express high levels of glutamate transporter proteins and are therefore very effective in removing glutamate from extracellular compartments, including synaptic regions (Bezzi et al., 2001). They also produce several different neurotrophic factors and cytokines that have been shown to affect neuronal vulnerability to excitotoxicity including bFGF (Mattson and Rychlik, 1990) and type 1 plasminogen activator (Gabriel et al., 2003). Gap junctions, intercellular channels that permit the passage of Ca^{2+}, nucleotides and even small peptides between cells, are expressed in astrocytes. Studies in which gap junctional communication in astrocytes is manipulated have provided evidence that these intercellular channels can modify neuronal vulnerability to different insults (Blanc et al., 1998a), although the underlying mechanism remains to be established. Although astrocytes have been shown to protect neurons against excitotoxicity, they can also release glutamate and oxyradicals that can promote excitotoxicity (McNaught and Jenner, 2000).

5.2 Oligodendrocytes

Oligodendrocytes are the myelinating cell in the central nervous system located in white matter. Early investigations showed that cells in white matter are damaged in animals subjected to severe epileptic seizures (Sperk et al., 1983), and subsequent studies of cultured oligodendrocytes revealed that this type of glial cell is very sensitive to being killed by glutamate. Glutamate kills oligodendrocytes by activating AMPA receptors resulting in excessive calcium influx (Yoshioka et al., 1996; McDonald et al., 1998; Matute et al., 2001). The damage to oligodendrocytes that occurs in AD likely involves an glutamate receptor activation because oligodendrocytes are vulnerable to amyloid β-peptide toxicity, and because mutations in presenilin-1 that cause early-onset inherited Alzheimer's disease increase the vulnerability of oligodendrocytes to excitotoxic death (Pak et al., 2003). Although oligodendrocytes are located in white matter and might therefore not be expected to modify the vulnerability of neurons to excitotoxicity, they do produce several trophic factors that are known to modify neuronal vulnerability to excitotoxicity including transforming growth factor-β and basic fibroblast growth factor (bFGF) (Du and Dreyfus, 2002).

5.3 Microglia

Microglia arise from the same monocyte lineage as peripheral macrophages. Microglia respond to tissue injury by moving to the sites of cellular damage where they produce several cytotoxic substances and can phagocytose apoptotic cells (Stoll et al., 1998). Toxic substances produced by microglia include nitric oxide and superoxide (Liu et al., 2002a), and excitotoxins such as quinolinic acid (Espey et al., 1997). In addition to promoting excitotoxic neuronal death, microglia may promote excitotoxic death of oligodendrocytes (Tahraoui et al., 2001).

6. Implications for the prevention and treatment of neurological disorders

Excitotoxicity has been implicated in the death of neurons that occurs in virtually every major acute and chronic neurodegenerative disorder. Analyses of patient tissues and animal and cell culture models strongly implicate activation of glutamate receptors in the neuronal deaths that occur in epilepsy (Cendes et al., 1995; Sagratella, 1995; Meldrum et al., 1999; Ben-Ari and Cossart, 2000; Kelsey et al., 2000; Armijo et al., 2002), ischemic stroke (Collins et al., 1989; Dirnagl et al., 1999) and traumatic brain and spinal cord injuries (Ikonomidou and Turski, 1996; Agrawal and Fehlings, 1997; Springer et al., 1997; Allen et al., 1999; McAdoo et al., 1999; Gaviria et al., 2000; Mills et al., 2002), AD (Mattson et al., 1992; Mark et al., 1995a; Guo et al., 1999a), PD (Turski et al., 1991; Miranda et al., 1997), HD (Brouillet et al., 1999) and ALS (Rothstein et al., 1995; Kruman et al., 1999). In each of these disorders the neurons that degenerate express glutamate receptors, and cell culture and in vivo studies have shown that equivalent populations of neurons are susceptible to excitotoxicity.

A sampling of some of the preventative/therapeutic approaches that have proven effective in animal and/or cell culture models of both acute and chronic neurodegenerative disorders is found in Figure 5.

ANTI-EXCITOTOXIC APPROACHES

Glutamate receptor antagonists: MK-801, memantine, kynurenate

Calcium channel antagonists: nimodipine, nifedipine, dantrolene

Antioxidants: tocopherol, lipoate, uric acid, lycopene

DNA damage response inhibitors: 3-AB, PFTα

Mitochondrial/energy stabilizers: cyclosporine A, creatine

Protease inhibitors: calpain inhibitors, caspase inhibitors

Cytoskeleton-modulating agents: cytochalasin D, taxol

Neurotrophic factors and cytokines: bFGF, BDNF, GDNF, ADNF, TGFβ, TNF, sAPP

Dietary manipulations: Fasting/caloric restriction, folate

Figure 5. Examples of anti-excitotoxic strategies. Data supporting the excitoprotective effects of each approach can be found in the following references: Akins and Atkinson, 2002; Albensi et al., 2000; Allen et al., 2000; Andreassen et al., 2001a, 2001b; Beal et al., 1998; Brustovetsky et al., 2001; Caba et al., 2002; Decker and Muller, 2002; Duan et al., 2002a, 2002b, 2003; Eldadah and Faden, 2000; Endres et al., 1999; Ferrante et al., 2002; Furukawa and Mattson, 1995; Furukawa et al., 1995; Gaviria et al., 2000; Goodman and Mattson, 1996; Guo et al., 1999a, 1999b; Gurney et al., 1996; Hadley et al., 1989; Kruman et al., 2002; Liu et al., 2002b; Loetscher et al., 2001; Martinou et al., 1994; Mattson et al., 1989; Mossakowski and Gadamski, 1990; Muller and Krieglstein, 1995; Nozaki et al., 1993; Shoulson, 1998; Tymianski et al., 1993; Yu et al., 1998; Yu and Mattson, 1999.

While effective in reducing neuronal death in various models (Boxer et al., 1990; Turski et al., 1991; Akins and Atkinson, 2002), most glutamate receptor antagonists suffer from serious side-effects due to the fact that they block the normal functions of neuronal circuits. Calcium channel antagonists may also reduce excitotoxic neuronal degeneration (Nuglisch et al., 1990; Berg et al., 1995; Guo et al., 1999a). Because oxidative stress is both a consequence of and a promoter of excitotoxicity, antioxidants such as vitamin E, coenzyme Q10 and lipoate exhibit beneficial effects in models of neurodegenerative disorders (Andreassen et al., 2001a; Ferrante et al., 2002). Approaches aimed at enhancing energy metabolism, such as creatine, have exhibited neuroprotective actions in animal models of traumatic brain injury (Sullivan et al., 2000) and HD (Andreassen et al., 2001b). Anti-apoptotic agents such as p53 inhibitors (Culmsee et al, 2001; Duan et al., 2002a) and caspase inhibitors have reduced neuronal loss and improved functional outcome in some models (Loetscher et al., 2001). Additional excitoprotective approaches are listed in Figure 5.

REFERENCES

Agrawal SK, Fehlings MG. (1997) Role of NMDA and non-NMDA ionotropic glutamate receptors in traumatic spinal cord axonal injury. J Neurosci. 17:1055-1063.

Aizenman E, Hartnett KA, Reynolds IJ (1990) Oxygen free radicals regulate NMDA receptor function via a redox modulatory site. Neuron 5: 841-846.

Akins PT, Atkinson RP. (2002) Glutamate AMPA receptor antagonist treatment for ischaemic stroke. Curr Med Res Opin. 18:S9-13.

Albensi BC, Sullivan PG, Thompson MB, Scheff SW, Mattson MP. (2000) Cyclosporin ameliorates traumatic brain-injury-induced alterations of hippocampal synaptic plasticity. Exp Neurol. 162:385-389.

Allen JW, Ivanova SA, Fan L, Espey MG, Basile AS, Faden AI. Group II metabotropic glutamate receptor activation attenuates traumatic neuronal injury and improves neurological recovery after traumatic brain injury. J Pharmacol Exp Ther. 290:112-120.

Allsopp TE, Kiseley S, Wyatt S, Davies AM (1995) Role of Bcl-2 in the brain-derived neurotrophic factor survival response. Eur J Neurosci. 7:1266-1272.

Andreassen OA, Dedeoglu A, Ferrante RJ, Jenkins BG, Ferrante KL, Thomas M, Friedlich A, Browne SE, Schilling G, Borchelt DR, Hersch SM, Ross CA, Beal MF (2001a) Creatine increase survival and delays motor symptoms in a transgenic animal model of Huntington's disease. Neurobiol Dis. 8:479-491.

Andreassen OA, Dedeoglu A, Friedlich A, Ferrante KL, Hughes D, Szabo C, Beal MF (2001b) Effects of an inhibitor of poly(ADP-ribose) polymerase, desmethylselegiline, trientine, and lipoic acid in transgenic ALS mice. Exp Neurol. 168:419-424.

Ankarcrona M, Dypbukt JM, Bonfoco E, Zhivotovsky B, Orrenius S, Lipton SA, Nicotera P (1995) Glutamate-induced neuronal death: a succession of necrosis or apoptosis depending on mitochondrial function. *Neuron* 15: 961-973.

Ankarcrona M (1998) Glutamate induced cell death: apoptosis or necrosis? Prog Brain Res. 116:265-272.

Armijo JA, Valdizan EM, De Las Cuevas I, Cuadrado A. (2002) Advances in the physiopathology of epileptogenesis: molecular aspects. Rev Neurol. 34:409-429.

Barger SW, Horster D, Furukawa K, Goodman Y, Krieglstein J, Mattson MP. (1995) Tumor necrosis factors alpha and beta protect neurons against amyloid beta-peptide toxicity: evidence for involvement of a kappa B-binding factor and attenuation of peroxide and Ca2+ accumulation. Proc Natl Acad Sci U S A. 92:9328-9332.

Barger SW, Mattson MP (1996) Induction of neuroprotective kappa B-dependent transcription by secreted forms of the Alzheimer's beta-amyloid precursor. Mol Brain Res. 40:116-126.

Beal MF, Matthews RT, Tieleman A, Shults CW. (1998) Coenzyme Q10 attenuates the 1-methyl-4-phenyl-1,2,3,tetrahydropyridine (MPTP) induced loss of striatal dopamine and dopaminergic axons in aged mice. Brain Res. 783:109-114.

Ben-Ari Y, Cossart R. (2000) Kainate, a double agent that generates seizures: two decades of progress. Trends Neurosci. 23:580-587.

Berg M, Bruhn T, Frandsen A, Schousboe A, Diemer NH. (1995) Kainic acid-induced seizures and brain damage in the rat: role of calcium homeostasis. J Neurosci Res. 40:641-646.

Bezzi P, Domercq M, Vesce S, Volterra A. (2001) Neuron-astrocyte cross-talk during synaptic transmission: physiological and neuropathological implications. Prog Brain Res. 132:255-265.

Bi X, Chang V, Siman R, Tocco G, Baudry M (1996) Regional distribution and time-course of calpain activation following kainate-induced seizure activity in adult rat brain. Brain Res. 726:98-108.

Blanc EM, Bruce-Keller AJ, Mattson MP (1998a) Astrocytic gap junctional communication decreases neuronal vulnerability to oxidative stress-induced disruption of Ca2+ homeostasis and cell death. J Neurochem. 70:958-970.

Blanc EM, Keller JN, Fernandez S, Mattson MP (1998b) 4-hydroxynonenal, a lipid peroxidation product, inhibits glutamate transport in astrocytes. *Glia* 22: 149-160.

Boxer PA, Cordon JJ, Mann ME, Rodolosi LC, Vartanian MG, Rock DM, Taylor CP, Marcoux FW. (1990) Comparison of phenytoin with noncompetitive N-methyl-D-aspartate antagonists in a model of focal brain ischemia in rat. Stroke 21:III47-51.

Brouillet E, Conde F, Beal MF, Hantraye P (1999) Replicating Huntington's disease phenotype in experimental animals. Prog Neurobiol. 59:427-468.

Bruce AJ, Boling W, Kindy MS, Peschon J, Kraemer PJ, Carpenter MK, Holtsberg FW, Mattson MP (1996) Altered neuronal and microglial responses to brain injury in mice lacking TNF receptors. *Nature Medicine* 2: 788-794.

Bruce-Keller AJ, Begley JG, Fu W, Butterfield DA, Bredesen DE, Hutchins JB, Hensley K, Mattson MP (1998) Bcl-2 protects isolated plasma and mitochondrial membranes against lipid peroxidation induced by hydrogen peroxide and amyloid β-peptide. *J. Neurochem.* 70: 31-39.

Bruce-Keller AJ, Umberger G, McFall R, Mattson MP (1999) Food restriction reduces brain damage and improves behavioral outcome following excitotoxic and metabolic insults. *Ann. Neurol.* 45: 8-15.

Brustovetsky N, Brustovetsky T, Dubinsky JM (2001) On the mechanisms of neuroprotection by creatine and phosphocreatine. J Neurochem. 76:425-434.

Caba E, Brown QB, Kawasaki B, Bahr BA (2002) Peptidyl alpha-keto amide inhibitor of calpain blocks excitotoxic damage without affecting signal transduction events. J Neurosci Res. 67:787-794.

Cendes F, Andermann F, Carpenter S, Zatorre RJ, Cashman NR. (1995) Temporal lobe epilepsy caused by domoic acid intoxication: evidence for glutamate receptor-mediated excitotoxicity in humans. Ann Neurol. 37:123-126.

Chan PH, Chu L, Chen SF, Carlson EJ, Epstein CJ (1990) Reduced neurotoxicity in transgenic mice overexpressing human copper-zinc-superoxide dismutase. Stroke 21:III80-III82.

Chen RW, Chuang DM (1999) Long term lithium treatment suppresses p53 and Bax expression but increases Bcl-2 expression. A prominent role in neuroprotection against excitotoxicity. J Biol Chem. 274:6039-6042.

Cheng B, Mattson MP (1991) NGF and bFGF protect rat and human central neurons against hypoglycemic damage by stabilizing calcium homeostasis. *Neuron* 7: 1031-1041.

Cheng B, Mattson MP (1992) IGF-I and IGF-II protect cultured hippocampal and septal neurons against calcium-mediated hypoglycemic damage. *J. Neurosci.* 12: 1558-1566.

Cheng B, Mattson MP (1994) NT-3 and BDNF protect CNS neurons against metabolic/excitotoxic insults. *Brain Res.* 640: 56-67.

Cheng B, Goodman Y, Begley JG, Mattson MP (1994a) Neurotrophin 4/5 protects hippocampal and cortical neurons against energy deprivation- and excitatory amino acid-induced injury. *Brain Res.* 650: 331-335.

Cheng B, Christakos S, Mattson MP (1994b) Tumor necrosis factors protect neurons against excitotoxic/metabolic insults and promote maintenance of calcium homeostasis. *Neuron* 12: 139-153.

Cheng B, Mattson MP (1995) PDGFs protect hippocampal neurons against energy deprivation and oxidative injury: evidence for induction of antioxidant pathways. J Neurosci. 15:7095-7104.

Cheng B, Furukawa K, O'Keefe JA, Goodman Y, Kihiko M, Fabian T, Mattson MP (1995) Basic fibroblast growth factor selectively increases AMPA-receptor subunit GluR1 protein level and differentially modulates Ca^{2+} responses to AMPA and NMDA in hippocampal neurons. *J. Neurochem.* 65: 2525-2536.

Collins RC, Dobkin BH, Choi DW. (1989) Selective vulnerability of the brain: new insights into the pathophysiology of stroke. Ann Intern Med. 110:992-1000.

Conover JC, Yancopoulos GD (1997) Neurotrophin regulation of the developing nervous system: analyses of knockout mice. Rev Neurosci. 8:13-27.

Culcasi M, Lafon-Cazal M, Pietri S, Bockaert J (1994) Glutamate receptors induce a burst of superoxide via activation of nitric oxide synthase in arginine-depleted neurons. J Biol Chem. 269:12589-12593.

Culmsee C, Zhu X, Yu QS, Chan SL, Camandola S, Guo Z, Greig NH, Mattson MP (2001) A synthetic inhibitor of p53 protects neurons against death induced by ischemic and excitotoxic insults, and amyloid beta-peptide. J Neurochem. 77:220-228.

Dargusch R, Piasecki D, Tan S, Liu Y, Schubert D (2001) The role of Bax in glutamate-induced nerve cell death. J Neurochem. 76:295-301.

David JC, Yamada KA, Bagwe MR, Goldberg MP. (1996) AMPA receptor activation is rapidly toxic to cortical astrocytes when desensitization is blocked. J Neurosci. 16:200-209.

Dawson VL, Kizushi VM, Huang PL, Snyder SH, Dawson TM (1996) Resistance to neurotoxicity in cortical cultures from neuronal nitric oxide synthase-deficient mice. J Neurosci. 16:2479-2487.

Dawson VL, Dawson TM (1998) Nitric oxide in neurodegeneration. Prog Brain Res. 118:215-229.

Decker P, Muller S (2002) Modulating poly (ADP-ribose) polymerase activity: potential for the prevention and therapy of pathogenic situations involving DNA damage and oxidative stress. Curr. Pharm. Biotechnol. 3: 275-283.

Didier M, Bursztajn S, Adamec E, Passani L, Nixon RA, Coyle JT, Wei JY, Berman SA. (1996) DNA strand breaks induced by sustained glutamate excitotoxicity in primary neuronal cultures. J Neurosci. 16:2238-2250.

Dirnagl U, Iadecola C, Moskowitz MA (1999) Pathobiology of ischemic stroke: an integrated review. *Trends Neurosci* 22:391-397.

Doble A (1999) The role of excitotoxicity in neurodegenerative disease: implications for therapy. Pharmacol Ther. 81:163-221.

Du Y, Dreyfus CF (2002) Oligodendrocytes as providers of growth factors. J Neurosci Res. 68:647-654.

Du Y, Bales KR, Dodel RC, Hamilton-Byrd E, Horn JW, Czilli DL, Simmons LK, Ni B, Paul SM (1997) Activation of a caspase 3-related cysteine protease is required for glutamate-mediated apoptosis of cultured cerebellar granule neurons. *Proc. Natl. Acad. Sci. U.S.A.* 94: 11657-11662.

Duan W, Mattson MP (1999) Dietary restriction and 2-deoxyglucose administration improve behavioral outcome and reduce degeneration of dopaminergic neurons in models of Parkinson's disease. J Neurosci Res. 57:195-206.

Duan W, Rangnekar VM, Mattson MP (1999) Prostate apoptosis response-4 production in synaptic compartments following apoptotic and excitotoxic insults: evidence for a pivotal role in mitochondrial dysfunction and neuronal degeneration. J Neurochem. 72:2312-2322.

Duan W, Guo Z, Mattson MP (2001) Brain-derived neurotrophic factor mediates an excitoprotective effect of dietary restriction in mice. J Neurochem. 76:619-626.

Duan W, Zhu X, Ladenheim B, Yu QS, Guo Z, Oyler J, Cutler RG, Cadet JL, Greig NH, Mattson MP (2002a) p53 inhibitors preserve dopamine neurons and motor function in experimental parkinsonism. Ann Neurol. 52:597-606.

Duan W, Ladenheim B, Cutler RG, Kruman II, Cadet JL, Mattson MP (2002b) Dietary folate deficiency and elevated homocysteine levels endanger dopaminergic neurons in models of Parkinson's disease. J Neurochem. 80:101-110.

Duan W, Guo Z, Jiang H, Ware M, Li XJ, Mattson MP (2003) Dietary restriction normalizes glucose metabolism and BDNF levels, slows disease progression and increases survival in Huntingtin mutant mice. *Proc. Natl. Acad. Sci. U.S.A.* In press.

Dugan LL, Sensi SL, Canzoniero LM, Handran SD, Rothman SM, Goldberg MP, Choi DW (1995) Mitochondrial production of reactive oxygen species in cortical neurons following exsposure to N-methyl-D-aspartate. *J. Neurosci.* 15: 6377-6388.

Eldadah BA, Faden AI. (2000) Caspase pathways, neuronal apoptosis, and CNS injury. J Neurotrauma. 17:811-829.

Endres M, Fink K, Zhu J, Stagliano NE, Bondada V, Geddes JW, Azuma T, Mattson MP, Kwiatkowski DJ, Moskowitz MA (1999) Neuroprotective effects of gelsolin and cytochalasin D during murine stroke. *J. Clin. Invest.* In press.

Espey MG, Chernyshev ON, Reinhard JF Jr, Namboodiri MA, Colton CA. (1997) Activated human microglia produce the excitotoxin quinolinic acid. Neuroreport. 8:431-434.

Ferrante RJ, Andreassen OA, Dedeoglu A, Ferrante KL, Jenkins BG, Hersch SM, Beal MF (2002) Therapeutic effects of coenzyme Q10 and remacemide in transgenic mouse models of Huntington's disease. J Neurosci. 22:1592-1599.

Frandsen A, Schousboe A (1991) Dantrolene prevents glutamate cytotoxicity and calcium release from intracelluar stores in cultured cerebral cortical neurons. *J Neurochem* 56: 1075-1078.

Furukawa K, Mattson MP (1995) Taxol stabilizes $[Ca^{2+}]_i$ and protects hippocampal neurons against excitotoxicity. *Brain Res.* 689: 141-146.

Furukawa K, Smith-Swintosky VL, Mattson MP (1995) Evidence that actin depolymerization protects hippocampal neurons against excitotoxicity by stabilizing $[Ca^{2+}]_i$. *Exp. Neurol.* 133: 153-163.

Furukawa K, Barger SW, Blalock E, Mattson MP (1996) Activation of K^+ channels and suppression of neuronal activity by secreted β-amyloid precursor protein. *Nature* 379: 74-78.

Furukawa K, Fu W, Li Y, Witke W, Kwiatkowski DJ, Mattson MP (1997) The actin-severing protein gelsolin modulates calcium channel and NMDA receptor activities and vulnerability to excitotoxicity in hippocampal neurons. *J. Neurosci.* 17: 8178-8186.

Furukawa K, Mattson MP (1998) The transcription factor NF-κB mediates increases in calcium currents and decreases in NMDA and AMPA/kainate-induced currents in response to TNFα in hippocampal neurons. *J. Neurochem.* 70: 1876-1886.

Furukawa K, Wang Y, Yao P, Fu W, Mattson MP, Itoyama Y, Onodera H, D'Souza I, Poorkaj P, Bird D, Schellenberg GD (2003) Prolonged calcium channel activation is responsible for the neurotoxic action of a familial frontotemporal dementia tau mutation *J. Neurosci.* Submitted.

Gabriel C, Ali C, Lesne S, Fernandez-Monreal M, Docagne F, Plawinski L, MacKenzie ET, Buisson A, Vivien D. (2003) Transforming growth factor alpha-induced expression of type 1 plasminogen activator inhibitor in astrocytes rescues neurons from excitotoxicity. FASEB J. 17:277-279.

Gallo V, Russell JT (1995) Excitatory amino acid receptors in glia: different subtypes for distinct functions? J Neurosci Res. 42:1-8.

Gary DS, Mattson MP (2001) Integrin signaling via the PI3-kinase-Akt pathway increases neuronal resistance to glutamate-induced apoptosis. J Neurochem. 76:1485-1496.

Gary DS, Milhavet O, Camandola S, Mattson MP (2003) Essential role for integrin linked kinase in Akt-mediated integrin survival signaling in hippocampal neurons. J Neurochem. 84:878-890.

Gaviria M, Privat A, d'Arbigny P, Kamenka J, Haton H, Ohanna F. (2000) Neuroprotective effects of a novel NMDA antagonist, Gacyclidine, after experimental contusive spinal cord injury in adult rats. Brain Res. 874:200-209.

Gilman CP, Mattson MP (2002) Do apoptotic mechanisms regulate synaptic plasticity and growth-cone motility? *Neuromolecular Med.* 2: 197-214.

Glazner GW, Mattson MP (1999) Differential effects of BDNF, ADNF-9, and TNFalpha on levels of NMDA receptor subunits, calcium homeostasis, and neuronal vulnerability to excitotoxicity. *Exp. Neurol.* 161:442-452.

Glazner GW, Chan SL, Lu C, Mattson MP (2000) Caspase-mediated degradation of AMPA receptor subunits: a mechanism for preventing excitotoxic necrosis and ensuring apoptosis. *J. Neurosci.* 20: 3641-3649.

Goodman Y, Steiner MR, Steiner SM, Mattson MP (1994) Nordihydroguaiaretic acid protects hippocampal neurons against amyloid beta-peptide toxicity, and attenuates free radical and calcium accumulation. Brain Res. 654:171-176.

Goodman Y, Mattson MP (1996) K^+ channel openers protect hippocampal neurons against oxidative injury and amyloid β-peptide toxicity. *Brain Res.* 706: 328-332.

Goodman Y, Bruce AJ, Cheng B, Mattson MP (1996) Estrogens attenuate and corticosterone exacerbates excitotoxicity, oxidative injury, and amyloid beta-peptide toxicity in hippocampal neurons. J Neurochem. 66:1836-1844.

Gorgias N, Maidatsi P, Tsolaki M, Alvanou A, Kiriazis G, Kaidoglou K, Giala M. (1996) Hypoxic pretreatment protects against neuronal damage of the rat hippocampus induced by severe hypoxia. Brain Res. 714:215-225.

Gunasekar PG, Kanthasamy AG, Borowitz JL, Isom GE (1995) NMDA receptor activation produces concurrent generation of nitric oxide and reactive oxygen species: implication for cell death. J Neurochem. 65:2016-2021.

Guo Q, Fu W, Xie J, Luo H, Sells SF, Geddes JW, Bondada V, Rangnekar V, Mattson MP (1998) Par-4 is a mediator of neuronal degeneration associated with the pathogenesis of Alzheimer's disease. *Nature Med.* 4: 957-962.

Guo Q, Fu W, Sopher BL, Miller MW, Ware CB, Martin GM, Mattson MP (1999a) Increased vulnerability of hippocampal neurons to excitotoxic necrosis in presenilin-1 mutant knock-in mice. *Nature Med.* 5: 101-107.

Guo Q, Sebastian L, Sopher BL, Miller MW, Glazner GW, Ware CB, Martin GM, Mattson MP (1999b) Neurotrophic factors [activity-dependent neurotrophic factor (ADNF) and basic fibroblast growth factor (bFGF)] interrupt excitotoxic neurodegenerative cascades promoted by a presenilin-1 mutation. *Proc. Natl. Acad. Sci. U.S.A.* 96: 4125-4130.

Gurney ME, Cutting FB, Zhai P, Doble A, Taylor CP, Andrus PK, Hall ED (1996) Benefit of vitamin E, riluzole, and gabapentin in a transgenic model of familial amyotrophic lateral sclerosis. *Ann. Neurol.* 39: 147-157.

Guttmann RP, Sokol S, Baker DL, Simpkins KL, Dong Y, Lynch DR (2002) Proteolysis of the N-methyl-d-aspartate receptor by calpain in situ. J Pharmacol Exp Ther. 302:1023-1030.

Hadley MN, Zabramski JM, Spetzler RF, Rigamonti D, Fifield MS, Johnson PC. (1989) The efficacy of intravenous nimodipine in the treatment of focal cerebral ischemia in a primate model. Neurosurgery. 25:63-70.

Haydon PG (2001) GLIA: listening and talking to the synapse. Nat Rev Neurosci. 2:185-193.

Hayes RL, Jenkins LW, Lyeth BG (1992) Neurotransmitter-mediated mechanisms of traumatic brain injury: acetylcholine and excitatory amino acids. *J. Neurotrauma* 9: S173-S187.

Ikonomidou C, Turski L. (1996) Prevention of trauma-induced neurodegeneration in infant and adult rat brain: glutamate antagonists. Metab Brain Dis. 11:125-141.

Keller JN, Mark RJ, Bruce AJ, Blanc EM, Rothstein JD, Uchida K, Mattson MP (1997a) 4-hydroxynonenal, an aldehydic product of membrane lipid peroxidation, impairs glutamate transport and mitochondrial function in synaptosomes. *Neuroscience* 80: 685-696.

Keller JN, Kindy MS, Holtsberg FW, St Clair DK, Yen HC, Germeyer A, Steiner SM, Bruce-Keller AJ, Hutchins JB, Mattson MP (1998) Mitochondrial manganese superoxide dismutase prevents neural apoptosis and reduces ischemic brain injury: suppression of peroxynitrite production, lipid peroxidation, and mitochondrial dysfunction. J Neurosci. 18:687-697.

Kelsey JE, Sanderson KL, Frye CA. (2000) Perforant path stimulation in rats produces seizures, loss of hippocampal neurons, and a deficit in spatial mapping which are reduced by prior MK-801. Behav Brain Res. 107:59-69.

Khodorov B, Pinelis V, Vergun O, Storozhevykh T, Vinskaya (1996) Mitochondrial deenergization underlies neuronal calcium overload following a prolonged glutamate challenge. FEBS Lett. 397:230-234.

Koketsu N, Berlove DJ, Moskowitz MA, Kowall NW, Caday CG, Finklestein SP (1994) Pretreatment with intraventricular basic fibroblast growth factor (bFGF) decreases infarct size following focal cerebral ischemia in rats. *Ann. Neurol.* 35: 451-457.

Kruman II, Pedersen WA, Springer JE, Mattson MP (1999) ALS-linked Cu/Zn-SOD mutation increases vulnerability of motor neurons to excitotoxicity by a mechanism involving increased oxidative stress and perturbed calcium homeostasis. *Exp. Neurol.* 160: 28-39.

Kruman II, Culmsee C, Chan SL, Kruman Y, Guo Z, Penix L, Mattson MP. (2000) Homocysteine elicits a DNA damage response in neurons that promotes apoptosis and hypersensitivity to excitotoxicity. J Neurosci. 20:6920-6926.

Kruman II, Kumaravel TS, Lohani A, Pedersen WA, Cutler RG, Kruman Y, Haughey N, Lee J, Evans M, Mattson MP. (2002) Folic acid deficiency and homocysteine impair DNA repair in hippocampal neurons and sensitize them to amyloid toxicity in experimental models of Alzheimer's disease. J Neurosci. 22:1752-1762.

Kullmann DM, Asztely F, Walker MC (2000) The role of mammalian ionotropic receptors in synaptic plasticity: LTP, LTD and epilepsy. Cell Mol Life Sci. 57:1551-1561.

Lee J, Duan W, Mattson MP (2002) Evidence that brain-derived neurotrophic factor is required for basal neurogenesis and mediates, in part, the enhancement of neurogenesis by dietary restriction in the hippocampus of adult mice. J Neurochem. 82:1367-1375.

Leist M, Volbracht C, Kuhnle S, Fava E, Ferrando-May E, Nicotera P (1997) Caspase-mediated apoptosis in neuronal excitotoxicity triggered by nitric oxide. Mol Med. 3:750-764.

Lipscombe D, Madison DV, Poenie M, Reuter H, Tsien RW, Tsien RY (1988) Imaging of cytosolic Ca2+ transients arising from Ca2+ stores and Ca2+ channels in sympathetic neurons. Neuron. 1:355-365.

Lipsky RH, Xu K, Zhu D, Kelly C, Terhakopian A, Novelli A, Marini AM. (2001) Nuclear factor kappaB is a critical determinant in N-methyl-D-aspartate receptor-mediated neuroprotection. J Neurochem. 78:254-264.

Lipton SA, Choi YB, Pan ZH, Lei SZ, Chen HSV, Sucher NJ, Loscaizo J, Singel DJ, Stamler JS (1993) A redox-based mechanism for the neuroprotective and neurodestructive effects of nitric oxide and related nitroso-compounds. *Nature* 364: 626-632.

Liu B, Gao HM, Wang JY, Jeohn GH, Cooper CL, Hong JS. (2002a) Role of nitric oxide in inflammation-mediated neurodegeneration. Ann N Y Acad Sci. 962:318-331.

Liu D, Lu C, Wan R, Auyeung WW, Mattson MP. (2002b) Activation of mitochondrial ATP-dependent potassium channels protects neurons against ischemia-induced death by a mechanism involving suppression of Bax translocation and cytochrome c release. J Cereb Blood Flow Metab. 22:431-443.

Loetscher H, Niederhauser O, Kemp J, Gill R. (2001) Is caspase-3 inhibition a valid therapeutic strategy in cerebral ischemia? Drug Discov Today. 6:671-680.

Lowenstein DH, Chan P, Miles M (1991) The stress protein response in cultured neurons: characterization and evidence for a protective role in excitotoxicity. *Neuron* 7: 1053-1060.

Lu CB, Fu W, Mattson MP (2001) Caspase-mediated suppression of glutamate (AMPA) receptor channel activity in hippocampal neurons in response to DNA damage promotes apoptosis and prevents necrosis: implications for neurological side effects of cancer therapy and neurodegenerative disorders. *Neurobol. Dis.* 8:194-206.

Lu C, Chan SL, Fu W, Mattson MP (2002a) The lipid peroxidation product 4-hydroxynonenal facilitates opening of voltage-dependent Ca2+ channels in neurons by increasing protein tyrosine phosphorylation. J Biol Chem. 277:24368-24375.

Lu C, Fu W, Mattson MP (2002b) Direct cleavage of AMPA receptor subunit GluR1 and suppression of AMPA currents by caspase-3: implications for synaptic plasticity and excitotoxic neuronal death. *NeuroMolecular Med.* 1: 69-79.

Ludolph AC, Meyer T, Riepe MW. (2000) The role of excitotoxicity in ALS--what is the evidence? J Neurol. 247: I7-I16.

Marini AM, Paul SM (1992) N-methyl-D-aspartate receptor-mediated neuroprotection in cerebellar granule cells requires new RNA and protein synthesis. Proc Natl Acad Sci U S A. 89:6555-6559.

Marini AM, Rabin SJ, Lipsky RH, Mocchetti I. (1998) Activity-dependent release of brain-derived neurotrophic factor underlies the neuroprotective effect of N-methyl-D-aspartate. J Biol Chem. 273:29394-29399.

Mark RJ, Hensley K, Butterfield DA, Mattson MP (1995) Amyloid β-peptide impairs ion-motive ATPase activities: evidence for a role in loss of neuronal Ca^{2+} homeostasis and cell death. *J. Neurosci.* 15: 6239-6249.

Mark RJ, Lovell MA, Markesbery WR, Uchida K, Mattson MP (1997a) A role for 4-hydroxynonenal in disruption of ion homeostasis and neuronal death induced by amyloid β-peptide. *J. Neurochem.* 68: 255-264.

Mark RJ, Pang Z, Geddes JW, Uchida K, Mattson MP (1997b) Amyloid β-peptide impairs glucose uptake in hippocampal and cortical neurons: involvement of membrane lipid peroxidation. *J. Neurosci.* 17: 1046-1054.

Martinou JC, Dubois-Dauphin M, Staple JK, Rodriguez I, Frankowski H, Missotten M, Albertini P, Talabot D, Catsicas S, Pietra C (1994) Overexpression of Bcl-2 in transgenic mice protects neurons from naturally occurring cell death and experimental ischemia. Neuron 13:1017-1030.

Mattson MP, Kater SB (1989) Excitatory and inhibitory neurotransmitters in the generation and degeneration of hippocampal neuroarchitecture. *Brain Res.* 478: 337-348.

Mattson MP, Murrain M, Guthrie PB, Kater SB (1989) Fibroblast growth factor and glutamate: Opposing actions in the generation and degeneration of hippocampal neuroarchitecture. *J. Neurosci.* 9: 3728-3740.

Mattson MP, Rychlik B (1990) Glia protect hippocampal neurons against excitatory amino acid-induced degeneration: involvement of fibroblast growth factor. Int J Dev Neurosci. 8:399-415.

Mattson MP, Cheng B, Davis D, Bryant K, Lieberburg I, Rydel RE. (1992) beta-Amyloid peptides destabilize calcium homeostasis and render human cortical neurons vulnerable to excitotoxicity. J Neurosci. 12:376-389.

Mattson MP, Cheng B, Culwell A, Esch F, Lieberburg I, Rydel RE (1993a) Evidence for excitoprotective and intraneuronal calcium-regulating roles for secreted forms of β-amyloid precursor protein. *Neuron* 10: 243-254.

Mattson MP, Zhang Y, Bose S (1993b) Growth factors prevent mitochondrial dysfunction, loss of calcium homeostasis and cell injury, but not ATP depletion in hippocampal neurons deprived of glucose. *Exp. Neurol.* 121: 1-13.

Mattson MP, Kumar K, Cheng B, Wang H, Michaelis EK (1993c) Basic FGF regulates the expression of a functional 71 kDa NMDA receptor protein that mediates calcium influx and neurotoxicity in cultured hippocampal neurons. *J. Neurosci.* 13: 4575-4588.

Mattson MP, Lovell MA, Furukawa K, Markesbery WR (1995) Neurotrophic factors attenuate glutamate-induced accumulation of peroxides, elevation of $[Ca^{2+}]_i$ and neurotoxicity, and increase antioxidant enzyme activities in hippocampal neurons. *J. Neurochem.* 65:1740-1751.

Mattson MP (1997) Neuroprotective signal transduction: relevance to stroke. Neurosci Biobehav Rev. 21:193-206.

Mattson MP. (1997) Cellular actions of beta-amyloid precursor protein and its soluble and fibrillogenic derivatives. Physiol. Rev. 77:1081-1132.

Mattson MP, Keller JN, Begley JG (1998) Evidence for synaptic apoptosis. *Exp. Neurol.* 153: 35-48.

Mattson MP (2000) Apoptosis in neurodegenerative disorders. Nat. Rev. Mol. Cell Biol. 1:120-129.

Mattson MP, Zhu H, Yu J, Kindy MS (2000) Presenilin-1 mutation increases neuronal vulnerability to focal ischemia in vivo and to hypoxia and glucose deprivation in cell culture: involvement of perturbed calcium homeostasis. J Neurosci. 20:1358-1364.

Mattson MP, Camandola S (2001) NF-kappaB in neuronal plasticity and neurodegenerative disorders. J. Clin. Invest. 107: 247-254.

Mattson MP (2003) Excitotoxic and excitoprotective mechanisms: abundant targets for the prevention and treatment of neurodegenerative disorders. Neuromolecular Med. In press.

Matute C, Alberdi E, Domercq M, Perez-Cerda F, Perez-Samartin A, Sanchez-Gomez MV (2001) The link between excitotoxic oligodendroglial death and demyelinating diseases. Trends Neurosci. 24:224-230.

McAdoo DJ, Xu GY, Robak G, Hughes MG. (1999) Changes in amino acid concentrations over time and space around an impact injury and their diffusion through the rat spinal cord. Exp Neurol. 159:538-544.

McDonald JW, Althomsons SP, Hyrc KL, Choi DW, Goldberg MP (1998) Oligodendrocytes from forebrain are highly vulnerable to AMPA/kainate receptor-mediated excitotoxicity. Nat Med. 4:291-297.

McNaught KS, Jenner P. (2000) Extracellular accumulation of nitric oxide, hydrogen peroxide, and glutamate in astrocytic cultures following glutathione depletion, complex I inhibition, and/or lipopolysaccharide-induced activation. Biochem Pharmacol. 60:979-988.

Meldrum BS, Akbar MT, Chapman AG. (1999) Glutamate receptors and transporters in genetic and acquired models of epilepsy. Epilepsy Res. 36:189-204.

Michaelis EK (1998) Molecular biology of glutamate receptors in the central nervous system and their role in excitotoxicity, oxidative stress and aging. Prog Neurobiol. 54:369-415.

Mills CD, Johnson KM, Hulsebosch CE. (2002) Group I metabotropic glutamate receptors in spinal cord injury: roles in neuroprotection and the development of chronic central pain. J Neurotrauma. 19:23-42.

Miranda AF, Boegman RJ, Beninger RJ, Jhamandas K (1997) Protection against quinolinic acid-mediated excitotoxicity in nigrostriatal dopaminergic neurons by endogenous kynurenic acid. Neuroscience. 78:967-975.

Mossakowski MJ, Gadamski R (1990) Nimodipine prevents delayed neuronal death of sector CA1 pyramidal cells in short-term forebrain ischemia in mongolian gerbils. *Stroke* 21: IV120-122.

Muller U, Krieglstein J (1995) Prolonged pretreatment with alpha-lipoic acid protects cultured neurons against hypoxic, glutamate-, or iron-induced injury. *J. Cereb. Blood Flow Metab.* 5: 624-630.

Nellgard B, Wieloch T (1992) Cerebral protection by AMPA- and NMDA-receptor antagonists administered after severe insulin-induced hypoglycemia. Exp Brain Res. 92:259-266.

Nicholls DG, Budd SL, Ward MW, Castilho RF (1999) Excitotoxicity and mitochondria. Biochem Soc Symp. 66:55-67.

Novelli A, Reilly JA, Lysko PG, Henneberry RC (1988) Glutamate becomes neurotoxic via the N-methyl-D-aspartate receptor when intracellular energy levels are reduced. *Brain Res.* 451: 205-212.

Nozaki K, Finklestein SP, Beal MF (1993) Basic fibroblast growth factor protects against hypoxia-ischemia and NMDA neurotoxicity in neonatal rats. *J. Cereb. Blood Flow Metab.* 13: 221-228.

Nuglisch J, Karkoutly C, Mennel HD, Rossberg C, Krieglstein J. (1990) Protective effect of nimodipine against ischemic neuronal damage in rat hippocampus without changing postischemic cerebral blood flow. J Cereb Blood Flow Metab. 10:654-659.

Olney JW (1969) Brain lesions, obesity, and other disturbances in mice treated with monosodium glutamate. *Science* 164:719-721.

Pak K, Chan SL, Mattson MP (2003) Presenilin-1 mutation sensitizes oligodendrocytes to glutamate and amyloid toxicities and exacerbates which matter damage and memory impairment in mice. *Neuromolecular Med.* In press.

Pappas IS, Parnavelas JG (1997) Neurotrophins and basic fibroblast growth factor induce the differentiation of calbindin-containing neurons in the cerebral cortex. Exp Neurol. 144:302-314.

Prehn JHM, Backhaub C, Krieglstein J (1993) Transforming growth factor-β1 prevents glutamate toxicity in rat neocortical cell cultures and protects mouse from ischemic injury in vivo. *J. Cereb. Blood Flow Metab.* 13: 521-525.

Rabchevsky AG, Fugaccia I, Turner AF, Blades DA, Mattson MP, Scheff SW. (2000) Basic fibroblast growth factor (bFGF) enhances functional recovery following severe spinal cord injury to the rat. Exp Neurol. 164:280-291.

Rothstein JD, VanKammen BA, Levey BA, Martin LJ, Kuncl RW (1995) Selective loss of glial glutamate transporter GLT-1 in amyotrophic lateral sclerosis. *Ann. Neurol.* 38: 73-84.

Sagratella S. (1995) NMDA antagonists: antiepileptic-neuroprotective drugs with diversified neuropharmacological profiles. Pharmacol Res. 32:1-13.

Saroff D, Delfs J, Kuznetsov D, Geula C (2000) Selective vulnerability of spinal cord motor neurons to non-NMDA toxicity. Neuroreport 11:1117-1121.

Scheibel AB (1979) Dendritic changes in senile and presenile dementias. Res Publ Assoc Res Nerv Ment Dis. 57:107-124.

Sengpiel B, Preis E, Krieglstein J, Prehn JH (1998) NMDA-induced superoxide production and neurotoxicity in cultured rat hippocampal neurons: role of mitochondria. Eur J Neurosci. 10:1903-1910.

Shoulson I (1998) DATATOP: a decade of neuroprotective inquiry. Parkinson Study Group. Deprenyl And Tocopherol Antioxidative Therapy Of Parkinsonism. Ann Neurol. 44:S160-166.

Sieradzan KA, Mann DM. (2001) The selective vulnerability of nerve cells in Huntington's disease. Neuropathol Appl Neurobiol. 27: 1-21.

Smith-Swintosky VL, Pettigrew LC, Craddock SD, Culwell AR, Rydel RE, Mattson MP (1994) Secreted forms of beta-amyloid precursor protein protect against ischemic brain injury. *J. Neurochem.,* 63: 781-784.

Smith-Swintosky VL, Zimmer S, Fenton JW, Mattson MP (1995) Protease nexin-I and thrombin modulate neuronal Ca^{2+} homeostasis and sensitivity to glucose deprivation-induced injury. *J. Neurosci.* 15: 5840-5850.

Sommer B, Burnashev N, Verdoorn TA, Keinanen K, Sakmann B, Seeburg PH (1992) A glutamate receptor channel with high affinity for domoate and kainate. EMBO J. 11:1651-1656.

Sonsalla PK, Albers DS, Zeevalk GD. (1998) Role of glutamate in neurodegeneration of dopamine neurons in several animal models of parkinsonism. Amino Acids. 14: 69-74.

Sperk G, Lassmann H, Baran H, Kish SJ, Seitelberger F, Hornykiewicz O. (1983) Kainic acid induced seizures: neurochemical and histopathological changes. Neuroscience. 10:1301-1315.

Springer JE, Azbill RD, Mark RJ, Begley JG, Waeg G, Mattson MP. (1997) 4-hydroxynonenal, a lipid peroxidation product, rapidly accumulates following traumatic spinal cord injury and inhibits glutamate uptake. J Neurochem. 68:2469-2476.

Stoll G, Jander S, Schroeter M (1998) Inflammation and glial responses in ischemic brain lesions. Prog Neurobiol. 56:149-171.

Sullivan PG, Geiger JD, Mattson MP, Scheff SW. (2000) Dietary supplement creatine protects against traumatic brain injury. Ann Neurol. 48:723-729.

Swann JW, Al-Noori S, Jiang M, Lee CL (2000) Spine loss and other dendritic abnormalities in epilepsy. Hippocampus 10:617-625.

Tahraoui SL, Marret S, Bodenant C, Leroux P, Dommergues MA, Evrard P, Gressens P. (2001) Central role of microglia in neonatal excitotoxic lesions of the murine periventricular white matter. Brain Pathol. 11:56-71.

Turski L, Bressler K, Rettig KJ, Loschmann PA, Wachtel H. (1991) Protection of substantia nigra from MPP+ neurotoxicity by N-methyl-D-aspartate antagonists. Nature 349:414-418.

Tymianski M, Wallace MC, Spigelman I, Uno M, Carlen PL, Tator CH, Charlton MP (1993) Cell-permeant Ca^{2+} chelators reduce early excitotoxic and ischemic neuronal injury in vitro and in vivo. *Neuron* 11: 221-235.

Vergun O, Sobolevsky AI, Yelshansky MV, Keelan J, Khodorov BI, Duchen MR (2001) Exploration of the role of reactive oxygen species in glutamate neurotoxicity in rat hippocampal neurones in culture. J Physiol. 531:147-163.

Volbracht C, Fava E, Leist M, Nicotera P (2001) Calpain inhibitors prevent nitric oxide-triggered excitotoxic apoptosis. Neuroreport. 12:3645-3648.

Wang GJ, Randall RD, Thayer SA (1994) Glutamate-induced intracellular acidification of cultured hippocampal neurons demonstrates altered energy metabolism resulting from Ca2+ loads. J Neurophysiol. 72:2563-2569.

Wang Y, Mattson MP, Furukawa K (2002) Endoplasmic reticulum calcium release is modulated by actin polymerization. *J. Neurochem.* 82: 945-952.

Weiss JH, Hartley DM, Koh J, Choi DW (1990) The calcium channel blocker nifedipine attenuates slow excitatory amino acid neurotoxicity. *Science* 247: 1474-1477.

Werth JL, Thayer SA (1994) Mitochondria buffer physiological calcium loads in cultured rat dorsal root ganglion neurons. *J. Neurosci.* 14: 348-356.

Yoshioka A, Bacskai B, Pleasure D (1996) Pathophysiology of oligodendroglial excitotoxicity. J Neurosci Res. 46:427-437.

Yu ZF, Bruce-Keller AJ, Goodman Y, Mattson MP (1998) Uric acid protects neurons against excitotoxic and metabolic insults in cell culture, and against focal ischemic brain injury in vivo. *J. Neurosci. Res.* 53: 613-625.

Yu ZF, Mattson MP (1999) Dietary restriction and 2-deoxyglucose administration reduce focal ischemic brain damage and improve behavioral outcome: evidence for a preconditioning mechanism. J Neurosci Res. 57:830-839.

Yu Z, Luo H, Fu W, Mattson MP (1999) The endoplasmic reticulum stress-responsive protein GRP78 protects neurons against excitotoxicity and apoptosis: suppression of oxidative stress and stabilization of calcium homeostasis. Exp Neurol. 155:302-314.

Part II

Acute and Chronic Neurological Disorders

Chapter 7

EPILEPSY AND SEIZURES: EXCITOTOXICITY OR EXCITOTROPHICITY?

Karen Gale
Department of Pharmacology, Georgetown University, Washington, D.C., U.S.A.

Abstract: Excitotoxic neuronal death has been clearly documented in the adult brain following status epilepticus (SE), a state of uninterrupted seizure activity that may last for hours. The severity and duration of SE determine the extent of neuronal injury; in most animal models damage is observed with durations of one hour or longer. Severity, duration and rate of onset may also influence the extent to which the neuronal death is mediated by apoptotic or necrotic processes. In contrast to the adult brain, the immature brain is resistant to SE-induced damage; nevertheless long-term deleterious effects of SE have been observed in animals that experienced SE in infancy without exhibiting neuronal loss. Likewise, in adult animals, long-term abnormalities such as spontaneous seizures have been documented following SE even under conditions in which the brain was protected from damage. This suggests that SE-induced excitotoxic injury is not required for the development of long-term disruption of neuronal function in the aftermath of SE. In contrast to SE, recurrent intermittent brief seizures such as those typically associated with epilepsy, do not necessarily cause neuronal injury. In patients with epilepsy it is often difficult to determine whether brain lesions are a cause or consequence of the seizure condition; little or no injury has been observed with chronic brief seizures in several animal models. In fact, exposure to repeated brief noninjurious seizures has been shown to exert a neuroprotective action, possibly as a consequence of induction of expression of neurotrophic factors. This "excitotrophic" effect of seizures extends beyond protection against SE-induced injury to include protection in models such as adrenalectomy-induced granule cell death. It is therefore possible that seizures may serve to protect against neurodegeneration and promote regrowth and remodeling in the face of insults to the nervous system. Thus, seizures can span a spectrum from trophic and adaptive to toxic and maladaptive, depending upon the conditions of their occurrence and the extent to which they are regulated. In this context, seizures may be analogous to fever: adaptive and protective in specific settings, with the capacity to become maladaptive and injurious in their own right if they develop into SE or a long lasting epileptic condition.

Keywords: seizures, limbic system, status epilepticus, neuroprotection, neurotrophic factors, apoptosis, electroconvulsive shock, immature brain.

INDEX

1. Introduction.
2. Status epilepticus (SE).
3. Neuronal death after prolonged SE: necrotic or apoptotic?
4. Nature of injury is dependent upon rate of onset and duration of SE.
5. Neuroprotective strategies.
6. "Epileptic tolerance".
7. The immature brain is resistant to SE-induced injury.
8. Long-term consequences of prolonged SE in adults: do they require injury?
9. Recurrent brief seizures: harmful or helpful?
10. Something beyond tolerance: excitotrophicity?
11. The immature brain: playing by different rules.

1. Introduction.

Epilepsy is a chronic neurological disorder characterized by recurring seizures. Although more than 10% of the general population is likely to experience a single isolated seizure episode, the incidence of epilepsy is closer to 3%. The term "epileps***ies***" more accurately reflects the fact that several different types of epileptic syndromes occur clinically and respond to distinct therapeutic interventions. Epileptic seizures, or paroxysmal discharges of neuronal ensembles, have been classified according to electrographic and behavioral criteria; these largely reflect the different neural pathways and networks that participate in the seizure onset and propagation. While different types of syndromes may arise from common etiologies, be they infectious, metabolic, vascular, genetic, or traumatic, certain syndromes are favored by specific etiological, especially genetic, factors. Because epilepsy typically appears in association with other neurological or developmental abnormalities, it is often impossible to determine which pathophysiological features are a result of the seizure condition, as distinguished from the underlying abnormality. Because of the complexity of the clinical syndromes associated with the epilepsies it is difficult to document a role for excitotoxicity in the progression or sequelae of these disorders. While there is ample documentation that neuronal damage occurs in association with certain types of epilepsy (Babb et al, 1990; Bernasconi et al, 2001; Blennow et al, 1978; Briellman et al, 2001; Cendes et al, 1995b; Fuerst et al, 2001; Glass and Dragunow, 1995; Houser, 1999; Mathern et al, 2002; Tasch et al, 1999; Theodore et al, 1999;

Wasterlain et al, 1996), the extent to which such damage is a cause or effect of seizures remains elusive in light of evidence that the damage may be a pre-existing etiological factor (Davies et al, 1996;Fernandez et al, 1998; Grunewald et al, 2001; Kuks et al, 1993). Furthermore, evidence of structural lesions in asymptomatic first-degree relatives of patients with epilepsy suggests that genetic predisposition may be a more fundamental than seizures as a determinant of hippocampal damage (Kobayashi et al, 2002). For this reason, studies in animal models are especially important because they allow for an analysis of the impact of recurrent seizures in the brains of animals that are otherwise neurologically normal. Another clinical condition closely related to epilepsy, is that of status epilepticus (SE). SE is an acute, often emergency, condition characterized by continuous uninterrupted seizure activity that becomes self-sustaining and can continue for hours in the absence of therapeutic intervention (DeLorenzo et al, 1992, 1996; Gilbert et al, 1999; Hauser et al, 1990; Shinnar et al, 1986; Shorvon, 2001). The brain regions most affected by these continuous seizures are those regions that participate in the particular type of seizure; it is at least theoretically possible for any type of seizure to become self-sustaining SE. The incidence of SE is higher in patients with epilepsy (between 10 and 20%, depending on the minimum seizure duration used to define SE) than in the general population (less than 0.1%). In patients with epilepsy, those with focal onset (partial) seizures have a higher risk of SE than patients with primary generalized seizures. SE occurs clinically in patients who may or may not have a history of epilepsy; in those without a history of epilepsy, an experience of SE may or may not be associated with a subsequent diagnosis of epilepsy. In some cases, SE has been caused in humans by intoxication with excitotoxic agents such as domoic acid (Cendes et al, 1995a). SE, especially if allowed to continue untreated past 30 min, can cause brain damage in adult humans and experimental animals, and it is likely that much of this damage is of an excitotoxic nature (DeGiorgio et al, 1992; Fountain and Lothman, 1995; Fujikawa et al, 2000a; Klitgaard et al, 2002; Meldrum, 1983, 1991; Treiman, 1992). However SE as a clinical disorder is relatively rare in comparison to the incidence of brief seizures associated with epilepsy. Therefore, although the term "seizure" is often used to refer to SE in the experimental animal literature, it is crucial to distinguish between SE and other epileptic seizures, both in the clinical arena and in experimental models. The present chapter will focus on evidence regarding excitotoxic damage in animal models in which the seizures predominantly affect limbic system and temporal lobe structures. These models most closely resemble human complex partial seizures. The vast majority of experimental studies of neural injury in the aftermath of seizures have been conducted in various models of SE in rodents (largely in rats) and these studies will be discussed first, followed by a discussion of studies of chronically recurring brief seizures not involving SE.

2. Status epilepticus (SE).

Most of the animal models used for studying seizure-induced injury are models of SE, predominantly in rodents. SE is induced by chemoconvulsants, many of which are themselves excitotoxins such as kainic acid, or by electrical stimulation of defined projections such as the perforant path (Bertram and Lothman, 1993; Lothman and Bertram, 1993; Sloviter and Damiano,1981; Sloviter et al, 1996) or the kindled amygdala (McIntyre et al, 1982). Because the condition of SE is self-sustaining, the seizure activity typically continues for 6 hr or longer if no attempt is made to terminate the seizure state pharmacologically. In some studies, the duration of SE is controlled by using diazepam or pentobarbital to stop SE at a defined time, while in other studies, the animals are allowed to continue in SE until they stop spontaneously. The former procedure allows for a careful time-dependent analysis but involves an additional pharmacologic treatment which may have neuroprotective actions. The latter procedure suffers from a significant rate of mortality as well as high variability in the duration of seizures across animals. Systemic and focally-induced SE exhibit overlapping features of brain pathology (see Goodman, 1998 for a review) and the extent of brain damage is influenced more by the severity and duration of the SE than by the technique employed for induction. When evaluating SE-induced excitotoxicity, it can be problematic to separate out the toxicity due to seizures from that due to direct effects of the drug (Sater and Nadler, 1988). The fact that drugs which block the seizures completely prevent the toxicity cannot be used as an argument for specificity because most of the drugs used to prevent seizures can also reduce the direct toxicity of many of the chemoconvulsant agents. The best argument in favor of SE-induced damage being derived from the prolonged seizure discharge rather than from the direct action of the chemoconvulsant drug itself, comes from studies of the time-dependent nature of the excitotoxic damage. Regardless of the mechanism employed of inducing SE, it is generally agreed that a minimum of 1 hr of continuous SE is required before significant neuronal damage can be detected either by histopathological methods or molecular methods, and that the damage becomes markedly enhanced by longer durations of SE (Kondratyev et al, 2001; Kondratyev and Gale, 2001; Lowenstein et al, 1991; McIntyre et al, 1982; Meldrum et al, 1973a,b,1975; Sater and Nadler, 1988; Shimosaka et al, 1992). If direct drug effects were a significant cause of the neuronal damage, these effects should be maximal within the first hour of SE (which is typically 0.5-2 hours following injection of the chemoconvulsant), when drug levels are at their peak. Additional evidence that neural injury in the aftermath of SE does not depend upon the presence of exogenous excitotoxic agents comes from studies of SE induced by focal electrical stimulation (e.g. McIntyre et al, 1982). The time course and pattern of neural injury after electrically-induced SE is quite comparable to that evoked by systemic or focal chemoconvulsants. The most frequently used models of chemoconvulsant-induced SE are: 1) systemic kainic acid (administered to rats i.p or i.v. in doses between 9 and 15 mg/kg); 2) systemic pilocarpine (350 – 400 mg/kg i.p. in rats); and 3) systemic pilocarpine (30-50 mg/kg i.p.) given after pretreatment with lithium chloride (Ben Ari, 1985; Jope et al, 1986; Turski at al, 1989). Differences between these models largely concern drug side effects (most problematic with pilocarpine because of the muscarinic-evoked secretions; these can however be ameliorated with repeated doses of a peripherally-acting muscarinic antagonist such as methyl-atropine), time course and profile of seizure manifestations (Morrisett et al 1987). At low to moderate doses in rats, kainic acid seizures are preferentially limbic-motor seizures with little or no hindbrain seizure manifestations (note that this selectivity is not present in most mouse strains), while pilocarpine seizures often include additional components such as "running-bouncing" which are typical of the activation of brain-stem seizure networks (Browning and Nelson, 1985, 1986; Browning et al 1990). While these models share common features in terms of regional profile of neural injury, certain seizure manifestations, and long-term consequences, it is also clear that they also possess unique attributes. For example, the susceptibility of specific mouse strains to neural injury evoked by kainic-acid induced SE does not correlate with susceptibility to injury induced by pilocarpine induced SE (Schauwecker, 2002b). SE and SE-induced injury can also be achieved with focal chemoconvulsant administration in selected brain regions. For kainic acid, one region from which focal administration evokes prolonged and injurious SE (resembling that evoked by systemic administration) is the substantia nigra in Sprague-Dawley rats, or in Wistar rats when the KA is given in combination with an NMDA antagonist (Maggio et al, 1990). This focally-evoked SE produced marked neuronal damage in forebrain limbic regions including piriform cortex, hippocampus, amygdala, and mediodorsal thalamus (Tanaka and Simon, 1994).

3. Neuronal death after prolonged SE: necrotic or apoptotic?

The mechanisms of SE-induced injury include many of the same molecular players that participate in standard excitotoxic insults. Considering the fact that continuous seizures represent highly synchronized repetitive discharge in large ensembles of neurons, many of which are glutamatergic, it is not surprising that excitotoxic damage would result with long durations of this type of activity. And it is also not surprising that the brain areas exhibiting the most consistent damage are those which are part of the core network through which the seizure propagates and which exhibit marked metabolic activation, including frontal, piriform, entorhinal, perirhinal and hippocampal cortices, amygdala and mediodorsal thalamus (Fig. 1) (Clifford et al, 1997; Handforth and Treiman, 1995; Kondratyev et al, 2001; Mohapel et al, 1996; Schwob et al, 1980; Sperk et al, 1983).

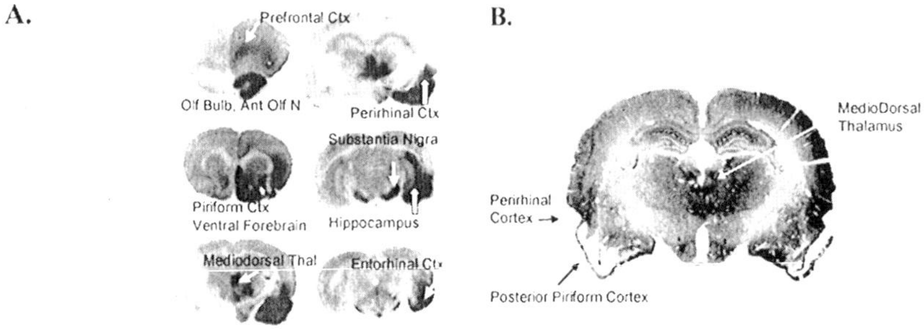

Figure 1. A. Mapping the seizure network: metabolic activation after a focally-evoked limbic motor seizure, assessed by ^{14}C 2-deoxyglucose accumulation. **B.** Silver stain for neuronal degeneration after status epilepticus.

With sufficient SE durations, the damage becomes so profound that complete disappearance of all tissue can be observed in the piriform cortex, a region that is especially susceptible to SE-induced damage. Oxidative stress, mitochondrial dysfunction and energy depletion (Bruce et al, 1995; Cock et al, 2002; Kunz et al, 1999, 2000), and increased production of nitric oxide (Lumme et al, 2000; Milatevic et al, 2002) have been documented in the aftermath of SE, typically occurring within the first 48 hours. In the case of limbic-SE (such as that induced by kainic acid, or perforant path stimulation), the increase in cerebral blood flow more than meets the increased oxygen demand created by the neural activity (Pinard et al, 1984), so that cerebral hypoxia is unlikely to account for the metabolic stress resulting from SE. Moreover, mild hypoxia does not appear to enhance the damage produced by SE (Sasahira et al, 1997). Furthermore, cellular stress and injury have been documented in artificially ventilated animals in which PCO_2 and pH is controlled during the period of SE (Meldrum et al, 1973b; Nevander et al, 1985). It has also been demonstrated that tetanic stimulation of the Shaffer collaterals in slices maintained *in vitro* can induce cell death in neurons in entorhinal cortex and in CA1 and CA3 of the hippocampus, but 12-14 hours of stimulation were required for this to occur (Pelletier et al, 1999). In this case, blocking increasing Ca++ availability either by blocking Ca++ entry or preventing Ca++ release from intracellular stores was sufficient to protect against the cell death. Markers of injury following SE in rats and mice have included factors that are characteristic of both apoptotic and necrotic cell death. Under certain conditions and in certain brain regions, necrotic patterns of cell death predominate after SE (Puig and Ferrer, 2002; Fujikawa et al, 1999, 2000c,d), while clear evidence of apoptotic profiles can be seen in other experimental conditions (Araki et al, 2002; Ebert et al, 2002; Faherty et al, 1999; Kondratyev et al, 2001; Liou et al, 2003; Pollard et al, 1994; Sankar et al, 1998; Sloviter et al, 1996). It appears that SE-induced injury is achieved via cell death mechanisms that combine biochemical and morphologic features of necrosis and apoptosis, and that an individual cell may show a mixed profile (Fujikawa et al, 2000b,c). Moreover, the process of death in a cell group is dynamic, changing over time from apoptotic to necrotic (e.g., secondary necrosis in the face of overwhelming apoptotic cell death), or vice versa, depending upon the energy state of the cells and the scavenging capacity in the immediate environment. Identification of necrosis is largely based on morphological signs (acute swelling, membrane lysis, release of intracellular debris followed by inflammatory responses) but may include molecular markers such as an increase in calpain or a smeared (non-ladder) pattern of DNA degradation. Confirmation of an apoptotic profile derives from both histological (cell shrinkage, membrane blebbing, apoptotic nuclei, chromatin condensation and pyknosis) and molecular markers such as caspase-3 activation, DNA fragmentation, a ladder-like pattern of internucleosomal DNA fragmentation, and the caspase-3 mediated cleavage of ICAD (or DFF-45), an inhibitor of "Caspase-Activated DNase" (CAD). Because it is not unusual to find necrotic features mixed with multiple apoptotic features, it has been argued that pure "apoptosis", as originally defined morphologically, does not occur after SE (Fujikawa, 2000; Fujikawa et al, 2000c). A mixed or variable profile of apoptosis and necrosis has been repeatedly described in various models of excitotoxic injury, so it is certainly not unique to SE-induced damage. The concept of an "apoptotic-necrotic continuum" (Covolon et al 2000, Portera-Cailliau et al, 1997a,b) creates a useful framework for interpreting the diversity of characteristics observed across experimental conditions, and obviates the controversy created by an "either-or" view. Subtle variations in the conditions under which the excitotoxic insult occurs may determine precisely where on the apoptotic-necrotic continuum an injured cell population will fall. For example, the severity of the insult is likely to be an important determinant, as it has been shown in cortical cell cultures that whereas a short duration (10 min) exposure to low concentrations of NMDA or peroxynitrite induce delayed (18 hour) neurotoxicity dominated by an apoptotic profile, higher doses of these neurotoxins trigger a more rapid and predominantly necrotic pattern of cell death (Bonfoco et al, 1995). In related studies it was shown that cultured cerebellar granule cells exposed to glutamate, exhibited a relatively rapid necrotic death in a subpopulation of neurons, while other neurons survived, recovered their energy levels, and then underwent apoptotic death after a delay (Ankarcrona et al, 1995). This suggests that the most vulnerable neurons succumb to necrosis early in the process, and that metabolically resilient neurons can survive to initiate the "cleaner" apoptotic death program at a later time. It stands to reason that the more severe and acute an insult is, the fewer neurons are likely to resist the necrotic phase and survive to exhibit apoptotic features. Understanding the mechanisms that contribute to cell death is a complex but worthy endeavor, as it has obvious implications for therapeutic strategies.

4. Nature of injury is dependent upon rate of onset and duration of SE.

In the case of SE, the severity of the insult is influenced by 3 major factors: 1) seizure severity; 2) rate of onset of SE; and 3) duration of SE. While seizure severity during SE is an important determinant of extent of injury (e.g., Zhang et al, 1997), to date there have been no systematic comparisons between the mechanisms of neural injury after mild and severe SE, largely owing to the difficulty in experimentally controlling seizure severity when employing convulsant drugs with very steep dose-response functions. On the other hand, *rate of onset* of SE has been shown to be an important determinant of the relative extent of apoptosis vs. necrosis, as evidenced by the fact that kainic acid-induced SE (duration of 2 hrs) with rapid onset (within 30-70 min following i.v. or i.p. injections) results in a predominantly necrotic profile of cell death, whereas the same duration of SE with a slower onset (SE starting > 90 min after kainic acid injection) results in a predominantly apoptotic profile (Kondratyev et al, in submission). It is noteworthy that in the latter study, the extent of total neuronal degeneration as confirmed by silver staining at 72 hr after slow-onset SE was equivalent to that seen after fast onset SE. It appears as if the slow-onset SE (which is typically preceded by several brief seizure events which may function as "warning" signals) allows the neurons to mount adaptive responses to the insult such that they are better equipped to survive the acute necrosis-triggering phase. This raises the possibility that therapeutic interventions limited to the early phase of SE may merely allow neurons to survive long enough to initiate apoptosis, underscoring the need for neuroprotective strategies that outlast the initial acute phase of the insult. Finally, *duration* of SE is a clear factor in determining the severity of insult, and it is possible to establish a defined "dose-response" relationship between duration of SE and extent of injury. Once the duration of SE reaches 4 hours or more, the integrity of tissue in many brain regions is so severely compromised that it would be difficult to evaluate changes in molecular markers against the background of tissue loss, and overwhelming necrosis has been documented after 3hr of SE (Fujikawa et al, 2000c). Accordingly, from both a practical and theoretical perspective, it would be more likely to find apoptotic profiles in tissues obtained from animals in which SE was limited to durations of 1-2 hr. With SE limited to this time frame, clear signs of apoptotic injury are evident (Becker et al, 1999; Kondratyev et al, 2001; Kondratyev and Gale, 1999) and the DNA fragmentation varies linearly in proportion to the duration of SE between 30 and 120 min (Kondratyev and Gale, 2001). The extent of DNA fragmentation in hippocampus and rhinal cortex was found to be comparable for 2 different models of SE: systemic kainic acid and focal application of bicuculline plus cyclothiazide into area tempestas (Kondratyev and Gale, 2001). Even with moderate to severe injury, the peak DNA fragmentation was not reached until 48 hours after SE (Kondratyev and Gale, 2001), confirming the delayed nature of the apoptotic response.

5. Neuroprotective strategies.

Several treatments have been documented to provide protection against SE-induced neuronal injury. However, a major complication in the interpretation of these studies is the fact that some of the treatments used can interfere with the seizure propagation itself. If the treatment is present while SE is in progress, then it becomes necessary to monitor SE onset, severity and duration to be sure that a reduction in injury is not simply a result of decreased seizure severity. Unfortunately, it is too often the case that seizure severity is neither monitored nor matched between treatment groups (e.g. by titrating the dose of chemoconvulsant to evoke an equivalent severity of seizures in the treated and control groups). This complication is especially problematic when evaluating treatments that interfere with excitatory transmission. Treatments which can ameliorate

SE-induced injury without significant effects on seizure severity include neurotrophic factors such as basic fibroblast growth factor (Liu and Holmes, 1996;) and nerve growth factor (Weiss et al, 1995), inhibitors of caspase-3 (Henshall et al, 2000; Kondratyev and Gale, 2000; Narkilahti et al, 2003), PARP inhibition (Filbert et al, 1999) and inhibitors of NO production (Jones et al, 1998; Montecot et al, 1998). Selective inhibition of caspase-3 and PARP will protect against the component of cell death due to apoptosis, so it would be expected that the effectiveness of these interventions will depend upon whether the conditions of SE favor an apoptotic response (see above). This may explain why some investigators fail to detect protection with caspase-3 inhibition (Fujikawa et al, 2002b) while others find robust protection (Henshall et al, 2000). Consistent with the observation that a 2 hr period of kainic acid-induced SE triggers a marked apoptotic response (Kondratyev and Gale, 2001) is the finding that caspase-3 inhibition protects against the injury under these same conditions (Kondratyev and Gale, 2000). In some situations, interventional treatments are given shortly following the termination of SE in an attempt to interfere with the evolving degenerative response. Here the expectation is that the delayed injury, which evolves over several hours or days, can be interrupted. Both intracellular molecular cascades as well as intercellular interactions contribute to the delayed injury, which can be necrotic and/or apoptotic. In the case of necrotic injury, treatments that block inflammatory responses (Vezzani et al, 2002) may prevent secondary injuries, whereas in the case of apoptotic processes, blockade of mediators such as caspase-3 or PARP (see above) may serve to interrupt the cell death machinery. One approach to identify potential molecular targets for neuroprotective therapeutic intervention in SE is the use of genetically engineered mouse strains. In particular, mice with a knockout of the p53 tumor suppressor gene were reported to be protected against SE damage in multiple brain regions (Morrison et al, 1996), mice deficient in JNK3 showed an absence of SE-induced neuronal death in hippocampus (Yang et al, 1997), mice deficient in DFF-45 were found to have reduced injury in CA3 neurons following kainic acid induced SE (Zhang et al, 2001), and protection against SE-induced injury also has been reported in mice with overexpression of heat shock protein 27 (Akbar, 2003), overexpression of a naturally-occurring antagonist to interleukin-1 (Vezzani et al, 2002), and overexpression of mitochondrial superoxide dismutase (Liang et al, 2000). While at first glance these results in mice appear straightforward, they are complicated by considerable strain differences in vulnerability to SE-induced injury, especially for SE induced by kainic acid (Faherty et al, 1999; Schauwecker, 2002a). In some cases strain differences are more substantial than the effects obtained by knockouts or overexpression of specific genes, and in at least one case a protective effect was attributed to a knockout of p53 (Morrison et al, 1996) when it was actually due to the injury resistance of the background strain that was used for the knockout rather than to the absence of the p53 gene (Schauwecker, 2002b). In the other studies cited above, the fact that some of the genetic manipulations resulted in a reduced seizure response to kainic acid makes it impossible to determine whether a specifically *neuroprotective* action was achieved. The study of Zhang et al provided evidence that seizure susceptibility was unchanged, but in all other cases the genetic alteration was accompanied by a reduced seizure response. Only the study of Yang et al directly compensated for the reduced seizure response by testing a higher dose of kainic acid in the JNK3 deficient mice: a dose of 45 mg/kg in these mice produced a state of SE comparable to that induced by 30 mg/kg in the control mice, but yet only 2 out of 15 exhibited cell death with the high dose as compared with 13 out of 17 controls given the lower dose.

6. "Epileptic Tolerance".

A phenomenon akin to that of ischemic tolerance (Dirnagl et al, 2003) occurs with SE preconditioning. Thus, if animals are pre-exposed to brief periods of SE that are insufficient to cause marked damage, they then show resistance to injury when challenged with a prolonged exposure to SE (Najm et al, 1998; Sasahira et al, 1995). Moreover, cross-tolerance between ischemic or hypoxic preconditioning and SE has been demonstrated (Emerson et al, 1999; Plamondon et al, 1999; Towfighi et al, 1999) suggesting that the adaptive changes in responses to SE, ischemia and hypoxia share one or more common substrates. The mechanisms responsible for the protective effect of the SE preconditioning remain to be elucidated, but they may be related to the mechanisms responsible for the neuroprotective actions of repeated brief non-injurious seizures (see below).

7. The immature brain is resistant to SE-induced injury.

More than half of the clinical cases of SE occur in children under 5 years of age, so investigations of the impact of SE in neonatal and juvenile animals is especially important. Despite the fact that immature animals are more susceptible to seizures and to SE during certain developmental periods, the vulnerability to excitotoxic injury seen in the adult brain is not seen during early postnatal development. This observation has been made repeatedly in various animals models of SE (Cavalheiro et al, 1987; Cilio et al, 2003; Druga et al, 2003; Gunderson et al, 1999;Haas et al, 2001; Holmes, 2002; Sperber et al, 1991,1999; Stafstrom et al, 1992) as well as in a specific clinical situations such as febrile seizures in human infants (Lewis, 1999; Moshe, 1998; Shinnar et al, 2001; Shinnar and Glauser, 2002; Tarkka et al, 2003). In the rat, a striking resistance to SE-induced injury is evident during the first 2 weeks postnatally, and this may be in part related to the effect of dietary fat on mitochondrial function (Sullivan et al, 2003). Nevertheless, an experience of SE in infancy, especially if it is sufficiently prolonged, may have deleterious long-term consequences, with or without detectable cell loss (Baram et al, 2002; Galanopoulou et al, 2002; Holmes and Ben Ari, 1998; Lado et al, 2000; McLachlan, 1995; Theodore et al, 2003). Some of the deleterious consequences that have been noted include an increased susceptibility to seizures in adulthood (Jensen and Baram, 2000; Schulz and Ebner, 2001; Sutula and Pitkanen, 2001), compromised cognitive function (Herman, 2002; Huang et al, 1999; Lynch et al, 2000; Stafstrom, 2002), and emotional deficits (Gunderson et al, 1999; Meletti et al, 2003). Considerable efforts are underway to identify the mechanisms responsible for these long-term deficits and to determine to what extent the seizures per se (apart from underlying abnormalities) contribute to the pathology (Mizrahy and Clancy, 2000; Moshe, 1998). At the same time that the immature brain exhibits little or no injury in response to SE, it exhibits a marked apoptotic response to acute exposure to glutamate antagonists and other agents that depress neuronal activity such as anesthetics and anticonvulsants (Bittigau et al, 2002, 2003; Ikonomidou et al, 1999; Olney et al, 2002). This is the inverse of an excitotoxic response, or what could be called "inhibitoxic". Given the fact that activity-dependent trophic support and functional validation of neuronal connections are core features of the developing brain, blockade of glutamate transmission is likely to interrupt these processes, thereby enhancing ongoing programmed cell death. It should be noted that SE in the immature brain, while not sufficient in and of itself to induce marked cell loss, can exacerbate the neural injury induced by other concomitant insults such as hypoxia or ischemia (Dzhala et al, 2000; Yager et al, 2002). Recognizing that in clinical situations SE is often precipitated by other neurological insults, the danger of brain damage posed by prolonged SE during infancy may be substantial.

8. Long-Term consequences of prolonged SE in adults: do they require injury?

Widespread neural injury during the first several hours or days following an incident of SE is a striking short-term consequence of the insult in the adult brain, but there are also dramatic long-term consequences that have fascinated investigators for the past 2 decades. In particular, several weeks following an incident of prolonged SE in adult rats, recurrent spontaneous seizures emerge in a high proportion of the animals (Cavalheiro et al, 1991; Goodman, 1998; Sloviter et al, 1996). Given the widespread neural damage that these animals have sustained, it is a daunting task to identify which components of the injury are necessary for the emergence of spontaneous seizures, or even if injury is a prerequisite condition for the long-term seizure state. Despite the daunting nature of this task, it is worthy of pursuit, because the long-term spontaneous seizure state bears resemblance to certain forms of epilepsy.

The most informative studies have been those that have sought to systematically eliminate certain conditions as being necessary or sufficient for the emergence of the long-term seizure state. One approach has been to compare the cell loss that occurs in rats exhibiting spontaneous seizures with cohorts from the same treatment group that do not develop the seizure condition. Studies in which this analysis has been done have shown that cell loss within hippocampus appears to be equivalent in both subgroups of animals suggesting that this damage is not *sufficient* to account for the spontaneous seizure state. On the other hand, spontaneous seizures have been documented in animals that do not exhibit significant cell loss in hippocampus (Sloviter, 1992; Buckmaster and Dudek, 1997) indicating that this cell loss is also not *necessary* for the development of the long-term seizure condition. More recently, Corcoran and coworkers (Zhang et al, 2002) have investigated the long-term consequences of SE in rats that have been protected from cell death by preconditioning with short period of SE (i.e. rats made tolerant to SE). The preconditioned rats were protected against neuronal loss and mossy fiber sprouting, but nevertheless went on to develop spontaneous seizures. The only difference noted between the rats with and without injury was that the severity of the spontaneous seizures was greater in the rats exhibiting cell loss. This effect of preconditioning was observed in both the kainic-acid induced model and in the pilocarpine-induced model of SE. A similar failure of neuroprotection to prevent SE-evoked epileptogenesis was documented following focally evoked SE (Narkilahti et al, 2003). These findings argue against excitotoxic cell loss being a prerequisite for the development of the late spontaneous seizures in the aftermath of SE, and suggest that protection against neuronal damage may not be sufficient to prevent deleterious long-term consequences of SE. Thus, additional therapeutic strategies aimed at interrupting the synaptic reorganization responsible for the generation of spontaneous seizures will be needed, but must await the identification of the precise synapses and networks responsible for this maladaptive response. In the meantime, numerous investigations have identified changes in components of glutamate transmission (e.g., Aronica et al, 1997; Friedman, 1998; Gorter et al, 2002; Mayat et al, 1994; Sommer et al, 2001; Tang et al, 2001; Ulas et al, 2000) after SE that may contribute to long-term synaptic alterations. It is conceivable that even in the absence of an understanding of the precise localization of the necessary changes, pharmacological strategies targeted to interrupt these changes may prove effective in preventing deleterious long-term sequelae throughout the brain.

9. Recurrent brief seizures: harmful or helpful?

As mentioned earlier, epilepsy is characterized by recurring brief seizure episodes, and SE is a relatively rare occurrence. Consequently, the question of how repeated brief seizures that do not involve SE affect short-term and long-term neurological function is exceptionally important. While brief seizures typically are not in themselves sufficient to provoke excitotoxic injury (Dalby et al, 1996; Ende et al, 2000; Sheline et al, 1999; Masco et al, 1999; Steward et al, 1998), the possibility that they may alter the vulnerability to excitotoxic injury from other origins must be considered. There are a large number of techniques available for inducing experimental seizures, some of which are more amenable to chronic treatment paradigms. Repeated administrations of low doses of short-acting chemoconvulsants such as bicuculline, picrotoxin, phentylenetetrazol, or flurothyl, focally-administered tetanus toxin, focal deposits of alumina cream, repeated administration of electroconvulsive shock, and electrical kindling of amygdala, hippocampus or other forebrain limbic regions have been used to induce repetitive seizures in animal models. Repeated treatment with chemoconvulsants is complicated by the fact that animals are chronically exposed to the drugs and their metabolites, often without adequate controls for the effects of drug exposure per se (e.g. at subconvulsant doses). Thus, for example, while chronic GABA receptor blockade (a mechanism shared by all the chemoconvulsants listed above) may be a good technique for inducing chronic seizures, the impact of such a treatment cannot be specifically attributed to seizures as opposed to other widespread effects resulting form diminished GABA transmission. Accordingly, methods of seizure induction that avoid systemic exposure to drugs are more likely to identify responses that are specifically related to the seizure activity itself. Detailed examination of the brains of adult experimental animals that have experienced multiple repeated daily seizures induced by electrical kindling stimulation, focal tetanus toxin, or electroshock treatments, have indicated that the seizures induce little or no neuronal damage (Bertram and Lothman, 1993; Gombos et al, 1999;

Jefferys et al, 1992; Kotloski et al, 1992; Masco et al, 1999; McIntyre et al, 2002; Tuuanen and Pitkanen, 2000). On the other hand, these treatments do activate a host of genomic responses, ranging from immediate early genes, genes for neurotrophic factors and neuropeptides, as well as multiple alterations in the regulation of neurotransmitters and their receptors in various brain regions (Angelucci et al, 2002; Blumcke et al, 2000; Follesa et al, 1994; Gwinn et al, 2002; Hughes et al, 1999; Kondratyev et al, 2002; Mody, 1999; Naylor et al, 1996; Nibuya et al, 1995; Pekary et al, 2000; Racine et al, 2002; Reti and Baraban, 2000; Sakhi et al, 1996; Sato et al, 1996; Zhang et al, 1996). If exposure to recurrent brief seizures is not injurious, does this exposure alter the vulnerability to subsequent excitotoxic insults? Studies conducted to date indicate that the answer to this question is unequivocally "yes". The most consistent finding is that exposure to repeated brief seizures, with or without kindling, induces a dramatic neuroprotective state in which the animals become resistant to brain damage induced by SE as evidenced by histopathology and sensitive molecular markers of cell death (Kelly and McIntyre, 1994; Kondratyev et al, 2001; Penner et al, 2001) (Fig.2).

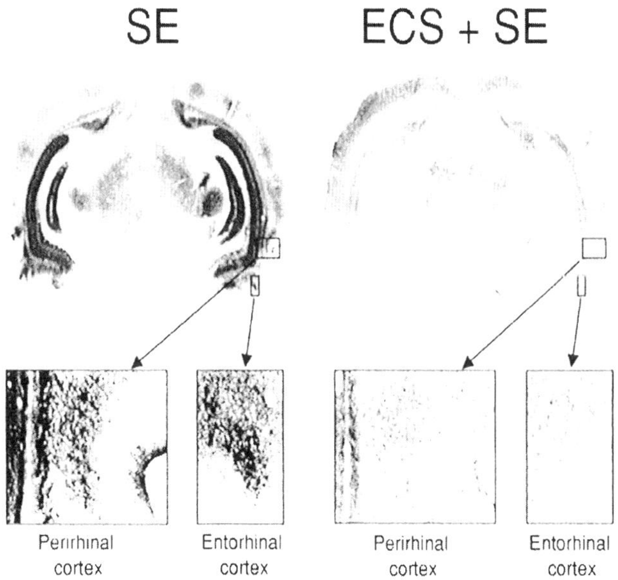

Figure 2. Pre-exposure to minimal noninjurious electroshock seizures (ECS) protects against neuronal death induced by prolonged seizures (SE).

The fact that neuroprotection requires multiple daily seizure treatments, and that a single treatment is not protective, indicates that acute responses to the seizures (e.g., postictal depression) cannot explain the injury-resistant state. Importantly, the neuroprotective effect of repeated non-injurious seizures is observed when seizure-exposed animals and controls are carefully matched for severity and duration of SE. It is important to recognize that the type of seizures which are effective in conferring a neuroprotective effect on forebrain regions are seizures that last only several seconds and engage limbic forebrain networks either through kindling (Kelly and McIntyre, 1994; Penner et al, 2001) or by electroshock administered via corneal electrodes (Kondratyev et al, 2001; Masco et al, 1999). These stimuli produce characteristic signs of limbic-motor seizures such as facial and forelimb clonus with rearing; typically without tonic-clonic manisfestations characteristic of generalized seizures. In sharp contrast, repeated exposure to seizures that selectively engage brainstem seizure circuitry and evoke tonic-clonic motor responses, as induced by electroshock administered via earclip electrodes (Browning and Nelson, 1985) fails to confer neuroprotection and may even worsen the outcome following SE (Andre et al, 2000). These observations emphasize the network-specificity of the protective responses. This also indicates that general responses such as stress associated with repeated seizures, various endocrine changes and other nonspecific physiological responses to seizures cannot account for the neuroprotection that is selective for seizures involving limbic forebrain activation. Instead, adaptive changes restricted to the network through which the seizures propagate appear to be required for the neuroprotective state. It is likely that the neuroprotective effect of repeated brief seizures is mediated by mechanisms that confer resistance to excitotoxicity, and it is likely to extend beyond SE to include protection against other types of trauma (Zai et al, 2001). Induction of neurotrophic factors is one of several potential mechanisms that may account for the decreased vulnerability to injury, as several studies have shown neurotrophic factors to attenuate excitotoxicity in vitro (Cheng and Mattson, 1994; Fernandez-Sanchez and Novelli, 1993; Finklestein et al, 1993; Freese et al, 1992; Glazner and Mattson, 200; Mattson et al, 1993,1995), and it is clear that repeated brief noninjurious seizures in adult animals increase the expression of bFGF (Gwinn et al, 2002), NGF and BDNF (Angelucci et al, 2002) in various brain regions. The fact that exogenous infusion of bFGF can protect against neuronal injury induced by SE (Liu et al, 1997; Liu et al, 1993) indicates that an increase in bFGF protein, as observed after several days of electroshock seizures (Gwinn et al, 2002) may be sufficient to account for the seizure-induced neuroprotective effects of this treatment. It is also likely that several additional neuroprotective adaptative responses are engaged by the repeated seizures and that the relative importance of any given factor may depend upon the cell type and the brain region examined.

10. Something beyond tolerance: seizure-induced 'excitotrophicity'?

The concept of "tolerance" implies that repeated exposure to a stimulus (such as a drug or physiological challenge) results in a diminished response to that stimulus, usually due to compensatory adaptations in the cellular elements (e.g. receptor) with which the stimulus interacts to produce an effect. In the case of tolerance to SE or ischemic tolerance, pre-exposure to brief periods of insult reduce the sensitivity to the injury produced by a more prolonged insult. Presumably, the pre-exposure to SE or ischemia in durations that are subthreshold for provoking injury, initiate early events in the injury cascade that serve to activate protective mechanisms, in the same way that muscle overload serves to enhance muscle strength and resistance to fatigue and injury (Gordon, 1995). While it is possible that repeated brief seizures confer tolerance, similar to that seen with pre-exposure to SE, it is also likely that a phenomenon distinct from tolerance is involved, given the fact that the brief seizures are remarkably protective even in the complete absence any evidence of injury or cellular stress. Moreover, the duration of the electroshock seizures are under 10 seconds, or at least 100 fold less than the seizure duration necessary for causing threshold signs of injury. In fact, it is possible that brief non-injurious seizure activity serves to trigger activity-dependent mechanisms of neuroplasticity which recapitulate the injury-resistant and resilient conditions characteristic of development. According to this view, seizures could serve to protect against neurodegenaration and promote regrowth and remodeling in the face of insults to the nervous system. This would predict that the 'excitotrophic' protection afforded by repeated seizures should not be limited to excitotoxic injury but should extend to other types of degeneration. The clearest evidence for a neuroprotective effect of repeated seizures in a non-excitotoxic neurodegenerative model, comes from a study in which electroshock seizures were administered to adrenalectomized rats (Masco et al, 1999). Removal of the adrenal glands in adult rats leads to a selective apoptotic degeneration of the dentate granule cells in the hippocampus (Liposits et al, 1997; Sloviter et al, 1989,1993), reflecting the fact that these neurons are directly or indirectly dependent upon adrenal cortisosteroids (Gould et al, 1990). Electroshock seizure treatment completely prevented the dentate granule cell degeneration, while sham treatments or daily exposure to restraint stress did not alter the profile of degeneration seen in the adrenalectomized animals. The facilitation of recovery of function following brain damage that is accomplished by daily exposure to pentylenetetrazol may also reflect a neuroprotective and plasticity-enhancing effect of seizures

(Hernandez and Schallert, 1988; Hamm et al, 1995). Thus, seizures that occur transiently in a clinical setting during the post-traumatic period may serve to attenuate the delayed or protracted components of neuronal damage. If this is the case, then the frequent procedure of placing patients on prophylactic anticonvulsant therapy immediately following either head injury or neurosurgical intervention may potentially retard or diminish functional recovery (Hernandez and Holling, 1994; Schallert et al, 1986). In this context, seizures may be analogous to fever — a symptom that may have adaptive and protective value in specific pathological settings; and like fever, seizures can become maladaptive and injurious in their own right if they develop into SE or a chronic epileptic condition. During nervous system development, a dynamic continuum between neuronal plasticity and neuronal death exists and synaptic stimulation can 'protect' or 'select' the population of neurons or synapses that will be maintained. In an analogous fashion, seizure evoked stimulation may allow the sparing of otherwise vulnerable populations of neurons in the injured adult brain, a phenomenon that could be called 'excitotrophic'. Thus, controlled administration of electroshock seizures may have therapeutic potential in the treatment of a variety of neurodegenerative disorders including Parkinson's and Huntington's Disease (Abrams, 1989; Balldin et al, 1980; Beale et al, 1997; Berger and deSoto, 1990; Birkett, 1990; Faber and Trimble, 1991; Kant et al, 1995; Kellner et al, 1994; Pridmore and Pollard, 1996; Rasmussen and Abrams, 1991; Sandyk, 1986; Stern, 1991), but it remains to be determined if the seizures slow the disease progression. Currently electroshock seizures are used in the treatment of bipolar affective disorders and their therapeutic impact may derive from neuroprotective actions (Gray et al, 2003; Manji et al, 2003). Further characterization of the mechanisms contributing to the neuroprotective impact of brief seizure episodes may generate novel strategies of neuroprotection and recovery of function following excitotoxic insults and other forms of injury to the central nervous system.

11. The immature brain: playing by different rules.

Accumulating evidence suggests that despite reduced neuronal vulnerability in immature animals, seizure activity during postnatal brain development may have substantial adverse effects. Children with epilepsy, especially those with a seizure onset within the first year of life, have an increased incidence of mental retardation, learning and behavioral problems as compared with normal children (Farwall et al, 1995; Holmes, 1991; Huttenlocher and Hapke,1990; Liu et al, 1999). However, in clinical studies it is difficult to separate the effects of the seizures themselves from the underlying neurological abnormalities that predispose to the seizures (Lewis, 1999). Therefore, analyses of the effects of seizures in immature animal models are especially important. Several reports have indicated that the occurrence of a series of recurrent brief seizures during early postnatal development can lead to significant long-term changes in seizure susceptibility, neuronal plasticity, vulnerability to SE in adulthood, dendritic spine density and hippocampal network properties, as well as learning and memory functions, even in the absence of significant neuronal cell loss (Baram et al, 2002; Schmid et al, 1999; Sogawa et al, 2001; Swann et al, 2001). The mechanisms underlying these structural and functional changes remain elusive. Whereas in the adult brain we think of neuronal cell death as generally a pathological and deleterious process, in the immature brain, the process of neuronal cell death is a necessary step in development of normal circuitry. During the development of the central nervous system, excessive neurons are produced and almost 50% of them are removed by programmed cell death (PCD) (Meier et al, 2000; Oppenheim,1991). Whereas apoptotic PCD of progenitor cells takes place mainly during the embryonic period, apoptosis of postmitotic neurons extends from prenatal into postnatal development (Kuan et al, 2000). A balance between the cell proliferation and PCD determines the number of neurons that survive in the mammalian brain (Kudryashov et al, 2001). The immature rat brain undergoes these incredibly dynamic but highly regulated developmental changes during the first 3 weeks postnatally, with postnatal PCD occurring according to distinct time courses in different brain regions (Naruse and Keino, 1995; Gould et al, 1991). In cases where the neurons removed by PCD have already established synapses, those synapses will likewise be eliminated leaving "appropriate" neuronal circuits to survive. Thus, a shift of apoptotic and anti-apoptotic signals during early postnatal development can have profound effects on postnatal neuronal organization. During development, neurotrophins serve to provide essential cell survival signaling as an influence against pro-apoptotic signals (Lowenstein and Arsenault, 1996), so that changes in neurotrophin levels represent one mechanism by which PCD can be controlled. To date, the only studies that have measured the effects of seizures on neurotrophic factors in immature animals have examined SE induced by pilocarpine or kainic acid (Berzaghi et al, 1993; Kornblum et al, 1997; Tandon et al, 1999), and the only study that has examined neuronal damage after brief neonatal seizures observed evidence of a transient necrotic neuronal death in amygdala and hippocampus after hyperthermia-induced seizures (Toth et al, 1998). It is important to recognize that some of the same seizure-induced increases in 'protective' factors that may be advantageous in the brain of the adult animal (e.g., to guard against further injury) may be *maladaptive* in the brains of immature, developing animals. One can imagine that repeated seizures in immature animals will cause a prolonged induction of expression of certain neuroprotective growth factors such as NGF and FGF-2, and that this could suppress the normally occurring PCD in certain brain regions during early postnatal development. Thus, one reason that early childhood seizures may have deleterious effects is because they might actually *prevent* the cell loss that is a necessary component of normal development. The highly dynamic state of the developing brain, coupled with the ongoing activity-dependent sculpting of synaptic connections, transforms seizure-induced 'excitotoxicity' from a phenomenon of injury by cell death, to injury by aberrant cell *surviv*al. Nevertheless, at the same time, the enhanced neurotrophic responsiveness in the immature brain is likely to contribute to the *beneficial* plasticity and decreased vulnerability to SE-induced damage, and it may also protect against excessive neuronal loss secondary to neurodevelopmental abnormalities.

REFERENCES

Abrams R. ECT for Parkinson's disease. Am J Psychiatry 1989; Nov;146(11):1391-3

Akbar M.T., Lundberg A.M., Liu K., Vidyadaran S., Wells K.E., Dolatshad H., Wynn S., Wells D.J., Latchman D.S., de Belleroche J. The neuroprotective effects of heat shock protein 27 overexpression in transgenic animals against kainate-induced seizures and hippocampal cell death. J Biol Chem 2003; May 30;278(22):19956-65

Andre V., Ferrandon A., Marescaux C., Nehlig A. Electroshocks delay seizures and subsequent epileptogenesis but do not prevent neuronal damage in the lithium-pilocarpine model of epilepsy. Epilepsy Res 2000; Nov;42(1):7-22

Angelucci F., Aloe L., Jimenez-Vasquez P., Mathe A.A. Electroconvulsive stimuli alter the regional concentrations of nerve growth factor, brain-derived neurotrophic factor, and glial cell line-derived neurotrophic factor in adult rat brain. J ECT 2002; Sep;18(3):138-43

Ankarcrona M., Dypbukt J.M., Bonfoco E., Zhivotovsky B., Orrenius S., Lipton S.A., Nicotera P. Glutamate- induced neuronal death: a succession of necrosis or apoptosis depending on mitochondrial function. Neuron 1995; 15:961-973

Araki T., Simon R.P., Taki W., Lan J.Q., Henshall D.C. Characterization of neuronal death induced by focally evoked limbic seizures in the C57BL/6 mouse. J Neurosci Res 2002; Sep 1;69(5):614-21

Aronica E.M., Gorter J.A., Paupard M.C., Grooms S.Y., Bennett M.V., Zukin R.S. Status epilepticus-induced alterations in metabotropic glutamate receptor expression in young and adult rats. J Neurosci 1997; Nov 1;17(21):8588-95

Babb T.L., Brown W.J., Pretorius J., Davenport C., Lieb J.P., Crandall P.H. Temporal lobe volumetric cell densities in temporal lobe epilepsy. Epilepsia 1990; 25:729-740.

Balldin J., Eden S., Granerus A.K, Modigh K., Svanborg A., Walinder J., Wallin L. Electroconvulsive therapy in Parkinson's syndrome with "on-off" phenomenon. J Neural Transm 1980; 47(1):11-21

Baram T.Z., Eghbal-Ahmadi M., Bender R.A. Is neuronal death required for seizure-induced epileptogenesis in the immature brain? Prog Brain Res 2002; 135:365-75

Beale M.D., Kellner C.H., Gurecki P., Pritchett J.T. ECT for the treatment of Huntington's disease: a case study. Convuls Ther 1997; Jun;13(2):108-12

Becker A.J., Gillardon F., Blumcke I., Langendorfer D., Beck H., Wiestler O.D. Differential regulation of apoptosis-related genes in resistant and vulnerable subfields of the rat epileptic hippocampus. Brain Res Mol Brain Res 1999; Apr 6;67(1):172-6

Ben-Ari Y. Limbic seizure and brain damage produced by kainic acid: mechanisms and relevance to human temporal lobe epilepsy. Neuroscience 1985; 14:375- 403

Berger M., de Soto D.A. The use of ECT for Parkinson symptoms in a nondepressed patient. Psychosomatics 1990; Fall;31(4):465-6

Bernasconi N., Bernasconi A., Caramanos Z., Dubeau F., Richardson J., Andermann F., Arnold D.L. Entorhinal cortex atrophy in epilepsy patients exhibiting normal hippocampal volumes. Neurology 2001; May 22;56(10):1335-9

Bertram E.H. 3rd, Lothman E.W. Morphometric effects of intermittent kindled seizures and limbic status epilepticus in the dentate gyrus of the rat. Brain Res 1993; Feb 12;603(1):25-31

Berzaghi M. P., Cooper J., Castren E., Zafra F., Sofroniew M., Thoenen H., and Lindholm D. Cholinergic regulation of brain-derived neurotrophic factor (BDNF) and nerve growth factor (NGF) but not neurotrophin-3 (NT-3) mRNA levels in the developing rat hippocampus. J Neurosci 1993; 13:3818-3826

Birkett D.P. Use of ECT in Parkinson's disease. Am J Psychiatry 1990; Jul;147(7):952

Bittigau P., Sifringer M., Ikonomidou C. Antiepileptic drugs and apoptosis in the developing brain. Ann N Y Acad Sci 2003; May;993:103-14; discussion 123-4

Bittigau P., Sifringer M., Genz K., Reith E., Pospischil D., Govindarajalu S., Dzietko M., Pesditschek S., Mai L., Dikranian K., Olney J., and Ikonomidou C. Antiepileptic drugs and apoptotic neurodegeneration in the developing brain. PNAS 2002; 99(23):15089-15094

Blennow G., Brierley J.B., Meldrum B.S., Siesjo B.K. Epileptic brain damage: the role of systemic factors that modify cerebral energy metabolism. Brain 1978; Dec;101(4):687-700

Blumcke I., Becker A.J., Klein C., Scheiwe C., Lie A.A., Beck H., Waha A., Friedl M.G., Kuhn R., Emson P., Elger C., Wiestler O.D. Temporal lobe epilepsy associated up-regulation of metabotropic glutamate receptors: correlated changes in mGluR1 mRNA and protein expression in experimental animals and human patients. J Neuropathol Exp Neurol 2000; Jan;59(1):1-10

Bonfoco E., Krainc D., Ankarcrona M., Nicotera P., Lipton S.A. Apoptosis and necrosis: two distinct events induced, respectively, by mild and intense insults with N-methyl-D-aspartate or nitric oxide/superoxide in cortical cell cultures. Proc Natl Acad Sci U S A 1995; Aug 1;92(16):7162-6

Briellmann R.S., Newton M.R., Wellard R.M., Jackson G.D. Hippocampal sclerosis following brief generalized seizures in adulthood. Neurology 2001; 57:315-7

Browning R.A., Nelson D.K. Modification of electroshock and pentylenetetrazol seizure patterns in rats after precollicular transections. Exp Neurol 1986; Sep;93(3):546-56

Browning R.A., Nelson D.K. Variation in threshold and pattern of electroshock-induced seizures in rats depending on site of stimulation. Life Sci 1985; Dec 9;37(23):2205-11

Browning R.A., Wang C., Lanker M.L., Jobe P.C. Electroshock- and pentylenetetrazol-induced seizures in genetically epilepsy-prone rats (GEPRs): differences in threshold and pattern. Epilepsy Res 1990; May-Jun;6(1):1-11

Bruce A.J., M. Baudry. Oxygen free radicals in rat limbic structures after kainate-induced seizures. Free Rad Biol Med 1995; 18:993-1002

Buckmaster P.S., Dudek F.E. Neuronal loss, granule cell axon reorganization, and functional changes in the dentate gyrus of epileptic kainate-treated rats. J Comp Neurol 1997; 385:385-404

Cassidy R.M., Gale K. Mediodorsal thalamus plays a critical role in the development of limbic motor seizures. J Neurosci 1998; Nov 1;18(21):9002-9

Cavalheiro E.A., Silva D.F., Turski W.A., Calderazzo-Filho L.S., Bortolotto Z.A., Turski L. The susceptibility of rats to pilocarpine is age-dependent. Exp Brain Res 1987; 37:43-58

Cavalheiro E.A., Leite J.P., Bortolotto Z.A., Turski W.A., Ikonomidou C., Turski L. Long-term effects of pilocarpine in rats: structural damage of the brain triggers kindling and spontaneous recurrent seizures. Epilepsia 1991; 32:778-782

Cendes F., Andermann F., Carpenter S., Zatorre R.J., Cashman N.R. Temporal lobe epilepsy caused by domoic acid intoxication: evidence for glutamate receptor-mediated excitotoxicity in humans. Ann Neurol 1995; Jan;37(1):123-6

Cendes F., Andermann F., Dubeau F., Gloor P., Evans A., Jones-Gotman M., Olivier A., Andermann E., Robitaille Y., Lopes-Cendes I., et al. Early childhood prolonged febrile convulsions, atrophy and sclerosis of mesial structures, and temporal lobe epilepsy: an MRI volumetric study. Neurology 1993; Jun;43(6):1083-7

Cheng B., Mattson M.P. NT-3 and BDNF protect CNS neurons against metabolic/excitotoxic insults. Brain Res 1994; Mar 21;640(1-2):56-67

Cilio M.R., Sogawa Y., Cha B.H., Liu X., Huang L.T., Holmes G.L. Long-term effects of status epilepticus in the immature brain are specific for age and model. Epilepsia 2003; Apr;44(4):518-28

Clifford D.B., Olney J.W., Maniotis A., Collins R.C., Zorumski C.F. The functional anatomy and pathology of lithium-pilocarpine and high-dose pilocarpine seizures. Neuroscience 1987; 23:953-968

Covolan L., Ribeiro L.T.C., Longo B.M., Mello L. Cell damage and neurogenesis in the dentate granule cell layer of adult rats after pilocarpine or kainate induced status epilepticus. Hippocampus 2000; 10:169-180

Dalby N.O., Tonder N., Wolby D.P., West M., Finsen B., Bolwig T.G. No loss of hippocampal hilar somatostatinergic neurons after repeated electroconvulsive shock: a combined stereological and in situ hybridization study. Biol Psychiatry 1996; Jul 1;40(1):54-60

Davies K.G., Hermann B.P., Dohan F.C. Jr., Foley K.T., Bush A.J., Wyler A.R. Relationship of hippocampal sclerosis to duration and age of onset of epilepsy, and childhood febrile seizures in temporal lobectomy patients. Epilepsy Res 1996; Jun;24(2):119-26

DeGiorgio C.M., Tomiyasu U., Gott P.S., Treiman D.M. Hippocampal pyramidal cell loss in human status epilepticus. Epilepsia 1992; 33:23-7

DeLorenzo R.J., Hauser W.A., Towne A.R., Boggs J.G., Pellock J.M., Penberthy L., *et al.* A prospective, population-based epidemiologic study of status epilepticus in Richmond,

Virginia. Neurology 1996; 46:1029-35

DeLorenzo R.J., Towne A.R., Pellock J.M., Ko D. Status epilepticus in children, adults, and the elderly. Epilepsia 1992; 33 Suppl 4:S15-25

Dirnagl U., Simon R.P., Hallenbeck J.M. Ischemic tolerance and endogenous neuroprotection. Trends Neurosci 2003; May;26(5):248-54

Druga R., Kubova H., Suchomelova L., Haugvicova R. Lithium/pilocarpine status epilepticus-induced neuropathology of piriform cortex and adjoining structures in rats is age-dependent. Physiol Res 2003; 52(2):251-64

Dzhala V., Ben-Ari Y., Khazipov R. Seizures accelerate anoxia-induced neuronal death in the neonatal rat hippocampus. Ann Neurol 2000; Oct;48(4):632-40

Ebert U., Brandt C., Loscher W. Delayed sclerosis, neuroprotection, and limbic epileptogenesis after status epilepticus in the rat. Epilepsia 2002; 43 Suppl 5:86-95

Emerson M.R., Nelson S.R., Samson F.E., Pazdernik T.L. A global hypoxia preconditioning model: neuroprotection against seizure-induced specific gravity changes (edema) and brain damage in rats. Brain Res Brain Res Protoc 1999; Dec;4(3):360-6

Ende G., Braus D.F., Walter S., Weber-Fahr W., Henn F.A. The hippocampus in patients treated with electroconvulsive therapy: a proton magnetic resonance spectroscopic imaging study. Arch Gen Psychiatry 2000; 57(10):937-943

Faber R., Trimble M.R. Electroconvulsive therapy in Parkinson's disease and other movement disorders. Mov Disord 1991; 6(4):293-303

Faherty C.J., Xanthoudakis S., Smeyne R.J. Caspase-3-dependent neuronal death in the hippocampus following kainic acid treatment. Brain Res Mol Brain Res 1999; Jun 18;70(1):159-63

Farwell J. R., Dodrill C. B., Batzel L. W. Neuropsychological abilities of children with epilepsy. Epilepsia 1985; 26:395-400

Fernandez G., Effenberger O., Vinz B., Steinlein O., Elger C.E., Dohring W., Heinze H.J. Hippocampal malformation as a cause of familial febrile convulsions and subsequent hippocampal sclerosis. Neurology 1998; Apr;50(4):909-17

Fernandez-Sanchez M.T., Novelli A. Basic fibroblast growth factor protects cerebellar neurons in primary culture from NMDA and non-NMDA receptor mediated neurotoxicity. FEBS Lett 1993; Nov 29;335(1):124-31

Finklestein S.P., Kemmou A., Caday C.G., Berlove D.J. Basic fibroblast growth factor protects cerebrocortical neurons against excitatory amino acid toxicity in vitro. Stroke 1993; Dec;24(12 Suppl):I141-3

Follesa P., Gale K., Mocchetti I. Regional and temporal pattern of expression of nerve growth factor and basic fibroblast growth factor mRNA in rat brain following electroconvulsive shock. Exp Neurol 1994; May;127(1):37-44

Fountain N.B., Lothman E.W. Pathophysiology of status epilepticus. J Clin Neurophysiol 1995; Jul;12(4):326-42

Freese A., Finklestein S.P., DiFiglia M. Basic fibroblast growth factor protects striatal neurons in vitro from NMDA-receptor mediated excitotoxicity. Brain Res 1992; Mar 20;575(2):351-5

Friedman L.K. Selective reduction of GluR2 protein in adult hippocampal CA3 neurons following status epilepticus but prior to cell loss. Hippocampus 1998; 8(5):511-25

Fuerst D., Shah J., Kupsky W.J., Johnson R., Shah A., Hayman- Abello B., *et al.* Volumetric MRI, pathological, and neuropsychological progression in hippocampal sclerosis. Neurology 2001; 57:184-8

Fujikawa D.G., Itabashi H.H., Wu A., Shinmei S.S. Status epilepticus-induced neuronal loss in humans without systemic complications or epilepsy. Epilepsia 2000; Aug;41(8):981-91

Fujikawa D.G., Ke X., Trinidad R.B., Shinmei S.S., Wu A. Caspase-3 is not activated in seizure-induced neuronal necrosis with internucleosomal DNA cleavage. J Neurochem 2002; Oct;83(1):229-40

Fujikawa D.G., Shinmei S.S., Cai B. Kainic acid-induced seizures produce necrotic, not apoptotic, neurons with internucleosomal DNA cleavage: implications for programmed cell death mechanisms. Neuroscience 2000c; 98(1):41-53

Fujikawa D.G., Shinmei S.S., Cai B. Lithium-pilocarpineinduced status epilepticus produces necrotic neurons with internucleosomal DNA fragmentation in adult rats. Eur J Neurosci 1999; 11:1605-14

Fujikawa D.G., Shinmei S.S., Cai B. Seizure-induced neuronal necrosis: implications for programmed cell death mechanisms. Epilepsia 2000d; 41 Suppl 6:S9-13

Fujikawa D.G. Confusion between neuronal apoptosis and activation of programmed cell death mechanisms in acute necrotic insults. Trends Neurosci 2000; Sep;23(9):410-1

Galanopoulou A.S., Vidaurre J., Moshe S.L. Under what circumstances can seizures produce hippocampal injury: evidence for age-specific effects. Dev Neurosci 2002; 24(5):355-63

Gilbert D.L., Gantside P.S., Glauser T.A. Efficacy and mortality in treatment of refractory generalized convulsive status epilepticus in children: a meta-analysis. J Child Neurol 1999; 14:602-609

Glass M., Dragunow M. Neurochemical and morphological changes associated with human epilepsy. Brain Res Brain Res Rev 1995; Jul;21(1):29-41

Glazner G.W., Mattson M.P. Differential effects of BDNF, ADNF9, and TNFalpha on levels of NMDA receptor subunits, calcium homeostasis, and neuronal vulnerability to excitotoxicity. Exp Neurol 2000; Feb;161(2):442-52

Gombos Z., Spiller A., Cottrell G.A., Racine R.J., McIntyre Burnham W. Mossy fiber sprouting induced by repeated electroconvulsive shock seizures. Brain Res 1999; Oct 9;844(1-2):28-33

Goodman J.H. "Experimental models of status epilepticus." In Neuropharmacological Methods in Epilepsy Research, S.L. Peterson and T.E. Albertson, eds. Boca Raton: CRC Press, 1998.

Gordon T. Fatigue in adapted systems. Overuse and underuse paradigms. Adv Exp Med Biol 1995; 384:429-56

Gorter J.A., Van Vliet E.A., Proper E.A., De Graan P.N., Ghijsen W.E., Lopes Da Silva F.H., Aronica E. Glutamate transporters alterations in the reorganizing dentate gyrus are associated with progressive seizure activity in chronic epileptic rats. J Comp Neurol 2002; Jan 21;442(4):365-77

Gould E., Woolley C.S., McEwen B.S. Short-term glucocorticoid manipulations affect neuronal morphology and survival in the adult dentate gyrus. Neuroscience 1990; 37(2):367-75

Gould E., Woolley C.S., McEwen B.S. Naturally occurring cell death in the developing dentate gyrus of the rat. J Comp Neurology 1991; 304:408-418

Gray N.A., Zhou R., Du J., Moore G.J., Manji H.K. The use of mood stabilizers as plasticity enhancers in the treatment of neuropsychiatric disorders. J Clin Psychiatry 2003; 64 Suppl 5:3-17

Grunewald R.A., Farrow T., Vaughan P., Rittey C.D., Mundy J. A magnetic resonance study of complicated early childhood convulsion. J Neurol Neurosurg Psychiatry 2001; Nov;71(5):638-42

Gunderson V.M., Dubach M., Szot P., Born D.E., Wenzel H.J., Maravilla K.R., Zierath D.K., Robbins C.A., Schwartzkroin P.A. Development of a model of status epilepticus in pigtailed macaque infant monkeys. Dev Neurosci 1999; Nov;21(3-5):352-64

Gwinn R.P., Kondratyev A., Gale K. Time-dependent increase in basic fibroblast growth factor protein in limbic regions following electroshock seizures. Neuroscience 2002; 114(2):403-9

Haas K.Z., Sperber E.F., Opanashuk L.A., Staton P.K., Moshe S.L. Resistance of immature hippocampus to morphologic and physiologic alteration following status epilepticus or kindling. Hippocampus 2001; 11(6):615-625

Hamm R.J., Pike B.R., Temple M.D., O'Dell D.M., Lyeth B.G. The effect of postinjury kindled seizures on cognitive performance of traumatically brain-injured rats. Exp Neurol 1995; Dec;136(2):143-8

Hauser W.A. Status epilepticus: epidemiologic considerations. Neurology 1990; 40(suppl 2):9-13.

Henshall D.C., Chen J., Simon R.P. Involvement of caspase-3 like protease in the mechanism of cell death following focally evoked limbic seizures. J Neurochem 2000; 74:1215-1223

Hermann B. Neurodevelopmental and Progressive Adverse Effects of Epilepsy on Higher Cognitive Functioning 2002 http://www.aesnet.org/edu_pub/PresidentialSymposium.cfm

Hernandez T.D., Holling L.C. Disruption of behavioral recovery by the anti-convulsant phenobarbital. Brain Res 1994; Jan 28;635(1-2):300-6

Hernandez T.D., Schallert T. Seizures and recovery from experimental brain damage. Exp Neurol 1988; Dec;102(3):318-24

Holmes G. L. Do seizures cause brain damage? Epilepsia 1991; 32:(suppl. 5) S14-S28

Holmes G. L. The long term effects of seizures on the developing brain: clinical and laboratory issues. Brain Dev 1991; 13:393-409

Holmes G.L., Khazipov R., Ben-Ari Y. New concepts in neonatal seizures. Neuroreport 2002; 13:A3-8

Holmes G.L., Ben-Ari Y. Seizures in the developing brain: perhaps not so benign after all. Neuron 1998; Dec;21(6):1231-4

Holmes G.L. Seizure-induced neuronal injury: animal data. Neurology 2002; Nov 12;59(9 Suppl 5):S3-6

Houser C.R. Neuronal loss and synaptic reorganization in temporal lobe epilepsy. Adv Neurol 1999; 79:743-761

Huang L., Cilio M. R., Silveira D.C., McCabe B.K., Sogawa Y., Stafstrom C.E., Holmes

G.L. Long-term effects of neonatal seizures: a behavioral, electrophysiological, and histological study, Dev Brain Res 1999; 118:99-107

Hughes P.E., Alexi T., Walton M., Williams C.E., Dragunow M., Clark R.G., Gluckman P.D. Activity and injury-dependent expression of inducible transcription factors, growth factors and apoptosis-related genes within the central nervous system. Prog Neurobiol 1999; Feb;57(4):421-50

Husseini M.K., Jorge Q.A., Sporn J., Payne J.L., Denicoff K., Gray N.A., Zarate Jr. C.A., Charney D.S. Enhancing neuronal plasticity and cellular resilience to develop novel, improved therapeutics for difficult-to-treat depression. Biol Psychiatry 2003; Apr 15; 53(8): 707-742

Huttenlocher P.R., Hapke R.J. A follow-up study of intractable seizures in childhood. Ann. Neurol 1990; 28:699-705

Ikononomidou C., Bosch F., Miksa M., Bittigua P., Vockler J., Dikranian K., Tenkova T., Stefovska V., Turski L., Olney J. Blockade of NMDA receptors and apoptotic neurodegeneration in the developing brain. Science 1999; 283:74

Jefferys J.G., Evans B.J., Hughes S.A., Williams S.F. Neuropathology of the chronic epileptic syndrome induced by intrahippocampal tetanus toxin in rat: preservation of pyramidal cells and incidence of dark cells. Neuropathol Appl Neurobiol 1992; Feb;18(1):53-70

Jensen F.E., Baram T.Z. Developmental seizures induced by common early-life insults: short- and long-term effects on seizure susceptibility. Ment Retard Dev Disabil Res Rev 2000; 6(4):253-7.

Jones P.A., Smith R.A., Stone T.W. Nitric oxide synthase inhibitors L-NAME and 7-nitroindazole protect rat hippocampus against kainate-induced excitotoxicity. Neurosci Lett 1998; Jun 19;249(2-3):75-8

Jope R.S., Morrisett R.A., Snead O.C. Characterization of lithium potentiation of pilocarpine-induced status epilepticus in rats. Exp Neurol 1986; Mar;91(3):471-80

Kant R., Bogyi A.M., Carosella N.W., Fishman E., Kane V., Coffey C.E. ECT as a therapeutic option in severe brain injury. Convuls Ther 1995; Mar;11(1):45-50

Kellner C.H., Beale M.D., Pritchett J.T., Bernstein H.J., Burns C.M. Electroconvulsive therapy and Parkinson's disease: the case for further study. Psychopharmacol Bull 1994; 30(3):495-500

Kelly M.E., McIntyre D.C. Hippocampal kindling protects several structures from the neuronal damage resulting from kainic acid-induced status epilepticus. Brain Res 1994; Jan 21;634(2):245-56

Kempermann G., Van Praag H., Gage F. H. Activity-dependent regulation of neuronal plasticity and self repair. Prog. Brain Res 2000; 127:35-48

Klitgaard H., Matagne A., Vanneste-Goemaere J., Margineanu D.G. Pilocarpine-induced epileptogenesis in the rat: impact of initial duration of status epilepticus on electrophysiological and neuropathological alterations. Epilepsy Res 2002; Sep;51(1-2):93-107

Kobayashi E., Li L.M., Lopes-Cendes I., Cendes F. Magnetic resonance imaging evidence of hippocampal sclerosis in asymptomatic, first-degree relatives of patients with familial mesial temporal lobe epilepsy. Arch Neurol 2002; Dec;59(12):1891-4

Kondratyev A., Ved R., Gale K. The effects of repeated minimal electroconvulsive shock exposure on levels of mRNA encoding fibroblast growth factor-2 and nerve growth factor in limbic regions. Neuroscience 2002; 114(2):411-6

Kondratyev A., Gale K. Temporal and spatial patterns of DNA fragmentation following focally or systemically-evoked status epilepticus in rats. Neurosci Lett 2001; 310:13-16

Kondratyev A., Sahibzada N., Gale K. Electroconvulsive shock exposure prevents neuronal apoptosis after kainic acid-evoked status epilepticus. Mol Brain Res 2001; 91:1-13

Kondratyev A., Gale K. Intracerebral injection of caspase-3 inhibitor prevents neuronal apoptosis after kainic acid-evoked status epilepticus. Brain Res Mol Brain Res 2000; 75: 216-224

Kornblum H.I., Sankar R., Shin D.H., Wasterlain C.G., Gall C.M. Induction of brain derived neurotrophic factor mRNA by seizures in neonatal and juvenile rat brain. Mol Brain Res 1997; 44:219-228

Kotloski R., Lynch M., Lauersdorf S., Sutula T. Repeated brief seizures induce progressive hippocampal neuron loss and memory deficits. Prog Brain Res 2002; 135:95-110

Kuan C.Y., Roth K.A., Flavell R.A., Rakic P. Mechanisms of programmed cell death in the developing brain. Trends Neurosci 2000; 23:291-297

Kudryashov I.E., Onufriev M.V., Kudryashova I.V., Gulyaeva N.V. Periods of postnatal maturation of hippocampus: synaptic modifications and neuronal disconnection. Dev Brain Res 2001; 132:113-120

Kuks J.B., Cook M.J., Fish D.R., Stevens J.M., Shorvon S.D. Hippocampal sclerosis in epilepsy and childhood febrile seizures. Lancet 1993; Dec 4;342(8884):1391-4

Kunz W.S., Goussakov I.V., Beck H., Elger C.E. Altered mitochondrial oxidative phosphorylation in hippocampal slices of kainate-treated rats. Brain Res 1999; 826:236-242

Kunz W.S., Kudin A.P., Vielhaber S., Blumcke I., Zuschratter W., Schramm J., Beck H., Elger C.E. Mitochondrial complex I deficiency in the epileptic focus of patients with temporal lobe epilepsy. Ann. Neurol 2000; 48:766-773

Lado F.A., Sankar R., Lowenstein D., Moshe S.L. Age-dependent consequences of seizures: relationship to seizure frequency, brain damage, and circuitry reorganization. Ment Retard Dev Disabil Res Rev 2000; 6(4):242-52

Lewis D.V., Barboriak D.P., MacFall J.R., Provenzale J.M., Mitchell T.V., VanLandingham K.E. Do prolonged febrile seizures produce medial temporal sclerosis? Hypotheses, MRI evidence and unanswered questions. Prog Brain Res 2002; 135:263-78

Lewis D.V. Febrile convulsions and mesial temporal sclerosis. Curr Opin Neurol 1999; Apr;12(2):197-201

Liang L.P., Ho Y.S., Patel M. Mitochondrial superoxide production in kainate-induced hippocampal damage. Neuroscience 2000; 101(3):563-70

Liou A.K., Clark R.S., Henshall D.C., Yin X.M., Chen J. To die or not to die for neurons in ischemia, traumatic brain injury and epilepsy: a on the stress-activated signaling pathways and apoptotic pathways. Prog Neurobiol 2003; Feb;69(2):103-42

Liposits Z., Kallo I., Hrabovszky E., Gallyas F. Ultrastructural pathology of degenerating "dark" granule cells in the hippocampal dentate gyrus of adrenalectomized rats. Acta Biol Hung 1997; 48(2):173-87

Liu Z., D'Amore P.A., Mikati M., Gatt A., Holmes G.L. Neuroprotective effect of chronic infusion of basic fibroblast growth factor on seizure-associated hippocampal damage. Brain

Res 1993; Oct 29;626(1-2):335-8

Liu Z., Holmes G.L. Basic fibroblast growth factor is highly neuroprotective against seizure-induced long-term behavioural deficits. Neuroscience 1997; Feb;76(4):1129-38

Liu Z., Yang Y., Silveira D.C., Sarkisian M.R., Tandon P., Huang L.T., Stafstrom C.E., Holmes G.L. Consequences of recurrent seizures during early brain development. Neuroscience 1999; 92:1443-1454

Lothman E.W., Bertram E.H. 3rd. Epileptogenic effects of status epilepticus. Epilepsia 1993; 34 Suppl 1:S59-70

Lowenstein D.H., Arsenault L. The effects of growth factors on the survival and differentiation of cultured dentate gyrus neurons. J Neurosci 1996; 16:1759-1769

Lumme A., Soinila S., Sadeniemi M., Halonen T., Vanhatalo S. Nitric oxide synthase immunoreactivity in the rat hippocampus after status epilepticus induced by perforant pathway stimulation. Brain Res 2000; 871:303-310

Lynch M., Sayin U., Bownds J., Janumpalli S., Sutula T. Long-term consequences of early postnatal seizures on hippocampal learning and plasticity. Eur J Neurosci 2000; Jul;12(7):2252-64

Maggio R., Liminga U., Gale K. Selective stimulation of kainate but not quisqualate or NMDA receptors in substantia nigra evokes limbic motor seizures. Brain Res 1990; Oct 1;528(2):223-30

Maher J., McLachlan R.S. Febrile convulsions. Is seizure duration the most important predictor of temporal lobe epilepsy? Brain 1995; Dec;118 (Pt 6):1521-8

Maher J., McLachlan R.S. Febrile convulsions. Is seizure duration the most important predictor of temporal lobe epilepsy? Brain 1995; Dec;118 (Pt 6):1521-8

Manji H.K., Quiroz J.A., Sporn J., Payne J.L., Denicoff K.A., Gray N., Zarate C.A. Jr., Charney D.S. Enhancing neuronal plasticity and cellular resilience to develop novel, improved therapeutics for difficult-to-treat depression. Biol Psychiatry 2003; Apr 15;53(8):707-42

Masco D., Sahibzada N., Switzer R., Gale K. Electroshock seizures protect against apoptotic hippocampal cell death induced by adrenalectomy. Neuroscience 1999; 91(4):1315-1319

Mathern G.W., Adelson P.D., Cahan L.D., Leite J.P. Hippocampal neuron damage in human epilepsy: Meyer's hypothesis revisited. Prog Brain Res 2002; 135:237-51

Mattson M.P., Kumar K.N., Wang H., Cheng B., Michaelis E.K. Basic FGF regulates the expression of a functional 71 kDa NMDA receptor protein that mediates calcium influx and neurotoxicity in hippocampal neurons. J Neurosci 1993; Nov;13(11):4575-88

Mattson M.P., Lovell M.A., Furukawa K., Markesbery W.R. Neurotrophic factors attenuate glutamate-induced accumulation of peroxides, elevation of intracellular Ca2+ concentration, and neurotoxicity and increase antioxidant enzyme activities in hippocampal neurons. J Neurochem 1995; Oct;65(4):1740-51

Mayat E., Lerner-Natoli M., Rondouin G., Lebrun F., Sassetti I., Reasens M. Kainate-induced status epilepticus leads to a delayed increase in various specific glutamate metabotropic receptor responses in the hippocampus. Brain Res 1994; May 9;645(1-2):186-200

McIntyre D.C., Nathanson D., Edson N. A new model of partial status epilepticus based on kindling. Brain Res 1982; Oct 28;250(1):53-63

McIntyre D.C., Poulter M.O., Gilby K. Kindling: some old and some new. Epilepsy Res 2002; Jun;50(1-2):79-92

Meier P, Finch A., Evan G. Apoptosis in development. Nature 2000; 407: 797-801

Meldrum B.S., Horton R.W., Brierley J.B. Epileptic brain damage in adolescent baboons following seizures induced by allylgycine. Brain 1974; Jun;97(2):407-18

Meldrum B.S., Papy J.J., Toure M.F., Brierley J.B. Four models for studying cerebral lesions secondary to epileptic seizures. Adv Neurol 1975; 10:147-61

Meldrum B.S., Vigouroux R.A., Brierley J.B. Systemic factors and epileptic brain damage. Prolonged seizures in paralyzed artificially ventilated baboons. Arch Neurol 1973a; Aug;29(2):82-7

Meldrum B.S., Vigouroux R.A., Rage P., Brierley J.B. Hippocampal lesions produced by prolonged seizures in paralyzed artificially ventilated baboons. Experientia 1973b; May 15;29(5):561-3

Meldrum B.S. Metabolic factors during prolonged seizures and their relation to nerve cell death. Adv Neurol 1983; 34:261-75

Meldrum B.S. Excitotoxicity and epileptic brain damage. Epilepsy Res 1991; 10:55-61

Meletti S., Benuzzi F., Rubboli G., Cantalupo G., Stanzani Maserati M., Nichelli P., Tassinari C.A. Impaired facial emotion recognition in early-onset right mesial temporal lobe epilepsy. Neurology 2003; Feb 11;60(3):426-31

Milatovic D., Gupta R.C., Dettbarn W.D. Involvement of nitric oxide in kainic acid-induced excitotoxicity in rat brain. Brain Research 2002; 957:330-337

Mizrahi E.M., Clancy R.R. Neonatal seizures: early-onset seizure syndromes and their consequences for development. Ment Retard Dev Disabil Res Rev 2000; 6(4):229-41

Mody I. Synaptic plasticity in kindling. Adv Neurol 1999; 79:631-43

Mohapel P., Dufresne C., Kelly M.E., McIntyre D.C. Differential sensitivity of various temporal lobe structures in the rat to kindling and status epilepticus induction. Epilepsy Res 1996; Apr;23(3):179-87

Montecot C., Rondi-Reig L., Springhetti V., Seylaz J., Pinard E. Inhibition of neuronal (type 1) nitric oxide synthase prevents hyperaemia and hippocampal lesions resulting from kainate-induced seizures. Neuroscience 1998; Jun;84(3):791-800

Morrisett R.A., Jope R.S., Snead O.C. III. Status epilepticus is produced by administration of cholinergic agonists to lithiumtreated rats: comparison with kainic acid. Exp Neurol 1987; 98:594-605

Morrison R.S., Wenzel H.J., Kinoshita Y., Robbins C.A., Donehower L.A., Schwartzkroin P.A. Loss of the p53 tumor suppressor gene protects neurons from kainate-induced cell death. J Neurosci 1996; Feb 5;16(4):1337-45

Moshe S.L. Brain injury with prolonged seizures in children and adults. J Child Neurol 1998; Oct;13 Suppl 1:S3-6

Najm I.M., Hadam J., Ckakraverty D., Mikuni N., Penrod C., Sopa C., Markarian G., Luders H.O., Babb T., Baudry M. A short episode of seizure activity protects from status epilepticus-induced neuronal damage in rat brain. Brain Res 1998; Nov 9;810(1-2):72-5

Narkilahti S., Nissinen J., Pitkanen A. Administration of caspase 3 inhibitor during and after status epilepticus in rat: effect on neuronal damage and epileptogenesis. Neuropharmacology 2003; Jun;44(8):1068-88

Naruse I., Keino H. Apoptosis in the developing CNS. Prog Neurobiol 1995; 47:135-155

Naylor P., Stewart C.A., Wright S.R., Pearson R.C., Reid I.C. Repeated ECS induces GluR1 mRNA but not NMDAR1A-G mRNA in the rat hippocampus. Brain Res Mol Brain Res 1996; Jan;35(1-2):349-53

Nevander G., Ingvar M., Auer R., Siesjo B.K. Status epilepticus in well-oxygenated rats causes neuronal necrosis. Ann Neurol 1985; 18:281-90

Nibuya M., Morinobu S., Duman R.S. Regulation of BDNF and trkB mRNA in rat brain by chronic electroconvulsive seizure and antidepressant drug treatments. J Neurosci 1995; Nov;15(11):7539-47

Olney J.W., Wozniak D.F., Jevtovic-Todorovic V., Farber N.B., Bittigau P., Ikonomidou C. Drug-induced apoptotic neurodegeneration in the developing brain. Brain Pathol 2002; Oct;12(4):488-98

Oppenheim R W. Cell death during development of the nervous system. Annu Rev Neurosci 1991; 14:453-501

Pekary A.E., Meyerhoff J.L, Sattin A. Electroconvulsive seizures modulate levels of thyrotropin releasing hormone and related peptides in rat hypothalamus, cingulate and lateral cerebellum. Brain Res 2000; Nov 24;884(1--2):174-83

Pelletier M.R., Wadia J.S., Mills L.R., Carlen P.L. Seizure-induced cell death produced by repeated tetanic stimulation in vitro: possible role of endoplasmic reticulum calcium stores. J Neurophysiol 1999; 81:3054-3064

Penner M.R., Pinaud R., Robertson H.A. Rapid kindling of the hippocampus protects against neural damage resulting from status epilepticus. NeuroReport 2001; 12(3): 453- 457

Pinard E., Tremblay E., Ben-Ari Y., Seylaz J. Blood flow compensates oxygen demand in the vulnerable CA3 regions of the hippocampus during kainate-induced seizures. Neuroscience 1984; 13:1039-1049

Plamondon H., Blondeau N., Heurteaux C., Lazdunski M. Mutually protective actions of kainic acid epileptic preconditioning and sublethal global ischemia on hippocampal neuronal death: involvement of adenosine A1 receptors and K(ATP) channels. J Cereb Blood Flow Metab 1999; Dec;19(12):1296-308

Pollard H., Cantagrel S., Charriaut-Marlangue C., Moreau J., Ben-Ari Y. Apoptosis associated DNA fragmentation in epileptic brain damage. NeuroReport 1994; 5:1053-1055

Portera-Cailliau C., Price D.L., Martin L.J. Excitotoxic neuronal death in the immature brain is an apoptosis-necrosis morphological continuum. J Comp Neurol 1997; Feb 3;378(1):70-87

Portera-Cailliau C., Price D.L., Martin L.J. Non-NMDA and NMDA receptor-mediated excitotoxic neuronal deaths in adult brain are morphologically distinct: further evidence for an apoptosis-necrosis continuum. J Comp Neurol 1997; Feb 3;378(1):88-104

Pridmore S., Pollard C. Electroconvulsive therapy in Parkinson's disease: 30 month follow up. J Neurol Neurosurg Psychiatry 1996; Jun;60(6):693

Puig B., Ferrer I. Caspase-3-associated apoptotic cell death in excitotoxic necrosis of the entorhinal cortex following intraperitoneal injection of kainic acid in the rat. Neurosci Lett 2002; Mar 22;321(3):182-6

Racine R.J., Adams B., Osehobo P., Fahnestock M. Neural growth, neural damage and neurotrophins in the kindling model of epilepsy. Adv Exp Med Biol 2002; 497:149-70

Rasmussen K., Abrams R. Treatment of Parkinson's disease with electroconvulsive therapy. Psychiatr Clin North Am 1991; Dec;14(4):925-33

Reti I.M., Baraban J.M. Sustained increase in Narp protein expression following repeated electroconvulsive seizure. Neuropsychopharmacology 2000; Oct;23(4):439-43

Roy M., Sapolsky R. Neuronal apoptosis in acute necrotic insults: why is this subject such a mess? Trends Neurosci 1999; Oct;22(10):419-22

Sakhi S., Sun N., Wing L.L., Mehta P., Schreiber S.S. Nuclear accumulation of p53 protein following kainic acid-induced seizures. NeuroReport 1996; 7:493-496

Sandyk R. Mechanisms of action of ECT in Parkinson's disease: possible role of pineal melatonin. Int J Neurosci 1990; Jan;50(1-2):83-94

Sankar R., Shin D.H., Liu H., Mazarati A., Pereira de Vasconcelos A., Wasterlain C.G. Patterns of status epilepticus-induced neuronal injury during development and long-term consequences. J Neurosci 1998; 18:8382-93

Sasahira M., Lowry T., Simon R.P., Greenberg D.A. Epileptic tolerance: prior seizures protect against seizure-induced neuronal injury. Neurosci Lett 1995; Feb 9;185(2):95-8

Sasahira M., Simon R.P., Greenberg D.A. Neuronal injury in experimental status epilepticus in the rat: role of hypoxia. Neurosci Lett 1997; Feb 7;222(3):207-9

Sasahira M., Lowry T., Simon R.P., Greenberg D.A. Epileptic tolerance: prior seizures protect against seizure-induced neuronal injury. Neurosci. Lett 1995; 185:95-98

Sater R.A., Nadler J.V. On the relation between seizures and brain lesions after intracerebroventricular kainic acid. Neurosci Lett 1988; Jan 11;84(1):73-8

Sato K., Kashihara K., Morimoto K., Hayabara T. Regional increases in brain-derived neurotrophic factor and nerve growth factor mRNAs during amygdaloid kindling, but not in acidic and basic fibroblast growth factor mRNAs. Epilepsia 1996; Jan;37(1):6-14

Schallert T., Hernandez T.D., Barth T.M. Recovery of function after brain damage: severe and chronic disruption by diazepam. Brain Res 1986; Jul 30;379(1):104-11

Schauwecker P.E. Complications associated with genetic background effects in models of experimental epilepsy. Prog Brain Res 2002; 135:139-48

Schauwecker P.E. Genetic background as a determinant of seizure-induced cell death 2002 http://www.aesnet.org/edu_pub/PresidentialSymposium.cfm

Schmid R., Tandon P., Stafstrom C.E., Holmes G.L. Effects of neonatal seizures on subsequent seizure-induced brain injury. Neurology 1999; 53(8):1754-1761

Schulz R., Ebner A. Prolonged febrile convulsions and mesial temporal lobe epilepsy in an identical twin. Neurology 2001; Jul 24;57(2):318-20

Schwob J.E., Fuller T., Price J.L., Olney J.W. Widespread patterns of neuronal damage following systemic or intracerebral injections of kainic acid: a histological study. Neuroscience 1980; 5:991-1014

Scott R.C., Gadian D.G., King M.D., Chong W.K., Cox T.C., Neville B.G., Connelly A. Magnetic resonance imaging findings within 5 days of status epilepticus in childhood. Brain 2002; Sep;125(Pt 9):1951-9. Erratum in: Brain 2002; Dec;125(Pt 12):2792

Sheline Y.I., Sanghavi M., Mintun M.A., Gado M.H. Depression duration but not age predicts hippocampal volume loss in medically healthy women with recurrent major depression. J Neurosci 1999; 19(12): 5034-5043

Shimosaka S., So Y.T., Simon R.P. Distribution of HSP72 induction and neuronal death following limbic seizures. Neurosci Lett 1992; Apr 27;138(2):202-6

Shinnar S., Berg A.T., Moshe S.L., O'Dell C., Alemany M., Newstein D., Kang H., Goldensohn E.S., Hauser W.A. The risk of seizure recurrence after a first unprovoked afebrile seizure in childhood: an extended follow-up. Pediatrics 1996; Aug;98(2 Pt 1):216-25

Shinnar S., Glauser T.A. Febrile seizures. J Child Neurol 2002; Jan;17 Suppl 1:S44-52

Shinnar S., Pellock J.M., Berg A.T., O'Dell C., Driscoll S.M., Maytal J., Moshe S.L., DeLorenzo R.J. Short-term outcomes of children with febrile status epilepticus. Epilepsia 2001; Jan;42(1):47-53

Shorvon S. Status Epilepticus. J Neurol Neurosurg Psychiatry 2001; Jun;70 Suppl 2:II22-7

Sloviter R.S. Possible functional consequences of synaptic reorganization in the dentate gyrus of kainate-treated rats. Neurosci Lett 1992; 137:91-96

Sloviter R.S., Dean E., Neubort S. Electron microscopic analysis of adrenalectomy-induced hippocampal granule cell degeneration in the rat: apoptosis in the adult central nervous system. J Comp Neurol 1993; Apr 15;330(3):337-51

Sloviter R.S., Dean E., Sollas A.L., Goodman J.H. Apoptosis and necrosis induced in different hippocampal neuron populations by repetitive perforant path stimulation in the rat. J Comp Neurol 1996; Mar 11;366(3):516-33

Sloviter R.S., Lowenstein D.H. Heat shock protein expression in vulnerable cells of the rat hippocampus as an indicator of excitation-induced neuronal stress. J Neurosci 1992; Aug;12(8):3004-9

Sloviter R.S., Valiquette G., Abrams G.M., Ronk E.C., Sollas A.L., Paul L.A., Neubort S. Selective loss of hippocampal granule cells in the mature rat brain after adrenalectomy. Science 1989; Jan 27;243(4890):535-8

Sloviter R.S., Damiano B.P. Sustained electrical stimulation of the perforant path duplicates kainate-induced electrophysiological effects and hippocampal damage in rats. Neurosci. Lett 1981; 24:279-284

Sogawa Y., Monokoshi M., Silveira D.C., Cha B.H., Cilio M.R., McCabe B.K., Liu X., Hu Y, Holmes G.L. Timing of cognitive deficits following neonatal seizures: relationship to histological changes in the hippocampus. Brain Res Dev Brain Res 2001; 26:73-83

Sommer C., Roth S.U., Kiessling M. Kainate-induced epilepsy alters protein expression of AMPA receptor subunits GluR1, GluR2 and AMPA receptor binding protein in the rat hippocampus. Acta Neuropathol (Berl) 2001; May;101(5):460-8

Sperber E.F., Haas K.Z., Stanton P.K., Moshe S.L. Resistance of the immature hippocampus to seizure-induced synaptic reorganization. Brain Res Dev Brain Res 1991; 60:88-93

Sperber E.F., Haas K.Z., Romero M.T., Stanton P.K. Flurothyl status epilepticus in developing rats: behavioral, electrographic histological and electrophysiological studies. Brain Res Dev Brain Res 1999; Aug 5;116(1):59-68

Sperk G., Lassmannv H., Baran H., Kish S.J., Seitelberger F., Hornykiewicz O. Kainic acid induced seizures: neurochemical and histopathological changes. Neuroscience 1983;

10:1301-1315

Stafstrom C.E. Assessing the behavioral and cognitive effects of seizures on the developing brain. Prog Brain Res 2002; 135:377-90

Stafstrom C.E., Thompson J.L., Holmes G.L. Kainic acid seizures in the developing brain: status epilepticus and spontaneous recurrent seizures. Brain Res Dev Brain Res 1992; Feb 21;65(2):227-36

Stern M.B. Electroconvulsive therapy in untreated Parkinson's disease. Mov Disord 1991; 6(3):265

Sullivan P.G., Dube C., Dorenbos K., Steward O., Baram T.Z. Mitochondrial uncoupling protein-2 protects the immature brain from excitotoxic neuronal death. Ann Neurol 2003; 53(6):711-7

Sutula T.P., Pitkanen A. More evidence for seizure-induced neuron loss: is hippocampal sclerosis both cause and effect of epilepsy? Neurology 2001; Jul 24;57(2):169-70

Swann J., Smith K., Lee C. Neuronal activity and the establishment of normal and epileptic circuits during brain development. Int'l Rev Neurobio 2001; 45:89-118

Tanaka K., Simon R.P. The pattern of neuronal injury following seizures induced by intranigral kainic acid. Neurosci Lett 1994; Aug 1;176(2):205-8

Tandon P., Yang Y., Das K., Holmes G.L., Stafstrom C.E. Neuroprotective effects of brain-derived neurotrophic factor in seizures during development. Neuroscience 1999; 91:293-303

Tang F.R., Lee W.L., Yang J., Sim M.K., Ling E.A. Expression of metabotropic glutamate receptor 1alpha in the hippocampus of rat pilocarpine model of status epilepticus. Epilepsy Res 2001; Aug;46(2):179-89

Tang F.R., Lee W.L., Yang J., Sim M.K., Ling E.A. Metabotropic glutamate receptor 8 in the rat hippocampus after pilocarpine induced status epilepticus. Neurosci Lett 2001; Mar 16;300(3):137-40

Tarkka R., Paakko E., Pyhtinen J., Uhari M., Rantala H. Febrile seizures and mesial temporal sclerosis: No association in a long-term follow-up study. Neurology 2003; Jan 28;60(2):215-8

Tasch E., Cendes F., Li L.M., Dubeau F., Andermann F., Arnold D.L. Neuroimaging evidence of progressive neuronal loss and dysfunction in temporal lobe epilepsy. Ann Neurol 1999; 45: 568-76

Theodore W.H., Bhatia S., Hatta J., Fazilat S., DeCarli C., Bookheimer S.Y., *et al.* Hippocampal atrophy, epilepsy duration, and febrile seizures in patients with partial seizures. Neurology 1999; 52:132-6

Theodore W.H., DeCarli C., Gaillard W.D. Total cerebral volume is reduced in patients with localization-related epilepsy and a history of complex febrile seizures. Arch Neurol 2003; Feb;60(2):250-2

Toth Z., Yan X., Haftoglou S., Ribak C., Baram T. Seizure-induced neuronal injury: Vulnerability to febrile seizures in an immature rat model. J. Neuroscience 1998; 18(11):4285-4294

Towfighi J., Housman C., Mauger D., Vannucci R.C. Effect of seizures on cerebral hypoxic-ischemic lesions in immature rats. Brain Res Dev Brain Res 1999; Mar 12;113(1-2):83-95

Treiman D.M., Walton N.Y., Gunawan S. Brain amino acid concentrations during specific electroencephalographic stages of status epilepticus in the rat. Epilepsy Res 1992 Suppl. 8:283-94

Turski L., Ikonomidou C., Turski W.A., Bortolotto Z.A., Cavalheiro E.A. Cholinergic mechanisms and epileptogenesis. The seizures induced by pilocarpine: a novel model of intractable epilepsy. Synapse 1989; 3:154-171

Tuunanen J., Pitkanen A. Do seizures cause neuronal damage in rat amygdala kindling? Epilepsy Res 2000; Apr;39(2):171-6

Ulas J., Satou T., Ivins K.J., Kesslak J.P., Cotman C.W., Balazs R. Expression of metabotropic glutamate receptor 5 is increased in astrocytes after kainate-induced epileptic seizures. Glia 2000; Jun;30(4):352-61

Vezzani A., Moneta D., Richichi C., Aliprandi M., Burrows S.J., Ravizza T., Perego C., De Simoni M.G. Functional role of inflammatory cytokines and antiinflammatory molecules in seizures and epileptogenesis. Epilepsia 2002; 43 Suppl 5:30-5

Wasterlain C.G., Shirasaka Y., Mazarati A.M., Spigelman I. Chronic epilepsy with damage restricted to the hippocampus: possible mechanisms. Epilepsy Res 1996; Dec;26(1):255-65

Yager J.Y., Armstrong E.A., Miyashita H., Wirrell E.C. Prolonged neonatal seizures exacerbate hypoxic-ischemic brain damage: correlation with cerebral energy metabolism and excitatory amino acid release. Dev Neurosci 2002; 24(5):367-81

Yang Y., Tandon P., Liu Z., Sarkisian M.R., Stafstrom C.E., Holmes G.L. Synaptic reorganization following kainic acid-induced seizures during development. Brain Res Dev Brain Res 1998; 107:169-177

Yang D.D., Kuan C.Y., Whitmarsh A.J., Rincon M., Zheng T.S., Davis R.J., Rakic P., Flavell R.A. Absence of excitotoxicity-induced apoptosis in the hippocampus of mice lacking the Jnk3 gene. Nature 1997; 389:865-870

Zhang J., Lee H., Agarwala A., Wen Lou D., Xu M. DNA fragmentation factor 45 mutant mice exhibit resistance to kainic acid-induced neuronal cell death. Biochem Biophys Res Commun 2001; Aug 3;285(5):1143-9

Zhang L.X., Smith M.A., Kim S.Y., Rosen J.B., Weiss S.R., Post R.M. Changes in cholecystokinin mRNA expression after amygdala kindled seizures: an in situ hybridization study. Brain Res Mol Brain Res 1996; Jan;35(1-2):278-84

Zhang X., Cui S.S., Wallace A.E., Hannesson D.K., Schmued L.C., Saucier D.M., Honer W.G., Corcoran M.E. Relations between brain pathology and temporal lobe epilepsy. J Neurosci 2002; Jul 15;22(14):6052-61

Zhang X., Gelowitz D.L., Lai C.T., Boulton A.A., Yu P.H. Gradation of kainic acid-induced rat limbic seizures and expression of hippocampal heat shock protein-70. Eur J Neurosci 1997; 9:760-769

Chapter 8
EXCITOTOXICITY IN CEREBRAL ISCHEMIA

Domenico E. Pellegrini-Giampietro, Elena Meli & Flavio Moroni
Dipartimento di Farmacologia Preclinica e Clinica, Viale Pieraccini 6, 50139 Firenze, Italy

Abstract: The hypothesis that cerebral ischemia may recognize an excitoxic pathophysiological component is supported by a number of consolidated arguments, including the neuroprotective activity of glutamate receptor antagonists in experimental animal models and the use of anti-excitotoxic agents in clinical trials for stroke. Despite the dramatic results in the preclinical setting, phase III clinical trials with neuroprotective drugs have been generally unsuccessful so far. Several complicating variables have been put forward to explain this discrepancy, including population heterogeneity, morphological and functional differences between human and animal brain, and side-effects of the tested compounds that prevent reaching effective plasma concentrations. Fine-tuning in the design of clinical trials, the use of imaging techniques for the evaluation of human brain injury, and the development of more appropriate experimental animal models are among the strategies that need to be utilized in future clinical studies. Also, drugs with a better therapeutic index and aimed at alternative targets in the excitotoxic cascade appear to be required. In our laboratory, we have recently investigated two alternative mechanisms promoted by extracellular glutamate that lead to post-ischemic neuronal damage: (1) the stimulation of mGlu1 receptors and (2) the overactivation of poly(ADP-ribose) polymerase.

Keywords: cerebral ischemia, stroke, glutamate receptors, free radicals, poly(ADP-ribose) polymerase

INDEX

1. Introduction
2. Cerebral ischemia
   2.1 Neuropathology
3. Excitotoxicity mechanisms in post-ischemic neurodegeneration: pre-clinical evidence
   3.1 Glutamate release
   3.2 Glutamate receptors
   3.3 Intracellular events
4. Clinical trials with anti-excitotoxic drugs

4.1 Reasons for negative results
4.2 The future
5. Alternative neuroprotective strategies
5.1 The role of mGlu1 receptors in cerebral ischemia
5.2 Poly(adp-ribose) polymerase inhibitors as neuroprotective agents in cerebral ischemia

1. Introduction

Neuronal injury resulting from glutamate receptor-mediated excitotoxicity has been implicated in a wide spectrum of neurological disease states, including cerebral ischemia, hypoglycemia, trauma, epilepsy and some types of neurodegenerative or psychiatric diseases. In particular, excitotoxicity appears to be involved in the pathogenesis of at least four types of *acute* injury to the central nervous system (CNS) (Choi, 1988): injury due to *i)* hypoxia, which can occur either alone (as in CO poisoning) or combined with reduced blood flow or ischemia (as in stroke or cardiac arrest), *ii)* hypoglycemia, as a component of ischemia or alone in insulin-induced diabetic coma, *iii)* mechanical trauma, or *iv)* prolonged seizures. According to the World Health Organization, stroke is the second leading cause of death and the most common cause of disability in the world. To alleviate the heavy burden of stroke, cardiac arrest and related cerebrovascular diseases, we need to understand the mechanisms that might form the basis for improved prevention and treatment.

The hypothesis that ischemia, hypoglycemia and trauma might share an excitotoxic pathophysiological component is supported by a number of arguments that have been accumulating over the last twenty years (see for reviews: Olney, 1990; Lipton and Rosenberg, 1994; Doble, 1999; Lipton, 1999; Meldrum, 2000). These include: neuropathology resembling the pattern of neurodegeneration induced by excitotoxic agents, accumulation of extracellular glutamate due to both increased release and/or decreased uptake, alteration in the number and/or subunit composition of glutamate receptors, abnormal and sustained increase in cytosolic free Ca^{2+}, dependence on excitatory afferents and delayed onset of neuronal damage. Probably, however, the most compelling pieces of evidence implicating excitotoxicity in cerebral ischemia, hypoglycemia and trauma are (1) the finding that glutamate receptor antagonists are neuroprotective in a variety of experimental models and (2) the fact that anti-excitotoxic neuroprotective agents have been and are currently used in clinical trials for stroke and head injury.

2. Cerebral ischemia

2.1 Neuropathology

Focal ischemia encompasses cerebrovascular disease (stroke), subarachnoid hemorrhage and traumatic brain injury (Choi, 1990). Stroke, being the most common and the prototype of this kind of disease, occurs when a cerebral artery is occluded and blood flow gradually but rapidly declines from the periphery to the core region of the affected vascular territory. This condition can be mimicked by surgical occlusion (transient or permanent) of the middle cerebral artery (MCA) in rats or mice. About 1 hr of focal ischemia causes cerebral infarction, characterized by pannecrosis involving neurons, glia, and endothelial cells; in this *core* region neuroprotection is impossible. However, there is a rim of moderately ischemic tissue between normally perfused brain and the evolving infarct, proximal to collateral vessels and known as the ischemic *penumbra*, in which metabolic changes are most dynamic, neuronal injury develops more slowly, and pharmacological intervention has been most successful (Ginsberg and Pulsinelli, 1994). In rat MCA occlusion models, the core of the infarct is located in the basal ganglia, and the penumbra extends to most of the parietal cortex.

Global ischemia is characterized by the transient reduction or elimination of blood flow to both cerebral hemispheres and in humans is typically a consequence of cardiac arrest following myocardial infarction or ventricular fibrillation. In animal models of severe global ischemia, as well as in patients successfully resuscitated from cardiorespiratory arrest, all forebrain areas are equally affected by oxygen and glucose deprivation but only selected neuronal populations degenerate and die. For example, the pattern of relative vulnerability in the rat four-vessel occlusion model (Pulsinelli et al., 1982) is, in decreasing order, pyramidal cells in the CA1 and CA4 subfields of the hippocampus and in the subiculum; neocortical layers 3, 5 and 6; and small to medium-sized striatal neurons. It is important to note that selective damage following global ischemia is also delayed: hippocampal and neocortical vulnerable neurons become irreversibly damaged after 10 min of ischemia, but histological signs of neurodegeneration are not apparent until 48-72 hrs after recirculation has been restored.

Until recently, the only possible therapeutic approach following a thrombotic stroke was to restore cerebral blood flow as soon as possible. In the last few years, experimental evidence with neuronal cultures and animal models of cerebral ischemia has modified our knowledge on the pathophysiology of post-ischemic neuronal damage and has suggested the possible utilization of more selective and efficient pharmacological interventions. It is now possible to discriminate two different pathogenetic processes that are thought to act in a sequential manner (Lipton, 1999; Dirnagl et al., 1999; Lee et al., 1999): (1) a *vascular* event, such as a thrombus or hemorrhage, which leads to a reduction in blood flow and in the delivery of oxygen and glucose to the brain; and (2) *parenchymal* events, such as overactivation of glutamate receptors and intracellular free Ca^{2+} overload, that are triggered by ischemia and are the direct cause of neuronal death. Accordingly, two general therapeutic approaches have emerged: (1) a reperfusion or thrombolytic approach, which has become established as an acute treatment for human stroke under selected conditions; and (2) a parenchymal or neuroprotective approach, focused mainly on directly or indirectly attenuating neuronal death triggered by glutamate and mediated by cytotoxic levels of intracellular Ca^{2+}, that has generated extremely encouraging results in preclinical experimental models. Because post-ischemic neuronal death begins at least a few hours, if not several days, following an ischemic insult (as is the case of neurons in the penumbra), it is not impossible nowadays to devise a rational strategy for neuroprotection at times that follow the occurrence of a stroke episode.

Although excitotoxic neuronal death has long been considered to represent necrosis, it now appears that many brain neurons undergo apoptosis after either global or focal ischemic insults (Choi, 1996; Leker and Shohami, 2002). Most investigators presently favor the view that programmed cell death and excitotoxicity are triggered in parallel in ischemic tissue, leading to a mosaic of morphological, and even biochemical features (Charriaut-Marlangue et al., 1996; Nicotera et al., 1999; Dirnagl et al., 1999; Lossi and Merighi, 2003). The concept of ischemic apoptosis has raised exciting prospects of combining anti-apoptotic with anti-excitotoxic treatments to achieve heightened therapeutic benefits in the brains of patiens traumatized by cardiac arrest or stroke. Interventions directed at blocking ischemic apoptosis seem to have two especially attractive features, the additivity with anti-excitotoxic therapy and a prolonged "window of opportunity".

3. Excitotoxicity mechanisms in post-ischemic neurodegeneration: pre-clinical evidence

A series of events, studied in neuronal primary cultures as well as in experimental models *in vivo*, are thought to be involved in the process linking brain ischemia to neuronal degeneration; some of them have been shown to be susceptible to modulation by pharmacological agents. These pathogenic mechanisms include glutamate release, glutamate receptor activation, and intracellular post-receptor events.

3.1 Glutamate release

An increased release of glutamate has been observed when brain tissue is exposed to hypoxic/hypoglycemic conditions both *in vitro* (Pellegrini-Giampietro et al., 1990) and *in vivo* (Benveniste et al., 1984). In stroke models, the magnitude of glutamate release varies in the different regions of the affected area, being positively correlated with infarct volume and the extent of the mean cerebral blood flow decrement. Hence, release is maximal in the core of the infarct and decreases towards the periphery of the penumbral area.

Ischemic release of excitatory amino acids (EAAs) may occur via both Ca^{2+}-dependent and Ca^{2+}-independent mechanisms (Szatkowski and Attwell, 1994). Increased glutamate release during and immediately after an ischemic episode is believed to be Ca^{2+}-independent and non-exocytotic, due to reversed operation of the cytoplasmic glutamate uptake carrier. The increase in $[Na^+]_i$ and $[K^+]_o$ that occurs during ischemia, together with anoxic membrane depolarization, tend to make the carrier run in reverse, in such a way that glutamate and Na^+ are pumped out of the cells into the extracellular space until a new equilibrium is reached at a neurotoxic level ($[\text{glutamate}]_o > 100\mu M$). Other mechanisms, such as a reduced uptake of glutamate into presynaptic terminals or glial cells, may also contribute to the increase in extracellular EAAs during ischemia. Within 15 min following transient global ischemia, ionic gradients and concentrations of ATP are restored, the carrier returns to operate in the forward direction, and glutamate release is again Ca^{2+}-dependent and at normal levels (Silver and Erecinska, 1992). Since the nature of neuronal death is delayed and neuroprotection can be obtained by administering drugs after the ischemic insult, it appears that although the ischemic increase in Ca^{2+}-independent glutamate release might be needed to trigger the neurotoxic chain of events, post-ischemic cell death requires the subsequent activation of perhaps supersensitive glutamate receptors by EAAs released during normal synaptic activity.

3.2 Glutamate receptors

Glutamate receptor antagonists have proven to be neuroprotective in models of cerebral ischemia *in vitro* and *in vivo*. In models of focal ischemia, both NMDA receptor AMPA/kainate receptor antagonists are protective against neocortical damage. Whereas NBQX is also effective in preventing delayed CA1 cell death induced by transient global ischemia, the issue of protection afforded by NMDA receptor antagonists in global ischemia has been quite more controversial in the past few years. It is now quite clear that drugs like MK-801 may still be protective if the ischemic insult is incomplete or moderate, but they are not effective, or their effect is mediated by hypothermia, in models of severe forebrain ischemia (Buchan and Pulsinelli, 1990; Nellgard and Wieloch, 1992).

Neuronal damage mediated by NMDA receptor is thought to be greater in the penumbra than in the core following focal ischemia because of the metabolic conditions of these regions. It has been shown that NMDA receptors need to be phosphorylated to be active and also that low pH inhibits the currents gated by these channels (Traynelis and Cull-Candy, 1990). Hence, NMDA receptors are thought to be inactive in the ischemic core or in regions exposed to severe global ischemia, in which the high energy phosphate and pH levels are extremely low. They could, however, be responsible for glutamate-mediated neurotoxicity in the penumbra, where these levels are reduced but still sufficient for the activity of NMDA receptors. As for AMPA receptors, our results suggest that following severe global ischemia AMPA receptors become permeable to Ca^{2+} in pyramidal cells of the CA1 hippocampal region, and that this modification may lead to delayed death of these neurons (Pellegrini-Giampietro et al., 1997).

3.3 Intracellular events

A glutamate-induced rise in intracellular Ca^{2+} is thought to play a critical role in neurodegeneration. Possible causes of a rise in intracellular Ca^{2+} include activation of Ca^{2+}-permeable ionotropic glutamate receptors, activation of metabotropic glutamate (mGlu) receptors of the I group positively linked to phosphoinositide breakdown, activation of voltage-sensitive Ca^{2+} channels, and/or de-activation of extrusion and/or sequestration systems. This rise in intracellular Ca^{2+} may lead to the onset of cell death by a number of mechanisms including activation of proteases, phospholipases and endonucleases, generation of free radicals that destroy cellular membranes by lipid peroxidation (Choi, 1995) and activation of pathways leading to apoptotic cell death (Takei and Endo, 1994).

A number of Ca^{2+}-activated enzymes exert their toxic effects through the formation of reactive oxygen radical species, such as the superoxide ion or the hydroxyl radical, or nitrogen-centered free radicals like NO and its derivative peroxynitrite. The direct measurement of free radicals in brain tissue following ischemia-reperfusion has been achieved by using magnetic resonance techniques and spin trap agents. Moreover, a number of studies have shown that free radical scavengers and/or lipid peroxidation inhibitors reduce brain damage following focal and global ischemia. Interestingly, one of these agents (the spin trap compound α-phenyl-tert-butyl-nitrone, α-PBN) appears to be neuroprotective in focal ischemia even when administered several hours (up to 12) after the occlusion (Cao and Phillis, 1994). In addition, the degree of cortical infarction induced by cerebral focal ischemia is reduced in transgenic mice overexpressing the CuZn-superoxide-dismutase-1 (CuZn-SOD-1) gene (Kinouchi et al., 1991) or deficient in neuronal NO synthase (Huang et al., 1994).

As discussed above, the occurrence or the prevalence of necrotic or apoptotic cell death in ischemic models depends on a number of factors, including the intensity and the duration of the stimulus, the brain region and the type of cell involved, and the developmental stage and functional status of the neurons. The mechanisms leading to apoptotic death in cerebral ischemia are intricate and involve several possible pathways including a NFκB-dependent pathway (Clemens et al., 1997), a p53-dependent pathway (Crumrine et al., 1994), activation of inducible proapoptotic members of the bcl family (Martinou et al., 1994), caspase activation (Namura et al., 1998), as well as a caspase-independent pathway (Cregan et al., 2002). Accordingly, pharmacological interventions that selectively block the apoptotic cascade have shown to afford neuroprotection *in vivo* (Martinou et al., 1994; Namura et al., 1998) and *in vitro* (Ray et al., 2000; Moroni et al., 2001).

4. Clinical trials with anti-excitotoxic drugs

In the past few years, clinical trials for stroke have been designed with drugs that have shown preclinical evidence for neuroprotection in experimental models of cerebral ischemia. Some of these drugs, that are able to block various glutamate receptor subtypes, reduce the concentrations of intracellular free Ca^{2+}, or attenuate the toxicity of free radicals have passed the preliminary phases and have been used in phase III clinical trials. However, despite the huge economical efforts and the active participation of many patients and medical and paramedical staff, phase III clinical trials for stroke with neuroprotective drugs have been generally unsuccessful so far (Lipton, 1999; Dirnagl et al., 1999; Lee et al., 1999). What are the reasons for this failure and what can be done in the future?

4.1 Reasons for negative results

There are numerous reasons that can explain the discrepancies between preclinical and clinical results with neuroprotective agents (Gladstone et al., 2002; Grotta, 2002). In addition to population heterogeneity, which alone may be sufficient to explain negative neuroprotective trials (Muir, 2002), it is becoming clear that existing animal models of cerebral ischemia are an imperfect representation of human stroke and may be relevant only to a minority of human stroke types: (1) Whereas in animal models it is possible to identify the exact time of onset of both ischemia and reperfusion, this is not always possible in humans. In some clinical trials, drugs were given outside of their therapeutic window of efficacy because it was difficult to determine this time frame in a reliable manner. This is complicated by the fact that stroke may have a nocturnal and asymptomatic onset and thus patients are transported to stroke units when it is too late. (2) Preclinical studies are performed in young and healthy animals, whereas human patients are often elderly and suffer from associated pathologies that may affect the efficacy and pharmacokinetics of the drug under study. In addition, the cerebral blood flow and metabolism, the vascular anatomy, and the volume of the penumbra can all be quite different in humans and animals. (3) In laboratory animals, the volume of the ischemic infarct is measured quantitatively and almost always within 7 days after the occlusion. Functional scores in humans are usually evaluated 3 to 5 months after the onset of the stroke and may be less sensitive to the action of drugs, as well as more difficult to be explored by statistical analysis. (4) A number of anti-excitotoxic compounds have psychotomimetic or cardiovascular side-effects. Therefore, some of them have been used in clinical trials at a lower dosage than that demonstrated to be effective in experimental models and may not have reached effective plasma concentrations.

4.2 The future

Before definitely putting aside the strategy of neuroprotection in stroke, that has generated remarkable results in preclinical studies, these problems must be re-evaluated and some steps must be taken in order to proceed forward (Koroshetz and Moskowitz, 1996; De Keyser et al., 1999). For example, imaging techniques based on magnetic resonance should be used to determine the evolving stage of the infarct, the extension of the penumbra and to evaluate the efficacy of the drug at early time points. Moreover, it is necessary to develop experimental animal models that reflect the human condition more accurately as well as drugs with a better therapeutic index. So far, clinical trials have only examined neuroprotective drugs in monotherapies, and not in combination with one or more drugs with a different mechanism of action. It could be useful to combine a neuroprotective drug with a thrombolytic agent such as tissue plasminogen activator (TPA). This strategy may help in removing the blood flow obstruction and thus could increase the amount of drug that reaches the damaged tissue. However, it appears to be of particular importance to properly design the clinical trials, selecting appropriate outcome measures and accurately selecting the population sample. Indeed, positive results with neuroprotective strategies are possible if these criteria are met (Ehrenreich et al., 2002; Chauhan et al., 2003).

As the understanding increases on the pathophysiology of ischemic brain injury and on the mechanisms underlying functional recovery, newer strategies are emerging and new targets are currently under investigation. In particular, drugs aimed at alternative targets in the excitotoxic cascade appear to be required for the development of new and effective neuroprotective therapies for cerebral ischemia.

5. Alternative neuroprotective strategies

In our laboratory, we have recently investigated two alternative mechanisms promoted by extracellular glutamate that lead to post-ischemic neuronal damage and could offer novel targets for a pharmacological approach in cerebral ischemia: (1) the stimulation of mGlu1 receptors and (2) the overactivation of poly(ADP-ribose) polymerase (PARP).

5.1 The role of mGlu1 receptors in cerebral ischemia

Recent evidence suggests that mGlu receptors contribute to the neurotoxic effects of glutamate. In particular, group I mGlu receptors coupled to phosphoinositide hydrolysis (i.e. mGlu1 and mGlu5) have been extensively investigated for their possible role in neurodegeneration (Nicoletti et al., 1999). Receptors of the mGlu1 and mGlu5 subtypes are known to be associated with a number of potentially detrimental events such as the increase in cytosolic free Ca^{2+} and activation of protein kinase C, the potentiation of NMDA and AMPA receptor responses, and the enhancement of glutamate release. The results of neurotoxicity studies have shown that group I mGlu receptor antagonists are consistently neuroprotective, whereas the use of agonists has generated conflicting results. Discrepancies may arise from the poor selectivity of available mGlu receptor ligands. It has also been proposed that a number of alternative factors, including the subunit composition of NMDA receptors, the exposure time to drugs or toxic insults, and the function of astrocytes may affect the responses to mGlu receptor agonists.

We have shown that 1-aminoindan-1,5-dicarboxylic acid (AIDA) and (*S*)-(+)-2-(3'-carboxybicyclo[1.1.1]pentyl)-glycine (CBPG), two group I mGlu receptor antagonists displaying preferential activity on mGlu1 over mGlu5 receptors (Moroni et al., 1997; Mannaioni et al., 1999), reduce neuronal damage in *in vitro* and *in vivo* models of cerebral ischemia (Pellegrini-Giampietro et al., 1999a; Pellegrini-Giampietro et al., 1999b). A number of similar studies in other experimental models also point to mGlu1 receptors as a potential target for neuroprotection (Mukhin et al., 1996; Strasser et al., 1998; Bruno et al., 1999). Although mGlu5 receptors have been implicated in the potentiation of ionotropic glutamate receptor responses (Doherty et al., 1997; Alagarsamy et al., 1999; Ugolini et al., 1999; Attucci et al., 2001) and antagonism of mGlu5 receptors has been demonstrated to be neuroprotective against NMDA-induced excitotoxicity (Bruno et al., 2000; Movsesyan et al., 2001), the contribution of mGlu5 receptor blockade to neuroprotection in post-ischemic neuronal injury is still unclear. By using the potent and selective mGlu5 noncompetitive antagonist 2-methyl-6-(phenylethynyl)-pyridine (MPEP) (Gasparini et al., 1999) in comparison with available selective mGlu1 antagonists we have demonstrated that inhibition of mGlu1 receptors is neuroprotective in different models of cerebral ischemia *in vitro* and *in vivo*, whereas the selective blockade of mGlu5 receptor is ineffective (Figure 1) (Meli et al., 2002). These results point out the differential role of the two group I mGlu receptors subtypes, suggesting that mGlu1 rather than mGlu5 receptors are implicated in the pathways leading to post-ischemic neuronal injury. Therefore, mGlu1 receptors antagonists should be regarded as promising targets for the development of innovative drugs against cerebral ischemia.

5.2 Poly(ADP-ribose) polymerase inhibitors as neuroprotective agents in cerebral ischemia

Poly(ADP-ribosyl)ation is a covalent post-translational modification of proteins catalyzed by a family of enzymes involved in numerous cellular processes, including DNA repair and telomere replication (Smith, 2001). Poly(ADP-ribose) polymerase-1 (PARP-1, E.C. 2.4.2.30), a nuclear protein that is strongly activated by exposure to agents that produce DNA strand breaks, is probably the most studied member of this family. Its activation represents an immediate cellular response to DNA lesions and facilitates maintenance of genome integrity (Shall and de Murcia, 2000; Herceg and Wang, 2001). PARP-1 catalyzes the transfer of multiple ADP-ribose units from NAD^+ to a variety of substrates including PARP-1 itself, histones and other target proteins involved in DNA repair and in gene expression (de Murcia et al., 1994; Chiarugi, 2002).

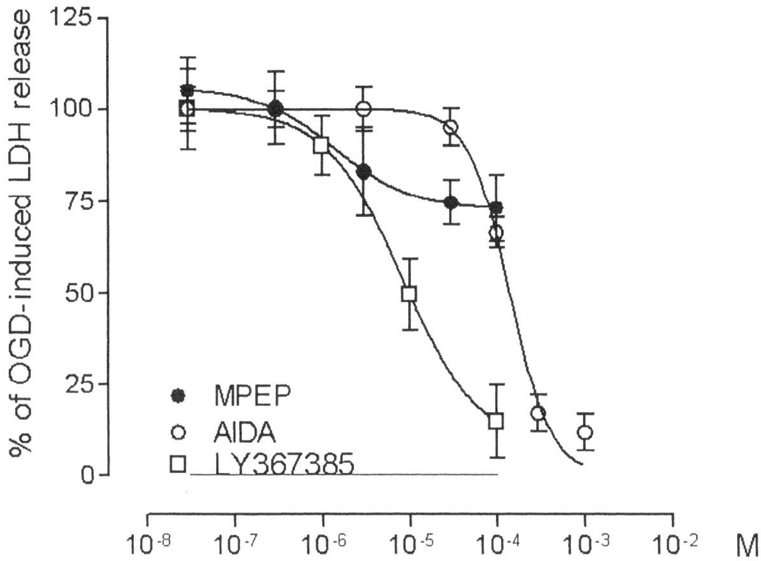

Figure 1. The mGlu1 receptor antagonists AIDA and LY367385 but not the mGlu5 receptor antagonist MPEP protect against glucose-oxygen deprivation (OGD) neurotoxicity in murine cortical cell cultures. Reprinted with permission from (Meli et al., 2002).

It has been suggested that an excessive activation of this enzyme, as occurs following glutamate receptor activation and the subsequent formation of free radicals like NO (Figure 2), may lead to intracellular NAD^+ and ATP deficiency, energy failure and cell death (Cosi et al., 1994; Zhang et al., 1994; Ha and Snyder, 1999). Accordingly, PARP-1 inhibitors have been shown to be of therapeutic value in a number of pathological conditions such as diabetes (Pieper et al., 1999; Virag and Szabo, 2002), inflammation (Szabó and Dawson, 1998) or brain injury (Endres et al., 1997; Ha and Snyder, 2000).The involvement of PARP activation in neuronal death following excitotoxicity and focal ischemia is further substantiated by the resistance to these types of insults observed in PARP-deleted gene mice (Eliasson et al., 1997; Endres et al., 1997) and by the finding that increased PAR immunoreactivity observed in focal ischemic animals is reduced by treatment with PARP inhibitors (Endres et al., 1997; Takahashi et al., 1997).

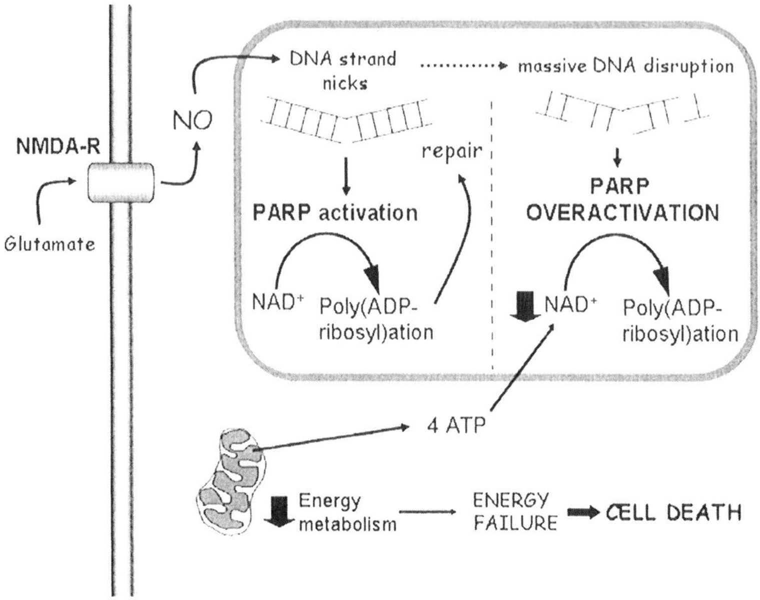

Figure 2. Role of PARP in glutamate-mediated necrosis: a schematic illustration of the "suicide hypothesis". When glutamate stimulates NMDA receptors there is an increase of NO and other free radicals, which are known to produce DNA strand breaks. When DNA damage is mild, poly(ADP-ribosyl)ation contributes to the complex process of DNA repair and manteinance of chromatin integrity *(physiological role).* Conversely, when DNA damage is massive, PARP overactivation can lead to cell death of the necrotic type, due to the marked depletion of NAD and ATP tissue stores and the resultant decrease of energy metabolism *(suicidal role)*. Reprinted with permission from (Meli et al., 2003).

We have shown that PARP inhibition attenuates neuronal injury in cortical cell cultures (Moroni et al., 2001; Chiarugi et al., 2003) but not in organotypic hippocampal slices exposed to oxygen-glucose deprivation (Moroni et al., 2001). By examining the necrotic and apoptotic features of neurodegeneration in both models, we have demonstrated that PARP overactivation appears to be involved in post-ischemic injury only when necrosis accounts for neuronal death (Moroni et al., 2001). These results confirm and extend previous observations showing that PARP overactivation leads to ATP intracellular depletion and energy deficit in necrotic cell death (Ha and Snyder, 1999). PARP may be transiently activated during the early phases of apoptosis (Simbulan-Rosenthal et al., 1998), but soon after the protein is cleaved and inactivated by caspase-3 (Wang, 2000). PARP cleavage in apoptosis is thought to prevent ATP depletion by PARP overactivation, preserving the energy required for the apoptotic active process. Along this line, PARP activation appears to be required for necrosis, suggesting that poly(ADP-rybosyl)ation may be regarded as an active step in this cell death pathway (Ha and Snyder, 1999). PARP inhibitors may therefore be of therapeutic importance in ischemic brain scenarios in which necrosis predominates (e.g. focal rather than global ischemia *in vivo*).

REFERENCES

Alagarsamy S, Marino MJ, Rouse ST, Gereau RW, IV, Heinemann SF, Conn PJ (1999) Activation of NMDA receptors reverses desensitization of mGluR5 in native and recombinant systems. Nat Neurosci 2:234-240.

Attucci S, Carlà V, Mannaioni G, Moroni F (2001) Activation of type 5 metabotropic glutamate receptors enhances NMDA responses in mice cortical wedges. Brit J Pharmacol 132:799-806.

Benveniste H, Drejer J, Schousboe A, Diemer N (1984) Elevation of extracellular concentrations of glutamate and aspartate in rat hippocampus during transient cerebral ischemia monitored by intracerebral microdialysis. J Neurochem 43:1369-1374.

Bruno V, Battaglia G, Kingston AE, O'Neill MJ, Catania MV, Di Grezla R, Nicoletti F (1999) Neuroprotective activity of the potent and selective mGlu1a metabotropic glutamate receptor antagonist, (+)-2-methyl-4-carboxyphenylglycine (LY367385): comparison with LY357366, a broader spectrum antagonist with equal affinity for mGlu1a and mGlu5 receptors. Neuropharmacology 38:199-207.

Bruno V, Ksiazek I, Battaglia G, Lukic S, Leonhardt T, Sauer D, Gasparini F, Kuhn R, Nicoletti F, Flor PJ (2000) Selective blockade of metabotropic glutamate receptor subtype 5 is neuroprotective. Neuropharmacology 39:2223-2230.

Buchan A, Pulsinelli WA (1990) Hypothermia but not the N-methyl-D-aspartate antagonist, MK-801, attenuates neuronal damage in gerbils subjected to transient global ischemia. J Neurosci 10:311-316.

Cao X, Phillis JW (1994) α-Phenyl-tert-butyl-nitrone reduces cortical infarct and edema in rats subjected to focal ischemia. Brain Res 644:267-272.

Charriaut-Marlangue C, Aggoun-Zouaoui D, Represa A, Ben-Ari Y (1996) Apoptotic features of selective neuronal death in ischemia, epilepsy and gp120 toxicity. Trends Neurosci 19:109-114.

Chauhan N, Zhao Z, Barber PA, Buchan AM (2003) Lessons in experimental ischemia for clinical stroke medicine. Curr Opin Neurol 16:65-71.

Chiarugi A (2002) Poly(ADP-ribose) polymerase: killer or conspirator? The 'suicide hypothesis' revisited. Trends Pharmacol Sci 23:122-129.

Chiarugi A, Meli E, Calvani M, Picca R, Baronti R, Camaioni E, Costantino G, Marinozzi M, Pellegrini-Giampiero DE, Pellicciari R, Moroni F (2003) Novel isoquinolinone-derived inhibitors of poly(ADP-ribose) polymerase-1: pharmacological characterization and neuroprotective effects in an in vitro model of cerebral ischemia. J Pharmacol Exp Ther

Choi D (1990) Cerebral hypoxia: some new approaches and unanswered questions. J Neurosci 10:2493-2501.

Choi DW (1988) Glutamate neurotoxicity and diseases of the nervous system. Neuron 1:623-634.

Choi DW (1995) Calcium: still center-stage in hypoxic-ischemic neuronal death. Trends Neurosci 18:58-60.

Choi DW (1996) Ischemia-induced neuronal apoptosis. Curr Opin Neurobiol 6:667-672.

Clemens JA, Stephenson DT, Dixon EP, Smalstig EB, Mincy RE, Rash KS, Little SP (1997) Global cerebral ischemia activates nuclear factor-kappa B prior to evidence of DNA fragmentation. Brain Res Mol Brain Res 48:187-196.

Cosi C, Suzuki H, Milani D, Facci L, Menegazzi M, Vantini G, Kanai Y, Skaper SD (1994) Poly(ADP-ribose) polymerase: early involvement in glutamate-induced neurotoxicity in cultured cerebellar granule cells. J Neurosci Res 39:38-46.

Cregan SP, Fortin A, MacLaurin JG, Callaghan SM, Cecconi F, Yu SW, Dawson TM, Dawson VL, Park DS, Kroemer G, Slack RS (2002) Apoptosis-inducing factor is involved in the regulation of caspase-independent neuronal cell death. J Cell Biol 158:507-517.

Crumrine RC, Thomas AL, Morgan PF (1994) Attenuation of p53 expression protects against focal ischemic damage in transgenic mice. J Cereb Blood Flow Metab 14:887-891.

De Keyser J, Sulter G, Luiten PG (1999) Clinical trials with neuroprotective drugs in acute ischemic stroke: are we doing the right thing? Trends Neurosci 22:535-540.

de Murcia G, Schreiber V, Molinete M, Saulier B, Poch O, Masson M, Niedergang C, Ménissier-de Murcia J (1994) Structure and function of poly(ADP-ribose) polymerase. Mol Cell Biochem 138:15-24.

Dirnagl U, Iadecola C, Moskowitz MA (1999) Pathobiology of ischaemic stroke: an integrated view. Trends Neurosci 22:391-397.

Doble A (1999) The role of excitotoxicity in neurodegenerative disease: implications for therapy. Pharmacol Ther 81:163-221.

Doherty AJ, Palmer MJ, Henley JM, Collingridge GL, Jane DE (1997) (RS)-2-Chloro-5-hydroxyphenylglycine (CHPG) activates $mGlu_5$, but not $mGlu_1$, receptors expressed in CHO cells and potentiates NMDA responses in the hippocampus. Neuropharmacology 36:265-267.

Ehrenreich H, Hasselblatt M, Dembowski C, Cepek L, Lewczuk P, Stiefel M, Rustenbeck HH, Breiter N, Jacob S, Knerlich F, Bohn M, Poser W, Ruther E, Kochen M, Gefeller O, Gleiter C, Wessel TC, De Ryck M, Itri L, Prange H, Cerami A, Brines M, Siren AL (2002) Erythropoietin therapy for acute stroke is both safe and beneficial. Mol Med 8:495-505.

Eliasson MJL, Sampei K, Mandir AS, Hurn PD, Traystman RJ, Bao J, Pieper A, Wang Z-Q, Dawson TM, Snyder SH, Dawson VL (1997) Poly(ADP-ribose) polymerase gene disruption renders mice resistant to cerebral ischemia. Nat Med 3:1089-1095.

Endres M, Wang Z-Q, Namura S, Waeber C, Moskowitz MA (1997) Ischemic brain injury is mediated by the activation of poly(ADP-ribose) polymerase. J Cereb Blood Flow Metab 17:1143-1151.

Gasparini F, Lingenhöhl K, Stoehr N, Flor PJ, Heinrich M, Vranesic I, Biollaz M, Allgeier H, Heckendorn R, Urwyler S, Varney MA, Johnson EC, Hess SD, Rao SP, Sacaan AI, Santori EM, Veliçelebi G, Kuhn R (1999) 2-Methyl-6-(phenylethynyl)-pyridine (MPEP), a potent, selective and systemically active mGlu5 receptor antagonist. Neuropharmacology 38:1493-1503.

Ginsberg MD, Pulsinelli WA (1994) The ischemic penumbra, injury thresholds, and the therapeutic window for acute stroke. Ann Neurol 36:553-554.

Gladstone DJ, Black SE, Hakim AM (2002) Toward wisdom from failure: lessons from neuroprotective stroke trials and new therapeutic directions. Stroke 33:2123-2136.

Grotta J (2002) Neuroprotection is unlikely to be effective in humans using current trial designs. Stroke 33:306-307.

Ha HC, Snyder SH (1999) Poly(ADP-ribose) polymerase is a mediator of necrotic cell death by ATP depletion. Proc Natl Acad Sci USA 96:13978-13982.

Ha HC, Snyder SH (2000) Poly(ADP-ribose) polymerase-1 in the nervous system. Neurobiol Dis 7:225-239.

Herceg Z, Wang ZQ (2001) Functions of poly(ADP-ribose) polymerase (PARP) in DNA repair, genomic integrity and cell death. Mutat Res 477:97-110.

Huang Z, Huang PL, Panahian N, Dalkara T, Fishman MC, Moskowitz MA (1994) Effects of cerebral ischemia in mice deficient in neuronal nitric oxid synthase. Science 265:1883-1885.

Kinouchi H, Epstein CJ, Mizui T, Carlson E, Chen SF, Chan PH (1991) Attenuation of focal cerebral ischemic injury in transgenic mice overexpressing CuZn superoxide dismutase. Proc Natl Acad Sci USA 88:11158-11162.

Koroshetz WJ, Moskowitz MA (1996) Emerging treatments for stroke in humans. Trends Pharmacol Sci 17:227-233.

Lee J-M, Zipfel GJ, Choi DW (1999) The changing landscape of ischaemic brain injury mechanisms. Nature 399 (supp.):A7-A14

Leker RR, Shohami E (2002) Cerebral ischemia and trauma-different etiologies yet similar mechanisms: neuroprotective opportunities. Brain Res Brain Res Rev 39:55-73.

Lipton P (1999) Ischemic cell death in brain neurons. Physiol Rev 79:1431-1568.

Lipton SA, Rosenberg PA (1994) Excitatory amino acids as a final common pathway for neurological disorders. N Engl J Med 330:613-622.

Lossi L, Merighi A (2003) In vivo cellular and molecular mechanisms of neuronal apoptosis in the mammalian CNS. Prog Neurobiol 69:287-312.

Mannaioni G, Attucci S, Missanelli A, Pellicciari R, Corradetti R, Moroni F (1999) Biochemical and electrophysiological studies on (*S*)-(+)-2-(3'-carboxybicyclo[1.1.1]pentyl)-glycine (CBPG), a novel $mGlu_5$ receptor agonist endowed with $mGlu_1$ receptor antagonist activity. Neuropharmacology 38:917-926.

Martinou J-C, Dubois-Dauphin M, Staple JK, Rodriguez I, Frankowski H, Missotten M, Albertini P, Talabot D, Catsicas S, Pietra C, Huarte J (1994) Overexpression of BCL-2 in transgenic mice protects neurons from naturally occurring cell death and experimental ischemia. Neuron 13:1017-1030.

Meldrum BS (2000) Glutamate as a neurotransmitter in the brain: review of physiology and pathology. J Nutr 130:1007S-1015S.

Meli E, Pangallo M, Baronti R, Chiarugi A, Cozzi A, Pellegrini-Giampietro DE, Moroni F (2003) Poly(ADP-ribose) polymerase as a key player in excitotoxicity and post-ischemic brain damage. Toxicol Lett 193:153-162.

Meli E, Picca R, Attucci S, Cozzi A, Peruginelli F, Moroni F, Pellegrini-Giampietro DE (2002) Activation of mGlu1 but not mGlu5 metabotropic glutamate receptors contributes to post-ischemic neuronal injury *in vitro* and *in vivo*. Pharmacol Biochem Behav 73:439-446.

Moroni F, Lombardi G, Thomsen C, Leonardi P, Attucci S, Peruginelli F, Albani Torregrossa S, Pellegrini-Giampietro DE, Luneia R, Pellicciari R (1997) Pharmacological characterization of 1-aminoindan-1,5-dicarboxylic acid, a potent mGluR1 antagonist. J Pharmacol Exp Ther 281:721-729.

Moroni F, Meli E, Peruginelli F, Chiarugi A, Cozzi A, Picca R, Romagnoli P, Pellicciari R, Pellegrini-Giampietro DE (2001) Poly(ADP-ribose) polymerase inhibitors attenuate necrotic but not apoptotic neuronal death in experimental models of cerebral ischemia. Cell Death Differ 8:921-932.

Movsesyan VA, O'Leary DM, Fan L, Bao W, Mullins PGM, Knoblach SM, Faden AI (2001) mGluR5 antagonists 2-methyl-6-(phenylethynyl)-pyridine and (*E*)-2-methyl-6-(phenylethynyl)-pyridine reduce traumatic neuronal injury in vivo and in vitro by antagonizing N-methyl-D-aspartate receptors. J Pharmacol Exp Ther 296:41-47.

Muir KW (2002) Heterogeneity of stroke pathophysiology and neuroprotective clinical trial design. Stroke 33:1545-1550.

Mukhin A, Fan L, Faden AI (1996) Activation of metabotropic glutamate receptor mGluR1 contributes to post-traumatic neuronal injury. J Neurosci 16:6012-6020.

Namura S, Zhu J, Fink K, Endres M, Srinivasan A, Tomaselli KJ, Yuan J, Moskowitz MA (1998) Activation and cleavage of caspase-3 in apoptosis induced by experimental cerebral ischemia. J Neurosci 18:3659-3668.

Nellgard B, Wieloch T (1992) Post ischemic blockade of AMPA but not NMDA receptors mitigates neuronal damage in the rat brain following transient severe forebrain ischemia. J Cereb Blood Flow Metab 12:1-11.

Nicoletti F, Bruno V, Catania MV, Battaglia G, Copani A, Barbagallo G, Ceña V, Sánchez-Prieto J, Spano PF, Pizzi M (1999) Group-I metabotropic glutamate receptors: hypotheses to explain their dual role in neurotoxicity and neuroprotection. Neuropharmacology 38:1477-1484.

Nicotera P, Leist M, Manzo L (1999) Neuronal cell death: a demise with different shapes. Trends Pharmacol Sci 20:46-51.

Olney JW (1990) Excitotoxic amino acids and neuropsychiatric disorders. Annu Rev Pharmacol Toxicol 30:47-71.

Pellegrini-Giampietro DE, Cherici G, Alesiani M, Carlà V, Moroni F (1990) Excitatory amino acid release and free radical formation may cooperate in the genesis of ischemia-induced neuronal damage. J Neurosci 10:1035-1041.

Pellegrini-Giampietro DE, Cozzi A, Peruginelli F, Leonardi P, Meli E, Pellicciari R, Moroni F (1999a) 1-Amino-1,5-dicarboxylic acid and (S)-(+)-2-(3'-carboxybicyclo[1.1.1]pentyl)-glycine, two mGlu1 receptor-preferring antagonists, reduce neuronal death in *in vitro* and *in vivo* models of cerebral ischemia. Eur J Neurosci 11:3637-3647.

Pellegrini-Giampietro DE, Gorter JA, Bennett MVL, Zukin RS (1997) The GluR2 (GluR-B) hypothesis: Ca^{2+}-permeable AMPA receptors in neurological disorders. Trends Neurosci 20:464-470.

Pellegrini-Giampietro DE, Peruginelli F, Meli E, Cozzi A, Albani Torregrossa S, Pellicciari R, Moroni F (1999b) Protection with metabotropic glutamate 1 receptor antagonists in models of ischemic neuronal death: time-course and mechanisms. Neuropharmacology 38:1607-1619.

Pieper AA, Brat DJ, Krug DK, Watkins CC, Gupta A, Blackshaw S, Verma A, Wang ZQ, Snyder SH (1999) Poly(ADP-ribose) polymerase-deficient mice are protected from streptozotocin-induced diabetes. Proc Natl Acad Sci U S A 96:3059-3064.

Pulsinelli WA, Brierley JB, Plum F (1982) Temporal profile of neuronal damage in a model of transient forebrain ischemia. Ann Neurol 11:491-498.

Ray AM, Owen DE, Evans ML, Davis JB, Benham CD (2000) Caspase inhibitors are functionally neuroprotective against oxygen glucose deprivation induced CA1 death in rat organotypic hippocampal slices. Brain Res 867:62-69.

Shall S, de Murcia G (2000) Poly(ADP-ribose) polymerase-1: what have we learned from the deficient mouse model? Mutat Res 460:1-15.

Silver IA, Erecinska M (1992) Ion homeostasis in rat brain in vivo: intra- and extracellular $[Ca^{2+}]$ and $[H^+]$ in the hippocampus during recovery from short- term, transient ischemia. J Cereb Blood Flow Metab 12:759-772.

Simbulan-Rosenthal CM, Rosenthal DS, Iyer S, Boulares AH, Smulson ME (1998) Transient poly(ADP-ribosyl)ation of nuclear proteins and role of poly(ADP-ribose) polymerase in the early stages of apoptosis. J Biol Chem 273:13703-13712.

Smith S (2001) The world according to PARP. Trends Biochem Sci 26:174-179.

Strasser U, Lobner D, Behrens MM, Canzoniero LMT, Choi DW (1998) Antagonists for group I mGluRs attenuate excitotoxic neuronal death in cortical cultures. Eur J Neurosci 10:2848-2855.

Szabó C, Dawson VL (1998) Role of poly(ADP-ribose) synthetase in inflammation and ischaemia-reperfusion. Trends Pharmacol Sci 19:287-298.

Szatkowski M, Attwell D (1994) Triggering and execution of neuronal death in brain ischemia: two phases of glutamate release by different mechanisms. Trends Neurosci 17:359-365.

Takahashi K, Greenberg JH, Jackson P, Maclin K, Zhang J (1997) Neuroprotective effects of inhibiting poly(ADP-ribose) synthetase on focal cerebral ischemia in rats. J Cereb Blood Flow Metab 17:1137-1142.

Takei N, Endo Y (1994) Ca2+ ionophore-induced apoptosis on cultured embryonic rat cortical neurons. Brain Res 652:65-70.

Traynelis S, Cull-Candy S (1990) Proton inhibition of N-methyl-D-aspartate receptors in cerebellar neurons. Nature 345:347-350.

Ugolini A, Corsi M, Bordi F (1999) Potentiation of NMDA and AMPA responses by the specific $mGluR_5$ agonist CHPG in spinal cord motoneurons. Neuropharmacology 38:1569-1576.

Virag L, Szabo C (2002) The therapeutic potential of poly(ADP-Ribose) polymerase inhibitors. Pharmacol Rev 54:375-429.

Wang KKW (2000) Calpain and caspase: can you tell the difference? Trends Neurosci 23:20-26.

Zhang J, Dawson VL, Dawson TM, Snyder SH (1994) Nitric oxide activation of poly(ADP-ribose) synthetase in neurotoxicity. Science 263:687-689.

Chapter 9
EXCITOTOXICITY AND TRAUMATIC BRAIN INJURY.
Pathophysiology, treatment approaches and controversies.

Niklas Marklund, John F. Stover and Tracy K. McIntosh
Department of Neurosurgery, University of Pennsylvania, Philadelphia, Pennsylvania 19104, USA.

Abstract: Traumatic brain injury (TBI) is a major health problem in the industrialized world where survivors suffer major physical, psychological and economical consequences. Despite numerous clinical trials investigating pharmaceutical compounds with promising pre-clinical efficacy, no pharmacological treatment has been developed with proven clinical efficacy. The present chapter reviews the pathophysiology of TBI with focus on the evidence of and mechanisms for glutamate-mediated excitotoxicity, which is considered to be a major contributor to the secondary injury process following both experimental and clinical TBI. Furthermore, we review the numerous pharmacological approaches targeting excitotoxicity employed in experimental TBI, and address the controversies that exist with regard to the importance of glutamate-mediated mechanisms following TBI. The impressive reductions in functional and structural impairment observed following experimental TBI after administration of compounds designed to attenuate glutamate excitotoxicity raised high expectations for improvement of outcome following clinical TBI, and we summarize the phase III clinical trials conducted thus far with glutamate receptor antagonists following clinical TBI. Finally, we address possible future directions in the search of a clinically useful treatment for human head injury.

Keywords: traumatic brain injury (TBI), excitotoxicity, glutamate, clinical trials, controversies, neuroprotection.

INDEX

1. Introduction.
2. Overview of the pathophysiology of traumatic brain injury (TBI).
3. Glutamate-mediated excitotoxicity in TBI.
   3.1 Evidence for increased glutamate release following TBI - experimental studies.

3.2 Evidence for increased glutamate release following TBI - clinical studies.
3.3 Glutamate-mediated damage following TBI.
3.4 Changes in glutamate and glutamate re-uptake receptors in TBI.
3.5 NMDA-receptor associated treatments in TBI.
3.5.1. NMDA receptor activation - an overview.
3.5.2. NMDA receptor channel blockers in TBI.
3.5.3. Polyamine site modulation in TBI.
3.5.4. Glycine site modulation in TBI.
3.5.5. Modulation of the NMDA receptor by zinc.
3.6 AMPA/KA receptor modulation in TBI.
3.7 Metabotropic receptor modulation in TBI.
3.8 Glutamate release inhibitors in TBI.
4. Clinical phase III trials in TBI with glutamate receptor antagonists.
5. Controversies.
6. Summary and future directions.

1. Introduction.

Traumatic brain injury (TBI) is a major health problem in the industrialized world; in the United States of America alone, 2.5 - 6.5 million individuals are living with the physical, psychological and economical consequences of TBI. Additionally, there are approximately 500.000 new cases per year (Consensus Conference, 1999). The outcome for TBI victims strongly depends on the severity of the injury and early prognostic factors include the level of consciousness early after injury (as assessed by early Glascow Coma Scale score), pupillary reactivity and age (Levin, 1995). The overall prognosis for TBI victims has improved during the last years, mainly attributed to improved monitoring and surveillance through the introduction of specialized neurointensive care units and aggressive treatment of potentially avoidable secondary injury factors. However, a large number of surviving TBI victims will suffer from significant long-term disabilities such as impairment of information processing, perceptual function and memory (Levin, 1995) and improved treatments are urgently needed. The numerous clinical trials investigating pharmaceutical compounds with promising pre-clinical efficacy all failed to identify a beneficial treatment reducing the disabling sequelae of head injury, and the search continues. Glutamate-mediated excitotoxicity is considered to be a major contributor to the secondary injury process following TBI, and the aim of this chapter is to highlight the importance of glutamate-mediated excitotoxicity in the pathophysiology of TBI, and summarize the successes, impediments and directions for possible future therapeutic interventions.

2. Overview of the pathophysiology of traumatic brain injury.

Traumatic brain injury is caused by the relative motions and displacements generated within the brain resulting from external forces (either contact or acceleration/deceleration) acting on the brain (Gennarelli, 1994). The majority of TBIs are closed head injuries which typically fall into two categories; a) focal injuries (intracranial hematomas, contusions and epidural- and subdural hematomas) and b) diffuse axonal injuries (McIntosh et al., 1996). Neuronal damage and cell loss have been documented in regions such as the cerebral cortex, hippocampus, and thalamus during the first few hours after TBI, and neurons in the cortex and the hippocampal CA3 and dentate gyrus regions appear to be the most vulnerable (McIntosh et al., 1998b). In particular, the commonly observed hippocampal damage is considered a major contributor to the cognitive disturbances seen following experimental and clinical TBI (Kotapka et al., 1994; Smith et al., 1994). The primary, mechanical insult causes a depolarization and disturbance of membrane ion homeostasis with uncontrolled potassium efflux and calcium influx (Katayama et al., 1990; Nilsson et al., 1993), which in turn activate energy-requiring ion pumps. This increased energy demand occurs in face of TBI-induced reduction in cerebral blood flow (CBF) and mitochondrial dysfunction (Marklund et al., 2002; Lifshitz et al., 2003) that may aggravate the underlying tissue damage. The acute phase of cell injury initiates a prolonged phase of neurodegeneration and reduced CBF lasting up to one year after the injury (Smith et al., 1997; Kochanek et al., 2002). The cortical tissue injury was shown to expand to 150-400 % or more of its initial volume within the first 24 hours after injury (see Stoffel et al., 2002), and a gradual and prolonged enlargement of the cerebral ventricles have been found in humans post-TBI (MacKenzie et al., 2002). This delayed ongoing and autodestructive process, the "secondary injury," appears to involve, among others, a cascade of events with alterations in CBF and metabolism, cell membrane damage, modifications of neurotransmitters and their receptors, changes in tissue content of cations, generation of excess free radicals and an inflammatory response (McIntosh et al., 1997; Laurer et al., 2000; Lenzlinger et al., 2001). There are numerous experimental approaches for the treatment of TBI (Gennarelli, 1993; Povlishock and Christman, 1995; Laurer and McIntosh, 2001), but one common denominator (and the topic of this chapter) to many of the secondary injury factors is the excessive release of excitatory amino acids (EAA), particularly glutamate, with concomitant activation of EAA receptors and influx of calcium. The following sections will focus on the evidence of and the mechanisms for glutamate-induced excitotoxicity.

3. Glutamate-mediated excitotoxicity in TBI.

There are three major classes of glutamate receptors characterized by their genetic profile and classified according to the binding of various pharmacological compounds: N-methyl-D-aspartate (NMDA; the subunits NR1, NR2A, NR2B, NR2C and NR2D), α-amino-3-hydroxy-5-methyl-4-isoxazole propionic acid (AMPA), kainate (GluR5-9, KA1 and KA2), and the metabotropic receptors, which again are subdivided into group I (mGluR1and mGluR5), II (mGluR2 and mGluR3) and III (mGluR4, mGluR6, mGluR7 and mGluR8). Importantly, functional glutamate receptors and transporters are expressed in both gray and white matter tracts (for recent reviews, see Matute et al., 2002; Danbolt, 2001). The basic neurochemistry and physiological aspects of glutamate synthesis, release, regulation and modes of action have been covered in detail in other parts of this volume.

3.1 Evidence for increased glutamate release following TBI: experimental studies

Although glutamate is ubiquitous in the CNS, only a fraction is present within the extracellular (EC) space or cerebrospinal fluid (CSF; 3-4 μM and 10 μM, respectively; Hamberger and Nystrom, 1984). After experimental TBI, EC glutamate levels are reversibly increased acutely after injury showing a strong dependency on the underlying injury severity and models employed (Faden et al., 1989; Nilsson et al., 1990; Palmer et al., 1993; Katayama et al., 1990; Bullock et al., 1991). For instance, Faden and colleagues (1989) found a 280-fold increase following lateral fluid percussion (LFP) injury, whereas Nilsson *et al.* (1990) found a more modest 8-fold increase following weight drop injury in the rat. There has been some debate (see controversies below) whether this surge in glutamate is sufficient to induce and maintain excitotoxic cell death. However, the facts that cortical injection of NMDA or AMPA produced remarkably similar histological lesions (cavitation and GFAP immunoreactions) as seen after controlled cortical impact (CCI; Palmer et al., 1993), and that cortical administration of glutamate through a microdialysis probe exacerbates LFP -induced histological damage (Di et al., 1999), support the notion that TBI, at least partly, is an excitotoxic injury. Initially, the surge in EC glutamate was thought to be transient as the pronounced early increase rapidly returned to normal levels within the first few hours post-injury measured at the site of the impact (Nilsson et al., 1990; Palmer et al., 1993). As an example, Nilsson and colleagues (1994) reported a peak of EC glutamate and aspartate during the first 5 min post-injury following mild weight drop injury. However, a recent study showed that EC glutamate was elevated in the injured cortex up to 7 days after CCI (Stover et al., 2003). Finally, there was a prolonged increase in CSF glutamate levels following CCI in rats with a peak at 24 hours (Stover and Unterberg, 2000).

3.2 Evidence for increased glutamate release following TBI: clinical studies

Following its adaptation to study biochemical changes in the extracellular fluid after human TBI more than a decade ago (Persson and Hillered, 1992), the microdialysis technique has revealed significant and persisting increases in extracellular glutamate in human TBI (Marion et al., 2002; Bullock et al., 1998) lasting up to 9 days following the injury (Vespa et al., 1998). These up to 50-fold increases were attributable to seizure activity, decreases in cerebral perfusion pressure (CPP) to <70 mm Hg (Vespa et al., 1998), secondary hypoxia/ischemia, focal contusions (Bullock et al., 1998) and increases in ICP (Koura et al., 1998). In the CSF, up to 30-fold increases of glutamate levels have been demonstrated (Baker et al., 1993), although no correlation between CSF glutamate levels and clinical outcome at 3 months was found (Brown et al., 1998).

3.3 Glutamate-mediated damage following TBI

The gradient of glutamate across the plasma membrane is several thousand-fold with the highest concentrations (100 mM) localized within the synaptic vesicles in nerve terminals (Storm-Mathisen et al., 1992). In general, the intracellular glutamate is considered non-toxic, but it is of crucial importance that the EC glutamate concentration is kept low, which is achieved by rapid removal of glutamate by glial and neuronal glutamate transporters and by a negative feedback mechanisms mediated by metabotropic and ionotropic glutamate receptors (for review see Danbolt, 2001). Excessive activation of the EAA receptor subtypes cause a massive influx of Na^+, efflux of K^+ and subsequent depolarization of the NMDA receptor, overcoming the voltage dependent block by Mg^{2+} (*vide infra*), allowing a large Ca^{2+} influx into the cell. Consequently, using atomic absorption spectrophotometry, total tissue concentrations of Na^+ and Ca^{2+} were increased following LFP, in contrast to decreased K^+, $Mg^{2+}$and Zn^{2+} concentrations (Soares et al., 1992; Smith et al, 1993b). Excessive intracellular accumulation of calcium may cause energetic collapse, activation of many cytoplasmatic intracellular enzymes, such as calcium-activated neutral protease (calpain) which targets the neuronal cytoskeleton and destroys the neuronal architecture and transport mechanisms (McIntosh et al., 1997). Consequently, mitochondrial depolarization, increased production of oxygen free radicals, and release of pro-apoptotic factors that activate caspases have all been reported to result from excitotoxic stimuli and are factors known to occur in TBI (Fiskum, 2000; Marklund et al., 2001).

Under conditions of post-traumatic metabolic compromise, glutamate may cause cell death even at normal physiological concentrations as seen after LFP (Di et al., 1999). Posttraumatic cognitive deficits due to glutamate-mediated hippocampal damage have been linked to the high density of glutamate receptors and the 3-4-fold elevation of EC glutamate levels seen in the hippocampus following CCI in the rat (Rose et al., 2002). Furthermore, blood-brain barrier (BBB) disturbances are well-characterized events after TBI (Soares et al., 1995; Hicks et al., 1997), and glutamate has been shown to contribute to vasogenic and cytotoxic edema by damaging the BBB (Mayhan and Didion, 1996). Although the role of inflammation in TBI is controversial (Lenzlinger et al., 2001), the TBI-induced inflammatory response may exacerbate the excitotoxic damage (Morimoto et al., 2002).

After TBI, the observed death of neurons and non-neuronal cells (glial cells, endothelial cells, etc.) occurs via necrosis and/or apoptosis (Colicos and Dash, 1996; Rink et al., 1995). In TBI, activation of cysteine proteases (caspases), in particular the caspase-3 family, is associated with apoptotic cell death after TBI (Yakovlev et al., 1997). Both apoptosis and necrosis have been reported as a mechanism of excitotoxic cell death of neurons (Tenneti and Lipton, 2000; Chihab et al., 1998).

3.4 Changes in glutamate re-uptake and receptors in TBI

The glial and neuronal glutamate transporters GLT and GLAST are abundantly distributed within the mammalian brain and are highly concentrated in areas of glutamatergic transmission. The importance of the astrocytic uptake system in the protection of neurons has been elucidated in numerous *in vitro* and *in vivo* reports (see Danbolt, 2001). Under *in vivo* conditions, genetically engineered GLT-1 or GLAST-deficient mice showed an increased susceptibility to secondary damage following cold injury (Tanaka et al., 1997; Watase et al., 1998). An early reduction in GLAST/GLT-2 and GFAP co-expressing astrocytes was seen after CCI in the rat, reaching minimum levels at 8 hours post-injury (Rao et al., 1998), which coincided with increased CSF glutamate concentrations (van Landeghem et al., 2001). Microglia are capable of de novo expression of glutamate transporters, indicating that ramified microglia may play an important compensatory role in the early regulation of posttraumatic increases in extracellular glutamate (van Landeghem et al., 2001). Recently, in fatal human TBI, survival of more than 24 hours post-injury was associated with a widespread decrease in the expression of the EAA transporter II (Ikematsu et al., 2002). In addition, changes in glutamate receptor expression and binding capacity has been found in TBI, and a decrease in NMDA but not AMPA/KA receptor binding was seen following central FPI at 3 hours in the hippocampus and at 5 minutes post-injury in the cortex (Miller et al., 1990). Recently, decreases in NMDA receptor binding were found in bilateral hippocampi and parietal cortices 2 and 12 hours following LFP (Sihver et al., 2001). These findings have been confirmed, showing a decrease in the affinity for MK-801 up to 7 days post-injury following CCI in the rat (Stover et al., unpublished observations). Furthermore, there was a decrease at 6 and 12 hours of the NMDA-receptor subunits NR1/NR2A/NR2B protein levels after CCI that recovered by 24 h post-injury without concomitant changes in mRNA levels (Kumar et al., 2002), and amplification of mRNA from TUNEL-positive cortical neurons following LFP revealed that NR2A and NR2C mRNA were increased and GluR2 reduced at 12 hours post-injury (O'Dell et al., 2000). Finally, metabotropic mGluR2 and 3 protein expression was reduced in the ipsilateral hippocampus and cortex at 7 but not 15 days following LFP (Gong et al., 1999).

3.5 NMDA-Receptor-associated treatments in TBI

3.5.1 NMDA Receptor Activation: an overview

The NMDA receptor is a transmembrane receptor channel complex with allosteric modulation allowing voltage-dependent entry of divalent and monovalent cations. Although glutamate will open the channel, an internal voltage-dependent blockade by magnesium must be overcome to allow Ca^{2+} influx. In addition, the activation of the NMDA receptor is facilitated by the binding of glycine to a specific binding site on the NMDA receptor (Fields et al., 1991). The NMDA receptor activity is also influenced by the binding of polyamines and zinc that may influence the glycine binding site. In line with the evidence for glutamate-induced neurotoxicity, blockade of the NMDA-receptor targeting these sites has been evaluated in numerous studies in experimental TBI summarized below and in Table 1.

3.5.2 NMDA receptor channel blockers in TBI

Examples of NMDA channel blockers evaluated in experimental TBI include dextromethorphan and dextrorphan, ketamine, MK-801, magnesium, HU-211 and remacemide hydrochloride which all have been shown to decrease neuronal death, edema and/or neurological dysfunction following experimental TBI and acute subdural hematoma (McIntosh et al., 1998a; Uchida et al., 2001). However, compounds such as MK-801, phencyclidine (PCP), and ketamine have strong psychomimetic effects and cause vacuolization in neurons (Olney et al., 1989), thus preventing their introduction into clinical practice. However, due to promising pre-clinical documentation the compounds CP-101,606, Selfotel and Cerestat (*vide infra*) were all investigated in phase III clinical trials. Newer noncompetitive NMDA receptor antagonists designed to have low toxicity but maintained efficacy include NPS 1506 and its analog NPS 846, and have been shown to improve neurological and cognitive deficits and delay edema following experimental TBI (Leoni et al., 2000; Gurevich et al., 1998). Of particular interest is the novel non-psychotropic cannabinoid HU-211 (Dexanabinol) that was found to improve neuromotor and cognitive outcome following brain trauma with a therapeutic window of 2-3 hours post-injury (Shohami et al., 1993; 1995). HU-211, a weak NMDA receptor antagonist with multiple mechanisms of action, including TNF-α inhibition, is currently investigated in a phase III clinical trial. In addition, the open-channel, non-competitive NMDA blocker memantine can easily cross the blood-brain barrier and shows minimal adverse effects on neuronal ultrastructure, learning, long-term potentiation (LTP) and synaptic transmission (Parsons et al., 1998).

Following CCI in the rat, post-injury treatment with memantine decreased CA2 and CA3 cell death at 7 days post-injury (Rao et al., 2001). Memantine is currently evaluated in clinical trials for dementia, Alzheimer's and Parkinson's disease, and is void of any apparent neuropyschiatric adverse effects.

NMDA-mediated receptor activation has a slow onset and long decay phase relative to AMPA/ KA receptors. This is though to result from the voltage-sensitive block within the NMDA receptor by Mg^{2+}, which is strongest between the membrane potential of – 80 to – 40 mV (Nowak et al., 1984). Mg^{2+} functions as a key endogenous modulator of the NMDA receptor, and TBI-induced loss of intracellular magnesium may influence NMDA-mediated neurotransmission and affect post-traumatic repair mechanisms. Decreased intracellular magnesium is a consistent finding in experimental TBI models (Vink et al., 1987; Heath and Vink, 1996; Vink et al., 1996; Heath and Vink, 2001; Bareyre et al., 1999) and decreased free intracellular magnesium concentrations were correlated with poor neurological recovery after TBI (Heath and Vink, 1999a). Consequently, pre- and post-treatment with magnesium salts ($MgCl_2$ or $MgSO_4$) improved neuromotor and cognitive deficits, decreased regional edema (McIntosh et al., 1988; McIntosh et al., 1989; Smith et al., 1993a; Okiyama et al., 1995; Heath and Vink, 1998), reduced cytoskeletal alterations, and attenuated cell death following LFP (Bareyre et al., 2000; Saatman et al., 2001). Finally, magnesium improved neuromotor function when administered up to 24 hours after closed head injury in rats (Heath and Vink, 1999b), and when administered 30 minutes post-injury, an improvement in motor and cognitive function was seen up to 28 days after injury (Vink et al., 2003).

3.5.3 Polyamine site modulation in TBI

Endogenous amines, spermidine and spermine, enhance NMDA receptor activation while the synthetic competitive antagonists of the polyamine site (eliprodil and ifenprodil) reduce NMDA receptor activation, and may have potential advantages over other noncompetitive antagonists due to less potential psychomimetic and neurotoxic side effects (Dempsey et al., 2000). When administered immediately following CCI in the rat, ifenprodil attenuated brain edema, BBB breakdown and reduced cortical injury volume (Dempsey et al., 2000). Eliprodil attenuated cortical tissue loss when delivered up to 18 hours after lateral fluid percussion injury, making this compound clinically interesting (Toulmond et al., 1993).

3.5.4 Glycine site modulation in TBI

Activation of the glycine site potentiates NMDA receptor binding, and inhibition of the glycine site has been shown to block the current evoked by NMDA (Huettner, 1989). When administered 15 minutes after injury, the glycine site antagonists kynurenate and I2CA both reduced cognitive

Action	Drug	Time and route of admin.	TBI Model	Outcome
Non-comp. NMDA RA Comp. NMDA RA Poly-amine site Glycine site Mg^{2+} site NMDA Agonist NMDA subunits ↓ Glutamate release ↓ ODC ↓ AMPA/ KA RA Selective $mGluR_1A$ $mGluR_5A$ Group II - III R agonists	MK-801 [1-14] Dextror.[15,18,81] Dextromtr.[16-17] Memantine[19] Ramacemide[20] Cerestat® [21] Gacyclidine[22] Ketamine[14,23-25] HU-211[26-27] NPS 1506[28-29] NPS 846 [30] PCP[31] CPP [1,13,15,32-35] APV [36] Selfotel®[13,16,81] CP-101,606 [37,39] CP-101,581[37] CP-98,113 [37-38] Eliprodil [6,40] Ifenprodil[41] KYNA [36,42-45] I2CA[44] ACEA1021[46] Mg [23,47-58] Cycloserine[59] Antisense oligos[60] BW1003C87[47] Riluzole[61-65] CNQX [36] 619C89[66-67] Lubeluzole[68] DFMO[69] Talampanel [70] RPR117824[71] NBQX [10,33] AIDA[72,75] MCPG[72-74] BAY 36-7620[76] MPEP[77] DCG-IV[78] and 4-phosphono-phenylglycine[78] LY354740[79] LY379268 [80]	**PRE-INJURY** >15 min [1,7,11,32-33,35, 37-38, 40,42, 46,60, 70,73] 0-15 min [2-3, 8-9, 12, 15(CPP)-16,31, 34,48, 67,72,74, 77,81] **POST-INJURY** 0-30 min [2-4, 6,10,13-15,17-21, 23, 26,29,37-41,43-47,49,51-56,58, 61-66,68-71,73, 75-76,78-80] >30 min – 2h [1,5,12,26-28,30,32-33,46, 50,53,57,63] >2-6 h [5-6,21-22,24-27,32-33,40,53,70] >6-24 h [6,32-33,40,53,59] **ROUTE OF ADMIN.** I.v. [3-4,11,13-16, 20-23,27-29, 39-40, 43-52, 54-56,58,61-68, 70-71,73,76,79-81] I.p.[1-2,5-9,12, 17-19,24-26,30-35,37-38,41,59, 69] I.m. [53-55] S.c.[20,38,57, 61-64,71] p.o.[17] Icv [15(CPP),36, 60,72-74,77] Intracerebr. [36,42,75,78] Intrathec. [10]	**LFP** [2-4, 6,15-16,18, 20,23,29, 36-38,40,42-45, 47-51,58-64,67, 70-75,77-79] **CHI + O_2↓**[7] **Bifrontal Contusion** [22] **CHI/IA** [1,5,10, 24-28,30, 52-57] **CFP** [31] **aSDH** [13-14, 17,34-35, 39,46, 66,76] **WD** [11,32-33] **CCI** [19,21,41, 65,68-69,80] **CFP+ deaf** [8-9] **CFP+ ischemia** [12]	**Lesion volume/ neuronal cell loss** d.0-7 ↓ [1(pre),6 (eliprodil),13-14 (MK-801),17,20-22,24,32-33(pre),35, 39,41,46, 58,61,65-66,70-71,76,80] d.0-7 No effect[5,6(MK-801),11,14(ketamine),29,32(post), 51,62, 64,68,70] d. 0-7 ↑ [1(post)] d. 8-30 ↓[10,50,63,72,77] d. 8-30 No effect [11] **Hippocampal damage** ↓ [9,19,29,32-33,45,67,70,73,75,78(DCG-IV)] ↑ [32 (post,CPP)] No change [11-12,32(NBQX), 50,60,63,78] **BBB breakdown** ↓[26,41] **Glucose utilization**↓[34,36] **Microdialys** [7,42.66,81] e.g. glutamate↓, lactate ↓ **FFA/DAG accum.** ↓ [43] **Ion homeostasis and/ or bioenergetic state** ↑[4, 15-16,25,28, 30,44, 48,52,55,57] No effect [16(Selfotel)] **Edema** ↓24-48h [4-5,21,26, 28,30, 37(CP-98,113)-38, 41,44, 47,51,57,62,64-65,69] No effect [24,27,35,68] **ICP** No effect[13-14,21,68,80] **Mortality** ↓ [31,60,72] **Neurological motor** d.1-7 ↑ [3(pre),5, 8,13,15,18,24, 26-28, 30-31,38,49,51-57,60,67,74] >d.7-6 wks ↑ [3(pre),15,18,27,31, 38,44,47-49,56, 60-61,64,72-73,75,77,79] No effect [3 (post),8,27,31] **Cognitive outcome** 48h -d.7 ↑ [22-23,26-27,29, 37-38,40, 44,64] 48h –d.7 No effect [20] >d.7-d.64 ↑ [2(pre),8-9, 11,27,38,47,56,59,72,74-75,77] >d.7-d.64 No effect [2(post),8 (d.61-64)]

Table 1. Pharmacological approaches targeting excitotoxicity in TBI. ***Abbreviations***: Comp; competitive, RA; Receptor Antagonist, ODC; ornithine decarboxylase synthesis inhibitor, admin; administration, oligos; oligonucleotides, Icv; intracerebro-ventricularly, Intracerebr; intracerebral, Intrathec; intrathecal LFP; lateral fluid percussion (FP) injury, WD; weight drop injury, $O_2\downarrow$; secondary hypoxia, CHI; closed head injury, CFP; central FP, CCI; controlled cortical impact, aSDH; acute subdural hematoma, deaf; de-afferentation, IA; Impact-acceleration, d.; day post-injury, BBB; blood brain barrier, GABA; γ-aminobutyrate, FFA; free fatty acids, DAG; diacylglycerol, ICP; intracranial pressure. All studies cited were performed in the rat.

REFERENCES TO TABLE 1

1) Pohl et al. PNAS, 1999 2) Hamm et al. Brain Res Cogn Brain Res, 1993 3) McIntosh et al. J Neurotrauma, 1989 4) McIntosh et al. J Neurochem, 1990 5) Shapira et al. J Neurotrauma, 1990 6) Toulmond et al. Brain Res, 1993 7) Katoh et al. Brain Res, 1997 8) Phillips et al. J Neurosci Res, 1997 9) Phillips et al. Hippocampus, 1998 10) Goda et al. J Neurotrauma, 2002 11) Lewen et al. Acta Neurochir, 1999 12) Jenkins et al. Brain Res, 1999 13) Kuroda et al. Neurosurgery, 1994 14) Uchida et al. Eur J Anaesthesiol, 2001 15) Faden et al. Science, 1989 16) Golding and Vink, Mol Chem Neuropathol, 1995 17) Duhaime et al. J Neurotrauma, 1996 18) Faden, J Neurotrauma, 1993 19) Rao et al. Brain Res, 2001 20) Smith et al. Neurosci Lett, 1997 21) Kroppenstedt et al. J Neurotrauma, 1998 22) Smith et al. Neural Plast, 2000 23) Smith et al. Neurosci Lett, 1993 24) Shapira et al. Stroke, 1994 25) Shapira et al. JCBF, 1993 26) Shohami et al. J Neurotrauma, 1993 27) Shohami et al. Brain Res, 1995 28) Lorber et al. J Neurosurg Anaesthesiol, 2000 29) Leoni et al. Exp Neurol, 2001 30) Gurevich et al. J Neurosurg, 1998 31) Hayes et al. J Neurotrauma, 1988 32) Ikonomidou et al. PNAS, 2000 33) Bernert and Turski, PNAS, 1996 34) Inglis et al. Acta Neurochir (Suppl), 1990 35) Bullock et al. Stroke, 1990 36) Kawamata et al. JCBF, 1992 37) Okiyama et al. J Neurotrauma, 1997 38) Okiyama et al. Brain Res, 1998 39) Tsuchida et al. J Neurotrauma, 1997 40) Hogg et al. J Neurotrauma, 1998 41) Dempsey et al. Neurosurgery, 2000 42) Kawamata et al. Brain res, 1995 43) Dhillon and Prasad, Brain Res, 1999 44) Smith et al. J Neurosci, 1999 45) Hicks et al. Brain Res, 1994 46) Tsuchida and Bullock, J Neurotrauma, 1995 47) Okiyama et al. J Neurochem, 1995 48) McIntosh et al. J Neurotrauma, 1988 49) McIntosh et al. Brain Res, 1989 50) Bareyre et al. J Neurotrauma, 2000 51) Guluma et al. J Neurotrauma, 1999 52) Heath and Vink, J Neurotrauma, 1998 53) Heath and Vink, J Neurosurg, 1999 54) Heath and Vink, Neurosci Lett, 1997 55) Heath and Vink, J Pharmacol Exp Ther, 1999 56) Vink et al. Neurosci Lett, 2003 57) Feldman et al. J Neurosurgery, 1996 58) Sathman et al. J Neuropath Exp Neurol, 2001 59) Temple and Hamm, Brain Res, 1996 60) Sun and Faden, Brain Res, 1995 (a, Sep.) 61) Wahl et al. Brain Res, 1997 62) Bareyre et al. J Neurotrauma, 1997 63) Zhang et al. J Neurosci Res, 1998 64) McIntosh et al. J Neurotrauma, 1996 65) Stover et al. J Neurotrauma, 2000 66) Tsuchida et al. J Neurosurgery, 1996 67) Sun and Faden, Brain Res, 1995 (b, Febr.) 68) Kroppenstadt et al. J Neurotrauma, 1999 69) Baskaya et al. J Neurotrauma, 1996 70) Belayev et al. J Neurotrauma, 2001 71) Mignani et al. Bioorg Medicinal Chem, 2002 72) Faden, Exp Neurol, 2001 73) Mukhin et al. J Neurosci, 1996 74) Gong et al. Brain Res, 1995 75) Lyeth et al. Exp Neurol, 2001 76) deVry et al. Eur J Pharmacol, 2001 77) Movsesyan et al. J Pharmacol Exp Therapy, 2001 78) Zwienenberg et al. Neurosurg, 2001 79) Allen et al. J Pharmacol Exp Therapy, 1999 80) Stover et al. J Neurotrauma, 2003 81) Panter and Faden, Neurosci Lett, 1992

and neurological motor deficits and reduced CA3 cell death at 2 weeks post-injury following LFP in the rat (Smith et al., 1993b; Hicks et al., 1994) and kynurenate reduced free fatty acid and diacylglycerol accumulation after LFP (Dhillon and Prasad, 1999). In addition, the compound ACEA1021 reduced ischemic damage in a model of acute subdural hematoma in the rat (Tsuchida and Bullock, 1995).

3.5.5 Modulation of the NMDA receptor by zinc

Zn^{2+} is present at high concentrations in mammalian brain, and is released in chelatable form after excitation of certain glutamatergic neurons. Recent observations suggest that it may play an important role in excitotoxic damage, where synaptically released zinc may enter postsynaptic neurons in toxic excess during ischemia and seizures. Prevention of the translocation of zinc has been shown to be neuroprotective in these conditions (for references see Weiss et al., 2000), and the translocation of zinc from presynaptic boutons into postsynaptic neurons was recently shown to occur following mechanical injury to the brain (Suh et al., 2000). However, zinc deficiency was associated with an increased cell death following a cortical stab wound in the rat (Yeiser et al., 2002). To date, however, the role and importance of Zn^{2+} in TBI has yet to be investigated.

3.6 AMPA/KA Receptor Modulation in TBI

While NMDA receptors have a high affinity to glutamate and slowly inactivate, the low affinity AMPA/ KA receptors desensitizes rather quickly. AMPA is thought to mediate the fast component while NDMA receptors mediate the slow component of fast excitatory neurotransmission (Dingledine et al., 1999). To date, only a few studies employing AMPA/ KA receptor antagonists in TBI have been published. However, post-injury treatment with the AMPA antagonist RPR117824 was recently shown to attenuate cortical lesion volume following LFP in the rat (Mignani et al., 2002). In addition, the AMPA antagonist Talampanel reduced CA1 damage and cortical lesion volume when administered 30 minutes after LFP (Belayev et al., 2001). Interestingly, the AMPA antagonist NBQX attenuated hippocampal damage when administered up to 7 h following a weight drop injury in the rat (Bernert and Turski, 1996).

3.7 Metabotropic Receptor Modulation in TBI

Abundant pre-and postsynaptically located metabotropic glutamate receptors (mGluRs) are thought to be important in modulating neurotransmission. Activation of group I receptors potentiates neuronal excitation and exacerbates excitotoxic cell death *in vitro* (Mukhin et al., 1996). In contrast, group II and III agonists reduce excitation, perhaps through presynaptic inhibition of glutamate release (Faden et al., 2001) or modulation of adenylate cyclase activity (Buisson and Choi, 1995). Thus, mGluR activation may contribute to both neurotoxic and neuroprotective processes depending on the activated receptor.

In vivo, intracerebral or intraventricular administration of the group 1 inhibitors MCPG, AIDA or MPEP has been used, starting pre-injury or up to 15 min following LFP, and consistent finding are improvements in neurological motor deficits, cortical lesion volume and cognitive outcome (Faden et al., 2001; Mukhin et al., 1996; Gong et al., 1995; Lyeth et al., 2001; Movsesyan et al., 2001). In addition, i.v. administration of the $mGluR_1$ antagonist BAY 36-7620 attenuated ischemic damage in a model of acute subdural hematoma in the rat (de Vry et al., 2001).

The activation of group-II and -III mGlu receptors is expected to attenuate neuronal degeneration without hampering the efficiency of fast excitatory synaptic transmission. *In vitro*, a clear neuroprotective effect of both group II and III mGluR agonists was demonstrated when neurons were exposed to mechanical injury (Allen et al., 1999). *In vivo* studies examining the role of Groups II and III mGluR in brain injury are rare, but administration of the Group II mGluR agonist LY 354740 after LFP injury in rats significantly improved neurological motor recovery (Allen et al., 1999). When the compounds DCG-IV (Group II) or (R,S)-4-phosphonophenylglycine (group III-agonist) were injected into the CA2 and CA3 area of the hippocampus immediately following LFP in the rat, only DCG-IV injections reduced the number of degenerating neurons in these regions (Zwienenberg et al., 2001). Finally, the specific group II mGluR agonist LY379268 decreased cortical contusion volume at 7 days following CCI but did not attenuate edema at 24 hours (Stover et al., 2003).

3.8 Glutamate Release Inhibitors in TBI

In addition to antagonism of EAA receptor function, modulation of EAA receptor activity may also be accomplished by inhibition of EAA release. BW1003C87 block glutamate release induced by the sodium channel agonist veratrine without blocking potassium-induced glutamate release, and when administered at 15 min after LFP in the rat, reduced edema in the ipsilateral cortex and hippocampus was found at 48 hours post-injury (Okiyama et al., 1995). In addition, post-injury treatment with the sodium channel blocker and glutamate release inhibitor riluzole was found to attenuate neurological motor, cognitive deficits and cortical lesion volume after fluid percussion injury in the rat (Wahl et al., 1997; Bareyre et al., 1997) and reduce edema after CCI (Stover et al., 2000). Finally, pre-treatment with 619C89 attenuated neurological deficits and reduced CA1 and CA3 cell loss following fluid percussion injury (Sun and Faden, 1995b). It should be mentioned that these compounds may all have other mechanisms of action besides inhibiting glutamate release contributing to their neuroprotective properties, including a primary effect on inactive sodium channels (riluzole and 619C89).

4 Clinical phase III trials with glutamate receptor antagonists.

The impressive reductions in functional and structural impairment following experimental TBI upon attenuating glutamate excitotoxicity have raised high expectations for neurological improvement following clinical TBI. Unfortunately, all clinical trials completed to date have failed to show any significant improvements in outcome (Teasdale and Bannan, 1997). Among others, reasons for these failures range from possibly ineffective drugs to improper patient selection, pre-clinical documentation, dosing and time of drug administration. In the following section, we will briefly summarize the phase III-trials which were completed to date (for details see Narayan et al. 2002).

Selfotel was the first glutamate antagonist to be tested in a phase III trial. Most of the pre-clinical evaluations with this competitive glutamate receptor antagonist were made using a pre-injury dosing paradigm in animal models of which only a few were related to TBI (e.g. Golding and Vink, 1995). A phase II trial showed an ICP-lowering effect, and four simultaneous trials in ischemia and two in TBI were launched. Due to higher mortality in the concomitant stroke trials, the TBI trials were terminated. On post-hoc analysis, an overall reduction in mortality was seen, but no effect on any of the TBI-related subgroups. The remarkable outcome in the placebo group found in this study suggests that the overall clinical care had been substantially improved during the last decades.

Cerestat is a noncompetitive glutamate receptor antagonist that attenuated post-traumatic edema after CCI in rats, although no effect on ICP was seen (Kroppenstedt et al., 1998). The basic idea was that this compound would dampen "the ionic storm" seen after TBI. An interim analysis of about 340 patients showed no benefit and no harmful effect, and a simultaneous stroke study showed neither a beneficial nor a harmful effect from the drug. The final evaluation of this trial has not been published or presented yet, although it appears that large inter-center variability interfered with the quality of the data as most of the 70 participating centers enrolled fewer than five patients.

D-CPP-ene (SDZ EAA 494; Saphir®), a competitive NMDA receptor antagonist that reduced e.g. ischemic damage after subdural hematoma in the rat (Bullock et al., 1990), was evaluated in 51 European Centers, enrolling 920 patients within 12 hours after injury. Unfortunately, a slightly, non-significant worse outcome at 6 months was seen and drug-treated patients required a prolonged ventilatory support and intensive care treatment. *Dexanabinol* (HU-211), a novel synthetic chemical analogue of marijuana has been shown to non-competetively inhibit the NMDA receptor and attenuated BBB damage, edema formation and neurological deficits with a therapeutic window of 3 hours or more following closed head injury in rats (Shohami et al., 1993; 1995). In a multicenter, randomized, double-blinded phase IIb study evaluating 101 patients with a 3 hour time-window, the compound was well tolerated, and tended to reduce mortality. Control of ICP was significantly ameliorated and hypotensive phases were significantly reduced, significantly improving outcome at one month post-injury. Currently, a phase III study is being conducted.

CP 101-606, a "second generation" NMDA antagonist targeting the NR2B subtype receptor showed promising pre-clinical effects in TBI (Okiyama et al., 1997; Tsuchida et al., 1997). This compound penetrates easily into the CNS, reaching a fourfold higher concentration as compared to plasma. No behavioral adverse side effects have been observed in humans, and in an "open-label" phase II study, a potential beneficial effect was seen (Narayan et al., 2002). The "phase IIb" trial with 400 enrolled patients has been completed, although the results have not yet been published.

5. Controversies

Under pathological conditions where glutamate induces excitotoxic injury, damage and/or dysfunction of the glutamate receptor network might impede normal LTP and disrupt normal cognitive processes. However, glutamate is involved in most aspects of normal brain function including cognition, memory and learning and glutamate is involved in synapse induction and elimination, cell migration and differentiation (see Danbolt, 2001a). Consequently, NMDA antagonists such as MK-801 and PCP inhibit LTP and cognitive processes (Davis et al., 1992), and MK-801 produced a profound amnesia in a passive avoidance task following LFP (Hamm et al., 1994). Treatments promoting the potentially beneficial effects of NMDA activation may be an alternative treatment strategy and beneficial effects on cognition were seen using the NMDA *agonist* D-cycloserine when administered at 24 h following LFP (Temple and Hamm, 1996). In addition, pre-treatment with the NMDA receptor antagonist CPP reduced but post-treatment *increased* the slowly progressing neurodegeneration (apoptosis) in the hippocampus following weight drop injury (Ikonomidou et al., 2000). Also, stimulation of NMDA receptors may be beneficial due to increases in the levels of putative neuroprotective neurotrophins (Rocamora et al., 1994), seen in the hippocampus after TBI (Hicks et al., 1999).

Although TBI causes a marked increase in extracellular glutamate and treatment directed against glutamate-mediated excitotoxicity improves outcome experimental TBI, it has been argued against the importance of high extracellular glutamate in the secondary injury process (summarized in (Obrenovitch and Urenjak, 1997). As one example, it has been suggested that the high extracellular glutamate levels seen after TBI merely reflects ongoing secondary ischemia and sustained anoxic depolarization as well as cellular lysis/ leakage in brain regions that can no longer be salvaged. In support of this theory, dialysate levels of aspartate increase synchronously with glutamate, with similar magnitude and kinetics (Nilsson et al., 1990; Nilsson et al., 1994) which suggests, as there is no vesicular aspartate, that most TBI-induced glutamate efflux is not due to exocytosis. In addition, many NMDA and non-NMDA receptor antagonists are effective even when administered well *after* the maximal increase of extracellular glutamate, suggesting that other mechanisms than reductions of glutamate-mediated excitotoxicity may be responsible for their neuroprotective properties. Furthermore, CSF glutamate was significantly elevated by 2-8-fold from 10 hours to days following clinical TBI. At the same time, CSF levels of the major *inhibitory* amino acid GABA was increased 56-317-fold compared to basal levels (Palmer et al., 1994), perhaps counteracting the effects of increased glutamate. In addition, in strains of mice susceptible or resistant to kainic acid-induced hippocampal damage, no differences in CA3 neuronal survival were seen at 4 days following FPI (Carbonell and Grady, 1999). In TBI, CA3 is preferentially damaged as compared to CA1, whereas the reverse situation is seen following glutamate, KA or NMDA exposure (Vornov et al., 1991). Finally, in rats, mice and man, CA3 has a predominance of AMPA receptors, whereas NMDA receptors are less concentrated (Watanabe et al., 1998), suggesting that other mechanisms may be involved in the reductions of CA3 neuronal damage seen using NMDA receptor antagonists following TBI (Rao et al., 2001; Hicks et al., 1994; Sun and Faden, 1995a).

6 Summary and future directions

Despite these controversies, there is a large amount of data stressing the importance of glutamate-mediated excitotoxicity following TBI. The large number of pre-clinical studies showing reduced neurological motor, cognitive and cellular damage with compounds targeting the effects of excessive glutamate receptor activation continues to nourish the hopes for a successful clinical translation. In addition, evolving knowledge concerning the role of physiological glutamate receptor activation stresses the importance to evaluate the time-window of initiation *and* termination of the treatment not to interfere with the attempt for regeneration. Thus, therapeutic strategies of EAA modulation may not be equally applicable to all central nervous system injuries. For clinical development, thorough pre-clinical documentation in TBI is crucial addressing toxicity, time-windows and dosing regimes in multiple clinically relevant models. Several new compounds with interesting profiles are currently being evaluated in various models of acute or chronic brain injuries in humans; these include memantine, HU-211 (Dexanabinol), currently in phase III-trial for TBI, and [1-(2-thienyl) cyclohexyl]piperidine (TCP) and its derivative gacyclidine, a potent noncompetitive NMDA receptor antagonists that has been suggested to be devoid of adverse side effects (Mitha and Maynard, 2001). One important lesson was learnt in a phase II study in stroke, where the AMPA antagonist ZK200775 transiently worsened the neurological condition and increased serum markers of glial toxicity (S-100B, Elting et al., 2002) suggesting the importance of closely monitoring adverse effects with the use of markers for cell toxicity. A novel approach to attenuate excess activation of e.g. the NMDA receptor was recently addressed; a prominent organizing protein of the multiprotein signaling complexes of ionotropic glutamate receptors is post-synaptic density (PSD)-95, which couples the NMDA receptor to intracellular proteins and signaling enzymes (Aarts et al., 2002), and PSD-95 deletion dissociates NMDA receptor activity from nitric oxide (NO) production (without blocking synaptic activity or calcium influx) and suppresses excitotoxicity (Sattler et al., 1999). This approach, however, has yet to be evaluated in TBI.

In summary, glutamate-mediated excitotoxicity seems to be an important contributor to the cognitive, neurological motor and cellular deficits following traumatic brain injury but to date, no clinically effective treatment has yet emerged targeting the effects of excess glutamate. It is hoped that better understanding of the role of excitotoxicity in the pathophysiology of TBI will lead to new and more effective treatments for the devastating sequelae of human TBI.

REFERENCES

Consensus conference. Rehabilitation of persons with traumatic brain injury. NIH Consensus Development Panel on Rehabilitation of Persons With Traumatic Brain Injury. JAMA 1999; 282:974-983.

Aarts M., Liu Y., Liu L., Besshoh S., Arundine M., Gurd J.W., Wang Y.T., Salter M.W., and Tymianski M. Treatment of ischemic brain damage by perturbing NMDA receptor- PSD-95 protein interactions. Science 2002; 298:846-850.

Allen J.W., Ivanova S.A., Fan L., Espey M.G., Basile A.S., and Faden A.I. Group II metabotropic glutamate receptor activation attenuates traumatic neuronal injury and improves neurological recovery after traumatic brain injury. J Pharmacol Exp Ther 1999; 290:112-120.

Baker A.J., Moulton R.J., MacMillan V.H., and Shedden P.M. Excitatory amino acids in cerebrospinal fluid following traumatic brain injury in humans. J Neurosurg 1993; 79:369-372.

Bareyre F., Wahl F., McIntosh T.K., and Stutzmann J.-M. Time course of cerebral edema after traumatic brain injury in rats: effects of riluzole and mannitol. J Neurotrauma 1997; 14:839-849.

Bareyre F.M., Saatman K.E., Helfaer M.A., Sinson G., Weisser J.D., Brown A.L., and McIntosh T.K. Alterations in ionized and total blood magnesium after experimental traumatic brain injury: Relationship to neurobehavioral outcome and neuroprotective efficacy of magnesium chloride. J Neurochem 1999; 73:271-280.

Bareyre F.M., Saatman K.E., Raghupathi R., and McIntosh T.K. Post-injury treatment with magnesium chloride attenuates cortical damage after traumatic brain injury in rats. J Neurotrauma 2000; 17:1029-1039.

Belayev L., Alonso O.F., Liu Y., Chappell A.S., Zhao W., Ginsberg M.D., and Busto R. Talampanel, a novel noncompetitive AMPA antagonist, is neuroprotective after traumatic brain injury in rats. J Neurotrauma 2001; 18:1031-1038.

Bernert H. and Turski L. Traumatic brain damage prevented by the non-N-methyl-D-aspartate antagonist 2,3-dihydroxy-6-nitro-7sulfamoylbenzo[f] quinoxaline. Proc Natl Acad Sci USA 1996; 93:5235-5240.

Brown J.I., Baker A.J., Konasiewicz S.J., and Moulton R.J. Clinical significance of CSF glutamate concentrations following severe traumatic brain injury in humans. J Neurotrauma 1998; 15:253-263.

Buisson A. and Choi D.W. The inhibitory mGluR agonist, S-4-carboxy-3-hydroxy-phenylglycine selectively attenuates NMDA neurotoxicity and oxygen-glucose deprivation-induced neuronal death. Neuropharmacol 1995; 34:1081-1087.

Bullock R., Butcher S.P., Chen M.H., Kendall L., and McCulloch J. Correlation of the extracellular glutamate concentration with extent of blood flow reduction after subdural hematoma in the rat. J Neurosurg 1991; 74:794-802.

Bullock R., McCulloch J., Graham D.I., Lowe D., Chen M.H., and Teasdale G.M. Focal ischemic damage is reduced by CPP-ene studies in two animal models. Stroke 1990; 21:III32-III36.

Bullock R., Zauner A., Woodward J.J., and et al. Factors affecting excitatory amino acid release following severe human head injury. J Neurosurg 1998; 89:507-518.

Carbonell W.S. and Grady M.S. Evidence disputing the importance of excitotoxicity in hippocampal neuron death after experimental traumatic brain injury. Ann NY Acad Sci 1999; 890:287-298.

Chihab R., Oillet J., Bossenmeyer C., and Daval J.L. Glutamate triggers cell death specifically in mature central neurons through a necrotic process. Mol Genet Metab 1998; 63:142-147.

Colicos M.A. and Dash P.K. Apoptotic morphology of dentate gyrus granule cells following experimental cortical impact injury in rats: possible role in spatial memory deficits. Brain Res 1996; 739:120-131.

Danbolt N.C. Glutamate uptake. Prog Neurobiol 2001; 65:1-105.

Davis S., Butcher S.P., and Morris R.G.M. The NMDA receptor antagonist D-2-amino-5-phosphonopentanoate (DAP5) impairs spatial learning and LTP in vivo at intracerebral concentrations comparable to those that block LTP in vitro. J Neurosci 1992; 12:21-34.

De Vry J., Horvath E., and Schreiber R. Neuroprotective and behavioral effects of the selective metabotropic glutamate mGlu(1) receptor antagonist BAY 36-7620. Eur J Pharmacol 2001; 428:203-214.

Dempsey R.J., kaya M.K., and an A. Attenuation of brain edema, blood-brain barrier breakdown, and injury volume by ifenprodil, a polyamine-site N-methyl-D-aspartate receptor antagonist, after experimental traumatic brain injury in rats. Neurosurg 2000; 47:399-404.

Dhillon H.S. and Prasad A.S. Kynurenate attenuates the accumulation of diacylglycerol and free fatty acids after experimental brain injury in the rat. Brain Res 1999; 832:7-12.

Di X., Gordon J., and Bullock R. Fluid percussion brain injury exacerbates glutamate-induced focal damage in the rat. J Neurotrauma 1999; 16:195-201.

Dingledine R., Borges K., Bowie D., and Traynelis S.F. The glutamate receptor ion channels. Pharmacol Rev 1999; 51:7-61.

Elting J.W., Sulter G.A., Kaste M., Lees K.R., Diener H.C., Hommel M., Versavel M., Teelken A.W., and De Keyser J. AMPA antagonist ZK200775 in patients with acute ischemic stroke: possible glial cell toxicity detected by monitoring of S-100B serum levels. Stroke 2002; 33:2813-2818.

Faden A.I., Demediuk P., Panter S.S., and Vink R. The role of excitatory amino acids and NMDA receptors in traumatic brain injury. Science 1989; 244:789-800.

Faden A.I., O'Leary D.M., Fan L., Bao W., Mullins P.G., and Movsesyan V.A. Selective blockade of the mGluR1 receptor reduces traumatic neuronal injury in vitro and improves outcome after brain trauma. Exp Neurol 2001; 167:435-444.

Fields R.D., Yu C., and Nelson P.G. Calcium, network activity, and the role of NMDA channels in synaptic plasticity in vitro. J Neurosci 1991; 11(1):134-146.

Fiskum G. Mitochondrial participation in ischemic and traumatic neural cell death. J Neurotrauma 2000; 17(10):843-55

Gennarelli T.A. Animate models of human head injury. J Neurotrauma 1994; 11:357-368.

Golding E.M. and Vink R. Efficacy of competitive vs noncompetitive blockade of the NMDA channel following traumatic brain injury. Mol Chem Neuropathol 1995; 24:137-150.

Gong Q.Z., Delahunty T.M., Hamm R.J., and Lyeth B.G. Metabotropic glutamate antagonist MCPG in the treatment of traumatic brain injury in rats. Brain Res 1995; 700(1-2):299-302.

Gong Q.Z., Phillips L.L., and Lyeth B.G. Metabotropic glutamate receptor protein alterations after traumatic brain injury in rats. J Neurotrauma 1999; 16:893-902.

Gurevich B., Artru A.A., Lam A.M., Mueller A.L., Merkind V., Talmor D., Katchko L., and Shapira Y. Neuroprotective effects of NPS 846, a novel N-methyl-D-aspartate receptor antagonist, after closed head trauma in rats. J Neurosurg 1998; 88:1066-1074.

Hamberger A. and Nystrom B. Extra- and intracellular amino acids in the hippocampus during development of hepatic encephalopathy. Neurochem Res 1984; 9:1181-1192.

Hamm R.J., Pike B.R., O'Dell D.M., and Lyeth B.G. Traumatic brain injury enhances the amnesic effect of an NMDA antagonist in rats. J Neurosurg 1994; 81:267-271.

Heath D.L. and Vink R. Traumatic brain axonal injury produces sustained decline in intracellular free magnesium concentration. Brain Res 1996; 738:150-153.

Heath D.L. and Vink R. Neuroprotective effects of $MgSO_4$ and $MgCl_2$ in closed head injury: a comparative phosphorus NMR study. J Neurotrauma 1998; 15:183-189.

Heath D.L. and Vink R. Brain free magnesium concentration is predictive of motor outcome following traumatic axonal brain injury in rats. Magnes Res 1999a; 12:269-277.

Heath D.L. and Vink R. Delayed therapy with magnesium up to 24 h following traumatic brain injury improves motor outcome. J Neurosurg 1999b; 90:504-509.

Heath D.L. and Vink R. Subdural hematoma following traumatic brain injury causes a secondary decline in brain free magnesium concentration. J Neurotrauma 2001; 18:465-469.

Hicks R.R., Baldwin S.A., and Scheff S.W. Serum extravasation and cytoskeletal alterations following traumatic brain injury in rats. Mol Chem Neuropathol 1997; 32:1-16.

Hicks R.R., Gennarelli T.A., and McIntosh T.K. Kynurenate is neuroprotective following experimental brain injury in the rat. Brain Res 1994; 655:91-96.

Hicks R.R., Martin V.B., Zhang L., and Seroogy K.B. Mild experimental brain injury differentially alters the expression of neurotrophin and neurotrophin receptor mRNAs in the hippocampus. Exp Neurol 1999; 160:469-478.

Huettner J.E. Indole-2-carboxylic acid: a competitive antagonist of potentiation by glycine at the NMDA receptor. Science 1989; 243:1611-1613.

Ikematsu K., Tsuda R., Kondo T., and Nakasono I. The expression of excitatory amino acid transporter 2 in traumatic brain injury. Forensic Sci Int 2002; 130:83-89.

Ikonomidou C., Stefovska V., and Turski L. Neuronal death enhanced by N-methyl-D-aspartate antagonists. Proc Natl Acad Sci USA 2000; 97:12885-12890.

Katayama Y., Becker D.P., Tamura T., and Hovda D.A. Massive increases in extracellular potassium and the indiscriminate release of glutamate following concussive brain injury. J Neurosurg 1990; 73:889-900.

Kochanek P.M., Hendrich K.S., Dixon C.E., Schiding J.K., Williams D.S., and Ho C. Cerebral blood flow at one year after controlled cortical impact in rats: assessment by magnetic resonance imaging. J Neurotrauma 2002; 19:1029-1037.

Kotapka M.J., Graham D.I., Adams J.H., and Gennarelli T.A. Hippocampal pathology in fatal human head injury without high intracranial pressure. J Neurotrauma 1994; 11:317-324.

Koura S.S., Doppenberg E.M., Marmarou A., Choi S., Young H.F., and Bullock R. Relationship between excitatory amino acid release and outcome after severe human head injury. Acta Neurochir (Suppl) 1998; 71:244-246.

Kroppenstedt S.-N., Schneider G.-H., Thomale U.-W., and Unterberg A.W. Protective effects of aptiganel HCl (Cerestat®) following controlled cortical impact injury in the rat. J Neurotrauma 1998; 15:191-197.

Kumar A., Zou L., Yuan X., Long Y., and Yang K. N-methyl-D-aspartate receptors: transient loss of NR1/NR2A/NR2B subunits after traumatic brain injury in a rodent model. J Neurosci Res 2002; 67:781-786.

Laurer H.L., Lenzlinger P.M., and McIntosh T.K. Models of traumatic brain injury. Eur J Trauma 2000; 26:95-110.

Laurer H.L. and McIntosh T.K. Pharmacologic therapy in traumatic brain injury: update on experimental treatment strategies. Curr Pharm Design 2001; 7:1505-1516.

Lenzlinger P.M., Morganti-Kossmann M.C., Laurer H.L., and McIntosh T.K. The duality of the inflammatory response to traumatic brain injury. Mol Neurobiol 2001; 24:169-181.

Leoni M.J., Chen X.-H., Cheney J.A., McIntosh T.K., and Smith D.H. NPS-1506 attenuates cognitive dysfunction and hippocampal neuron death following brain trauma in the rat. Exp Neurol 2000; 166:442-449.

Levin H.S. Prediction of recovery from traumatic brain injury. J Neurotrauma 1995; 12:913-922.

Lifshitz J., Friberg H., Neumar R.W., Raghupathi R., Welsh F.A., Janmey P., Saatman K.E., Wieloch T., Grady M.S., and McIntosh T.K. Structural and functional damage sustained by mitochondria after traumatic brain injury in the rat: evidence for differentially sensitive populations in the cortex and hippocampus. J Cereb Blood Flow Metab 2003; 23:219-231.

Lyeth B.G., Gong Q.Z., Shields S., Muizelaar J.P., and Berman R.F. Group I metabotropic glutamate antagonist reduces acute neuronal degeneration and behavioral deficits after traumatic brain injury in rats. Exp Neurol 2001; 169:191-199.

MacKenzie J.D., Siddiqi F., Babb J.S., Bagley L.J., Mannon L.J., Sinson G.P., and Grossman R.I. Brain atrophy in mild or moderate traumatic brain injury: a longitudinal quantitative analysis. Am J Neuroradiol 2002; 23:1509-1515.

Marion D.W., Puccio A., Wisniewski S.R., Kochanek P., Dixon C.E., Bullian L., and Carlier P. Effect of hyperventilation on extracellular concentrations of glutamate, lactate, pyruvate, and local cerebral blood flow in patients with severe traumatic brain injury.[comment]. Crit Care Med 2002; 30:2619-2625.

Marklund N., Clausen F., Lewander T., and Hillered L. Monitoring of reactive oxygen species production after traumatic brain injury in rats with microdialysis and the 4-hydroxybenzoic acid trapping method. J Neurotrauma 2001; 18:1217-1227.

Marklund N., Sihver S., Langstrom B., Bergstrom M., and Hillered L. Effect of traumatic brain injury and nitrone radical scavengers on relative changes in regional cerebral blood flow and glucose uptake in rats. J Neurotrauma 2002; 19:1139-1153.

Matute C., Alberdi E., Ibarretxe G., and Sanchez-Gomez M.V. Excitotoxicity in glial cells. Eur J Pharmacol 2002; 447:239-246.

Mayer M.L. and Westbrook G.L. The action of N-methyl-D-aspartic acid on mouse spinal neurones in culture. J Physiol (Lond) 1985; 361:65-90.

Mayhan W.G., Didion J.P.. Glutamate-induced disruption of the blood-brain barrier in rats. Role of nitric oxide. Stroke 1996; 27: 965-969.

McIntosh T.K., Juhler M., and Wieloch T. Novel pharmacologic strategies in the treatment of experimental traumatic brain injury: 1998. J Neurotrauma 1998a; 15:731-769.

McIntosh T.K., Saatman K.E., and Raghupathi R. Calcium and the pathogenesis of traumatic CNS injury: Cellular and molecular mechanisms. The Neuroscientist 1997; 3:169-175.

McIntosh T.K., Saatman K.E., Raghupathi R., Graham D.I., Smith, DH, Lee V.M., and Trojanowski J.Q. The Dorothy Russell Memorial Lecture. The molecular and cellular sequelae of experimental traumatic brain injury: pathogenetic mechanisms. Neuropath Appl Neurobiol 1998b; 24:251-267.

McIntosh T.K., Smith D.H., Meaney D.F., Kotapka M.J., Gennarelli T.A., and Graham D.I. Neuropathological sequelae of experimental brain injury: Relationship to neurochemical and biomechanical mechanisms. Lab Invest 1996; 74:315-342.

McIntosh T.K., Vink R., Yamakami I., and Faden A.I. Magnesium deficiency exacerbates and pretreatment improves outcome following traumatic brain injury in rats. ^{31}P Magnetic resonance spectroscopy and behavioral studies. J Neurotrauma 1988; 5:17-31.

McIntosh T.K., Vink R., Yamakami I., and Faden A.I. Magnesium protects against neurological deficit after brain injury. Brain Res 1989; 482:252-260.

Mignani S., Bohme G.A., Birraux G., Boireau A., Jimonet P., Damour D., Debono M.W., Pratt J., Vuilhorgne M., Wahl F., and Stutzmann J.-M. 9-Caroboxymethyl-5H,10H-imidazo[1,2-a]indeno[1,2-e]pyrazin-4-one-2-carbolic acid (RPR117824): selective anticonvulsive and neuroprotective AMPA antagonist. Bioorg Med Chem 2002; 10:1627-1637.

Miller L.P., Lyeth B.G., Jenkins L.W., Oleniak L., Panchision D., Hamm R., Phillips L.L., Dixon C.E., Clifton G., and Hayes R.L. Excitatory amino acid receptor subtype binding following traumatic brain injury. Brain Res 1990; 526:103-107.

Mitha A.P. and Maynard K.I. Gacyclidine (Beaufour-Ipsen). Curr Opin Invest Drugs 2001; 2:814-819.

Morimoto K., Murasugi T., and Oda T. Acute neuroinflammation exacerbates excitotoxicity in rat hippocampus in vivo. Exp Neurol 2002; 177:95-104.

Movsesyan V.A., O'Leary D.M., Fan L., Bao W., Mullins P.G., Knoblach S.M., and Faden A.I. mGluR5 antagonists 2-methyl-6-(phenylethynyl)-pyridine and (E)-2-methyl-6-(2-phenylethenyl)-pyridine reduce traumatic neuronal injury in vitro and in vivo by antagonizing N-methyl-D-aspartate receptors. J Pharmacol Exp Ther 2001; 296:41-47.

Mukhin A., Fan L., and Faden A.I. Activation of metabotropic glutamate receptor subtype mGluR1 contributes to post-traumatic neuronal injury. J Neurosci 1996; 16:6012-6020.

Nakanishi S. Metabotropic glutamate receptors: synaptic transmission, modulation, and plasticity. Neuron 1994; 13:1031-1037.

Narayan R.K., Michel M.E., Ansell B., Baethmann A., Biegon A., Bracken M.B., Bullock M.R., Choi S.C., Clifton G.L., Contant C.F., Coplin W.M., Dietrich W.D., Ghajar J., Grady S.M., Grossman R.G., Hall E.D., Heetderks W., Hovda D.A., Jallo J., Katz R.L., Knoller N., Kochanek P.M., Maas A.I., Majde J. , Marion D.W., Marmarou A., Marshall L.F., McIntosh T.K., Miller E., Mohberg N., Muizelaar J.P., Pitts L.H., Quinn P., Riesenfeld G., Robertson C.S., Strauss K.I., Teasdale G., Temkin N., Tuma R., Wade C., Walker M.D., Weinrich M., Whyte J., Wilberger J., Young A.B., and Yurkewicz L. Clinical trials in head injury. J Neurotrauma 2002; 19:503-557.

Nilsson B., Hillered L., Ponten U., and Ungerstedt U. Changes in cortical extracellular levels of energy-related metabolites and amino acids following concussive brain injury in rats. J Cereb Blood Flow and Metab 1990; 10:631-637.

Nilsson P., Hillered L., Olsson Y., Sheardown M., and Hansen A. Regional changes in interstitial K^+ and Ca^{2+} levels following cortical compression contusion trauma in rats. J Cereb Blood Flow Metab 1993; 13:183-192.

Nilsson P., Ronne-Engstrom E., Flink R., Ungerstedt U., Carlson H., and Hillered L. Epileptic seizure activity in the acute phase following cortical impact trauma in rat. Brain Res 1994; 637:227-232.

Nowak L., Bregestovski P., Ascher P., Herbelt A., and Prochiantz A. Magnesium gates glutamate-activated channels in mouse central neurons. Nature 1984; 307:462-465.

O'Dell D.M., Raghupathi R., Crino P.B., Eberwine J.H., and McIntosh T.K. Traumatic brain injury alters the molecular fingerprint of TUNEL-positive cortical neurons *in vivo*: A single-cell analysis. J Neurosci 2000; 20:4821-4828.

Obrenovitch T.P. and Urenjak J. Is high extracellular glutamate the key to excitotoxicity in traumatic brain injury? J Neurotrauma 1997; 14:677-698.

Okiyama K., Smith D.H., Gennarelli T.A., Simon R.P., Leach M., and McIntosh T.K. The sodium channel blocker and glutamate release inhibitor BW1003C87 and magnesium attenuate regional cerebral edema following experimental brain injury in the rat. J Neurochem 1995; 64(2):802-809.

Okiyama K., Smith D.H., White W.F., Richter K., and McIntosh T.K. Effects of the novel NMDA antagonists CP-98,113, CP-101,581, and CP-101,606 on cognitive function and regional cerebral edema following experimental brain injury in the rat. J Neurotrauma 1997; 14:211-222.

Olney J.W., Labruyere J., and Price M. Pathological changes induced in cerebrocortical neurons by phencyclidine and related drugs. Science 1989; 244:1360-1362.

Palmer A.M., Marion D.W., Botscheller M.L., Bowen D.M., and DeKosky S.T. Increased transmitter amino acid concentration in human ventricular CSF after brain trauma. NeuroReport 1994; 6:153-156.

Palmer A.M., Marion D.W., Botscheller M.L., Swedlow P.E., Styren S.D., and DeKosky S.T. Traumatic brain injury-induced excitotoxicity assessed in a controlled cortical impact model. J Neurochem 1993; 61:2015-2024.

Parsons C.G., Hartmann S., and Spielmanns P. Budipine is a low affinity, N-methyl-D-aspartate receptor antagonist: patch clamp studies in cultured striatal, hippocampal, cortical and superior colliculus neurones. Neuropharmacol 1998; 37:719-727.

Persson L. and Hillered L. Chemical monitoring of neurosurgical intensive care patients using intracerebral microdialysis. J Neurosurg 1992; 76:72-80.

Povlishock J.T. and Christman C.W. The pathobiology of traumatically induced axonal injury in animals and humans: a review of current thoughts. J Neurotrauma 1995; 12:555-564.

Rao V.L., Dogan A., Todd K.G., Bowen K.K., and Dempsey R.J. Neuroprotection by memantine, a non-competitive NMDA receptor antagonist after traumatic brain injury in rats. Brain Res 2001; 911:96-100.

Rao V.L., Kaya M.K., an A., Rothstein J.D., and Dempsey R.J. Traumatic brain injury down-regulates glial glutamate transporter (GLT-1 and GLAST) proteins in rat brain. J Neurochem 1998; 70:2020-2027.

Rink A.D., Fung K.-M., Trojanowski J.Q., Lee V., Neugebauer E., and McIntosh T.K. Evidence of apoptotic cell death after experimental traumatic brain injury in the rat. Am J Pathol 1995; 147:1575-1583.

Rocamora N., Massieu L., Boddeke H.W., Palacios J.M., and Mengod G. Differential regulation of the expression of nerve growth factor, brain-derived neurotrophic factor and neurotrophin-3 mRNAs in adult rat brain after intrahippocampal injection of quinolinic acid. Mol Brain Res 1994; 26:89-98.

Rose M.E., Huerbin M.B., Melick J., Marion D.W., Palmer A.M., Schiding J.K., Kochanek P.M., and Graham S.H. Regulation of interstitial excitatory amino acid concentrations after cortical contusion injury. Brain Res 2002; 943:15-22.

Saatman K.E., Bareyre F.M., Grady M.S., and McIntosh T.K. Acute cytoskeletal alterations and cell death induced by experimental brain injury are attenuated by magnesium treatment and excerbated by magnesium deficiency. J Neuropathol Exp Neurol 2001; 60:183-194.

Sattler R., Xiong Z., Lu W.Y., Hafner M., MacDonald J.F., and Tymianski M. Specific coupling of NMDA receptor activation to nitric oxide neurotoxicity by PSD-95 protein. Science 1999; 284:1845-1848.

Shohami E., Novikov M., and Bass R. Long-term effect of HU-211, a novel non-competitive NMDA antagonist, on motor and memory functions after closed head injury in the rat. Brain Res 1995; 674:55-62.

Shohami E., Novikov M., and Mechoulam R. A nonpsychotropic cannabinoid, HU-211, has cerebroprotective effects after closed head injury in the rat. J Neurotrauma 1993; 10:109-119.

Sihver S., Marklund N., Hillered L., Langstrom B., Watanabe Y., and Bergstrom M. Changes in mACh, NMDA and GABA(A) receptor binding after lateral fluid-percussion injury: in vitro autoradiography of rat brain frozen sections. J Neurochem 2001; 78:417-423.

Smith D.H., Chen X.-H., Pierce J.E.S., Wolf J.A., Trojanowski J.Q., Graham D.I., and McIntosh T.K. Progressive atrophy and neuronal death for one year following brain trauma in the rat. J Neurotrauma 1997; 14:715-727.

Smith D.H., Lowenstein D.H., Gennarelli T.A., and McIntosh T.K. Persistent memory dysfunction is associated with bilateral hippocampal damage following experimental brain injury. Neurosci Lett 1994; 168:151-154.

Smith D.H., Okiyama K., Gennarelli T.A., and McIntosh T.K. Magnesium and ketamine attenuate cognitive dysfunction following experimental brain injury. Neurosci Lett 1993a; 157:211-214.

Smith D.H., Okiyama K., Thomas M.J., and McIntosh T.K. Effects of the excitatory amino acid receptor antagonists kynurenate and indole-2-carboxylic acid on behavioral and neurochemical outcome following experimental brain injury. J Neurosci 1993b; 13:5383-5392.

Soares H.D., Hicks R.R., Smith D.H., and McIntosh T.K. Inflammatory leukocytic recruitment and diffuse neuronal degeneration are separate pathological processes resulting from traumatic brain injury . J Neurosci 1995; 15:8223-8233.

Soares H.D., Thomas M.J., Cloherty K., and McIntosh T.K. Development of prolonged focal cerebral edema and regional cation change following experimental brain injury in the rat. J Neurochem 1992; 58:1845-1852.

Stoffel M., Plesnila S., Eriskat J., Furst M., and Baethmann A. Release of excitatory amino acids in the penumbra of a focal cortical necrosis. J Neurotrauma 2000; 4:467-477.

Storm-Mathisen J., Danbolt N.C., Rothe F., Torp R., Zhang N., Aas J.E., Kanner B.I., Langmoen I., and Ottersen O.P. Ultrastructural immunocytochemical observations on the localization, metabolism and transport of glutamate in normal and ischemic brain tissue. Prog Brain Res 1992; 94:225-241.

Stover J.F., Beyer T.F., and Unterberg A.W. Riluzole reduces brain swelling and contusion volume in rats following controlled cortical impact injury. J Neurotrauma 2000; 17:1171-1178.

Stover J.F., Sakowitz O.W., Beyer T.F., Dohse N.K., Kroppenstedt S.-N., Thomale U.-W., Schaser K.-D., and Unterberg A.W. Effects of LY379268, a selective group II metabotropic glutamate receptor agonist on EEG activity, cortical perfusion, tissue damage, and cortical glutamate, glucose, and lactate levels in brain-injured rats. J Neurotrauma 2003; in press.

Stover J.F. and Unterberg A.W. Increased cerebrospinal fluid glutamate and taurine concentrations are associated with traumatic brain edema formation in rats. Brain Res 2000; 875:51-55.

Suh S.W., Chen J.W., Motamedi M., Bell B., Listiak K., Pons N.F., Danscher G., and Frederickson C.J. Evidence that synaptically-released zinc contributes to neuronal injury after traumatic brain injury. Brain Res 2000; 852:268-273.

Sun F.-Y. and Faden A.I. *N*-Methyl-D-aspartate receptors mediate post-traumatic increases of protein kinase C in rat brain. Brain Res 1995a; 661:63-69.

Sun F.-Y. and Faden A.I. Neuroprotective effects of 619C89, a use-dependent sodium channel blocker, in rat traumatic brain injury. Brain Res 1995b; 673:133-140.

Tanaka K., Watase K., Manabe T., Yamada K., Watanabe M., Takahashi K., Iwama H., Nishikawa T., Ichihara N. , Kikuchi T., Okuyama S., Kawashima N., Hori S., Takimoto M., and Wada K. Epilepsy and exacerbation of brain injury in mice lacking the glutamate transporter GLT-1. Science 1997; 276:1699-1702.

Teasdale GM and Bannan PE. 1997. Neuroprotection in head injury. In: Reilly P and Bullock R, editors. Head Injury. London: Chapman & Hall. p 423-438.

Temple M.D. and Hamm R.J. Chronic, postinjury administration of D-cycloserine, an NMDA partial agonist, enhances cognitive performance following experimental brain injury. Brain Res 1996; 741:246-251.

Tenneti L. and Lipton S.A. Involvement of activated caspase-3-like proteases in N-Methyl-D-Aspartate-induced apoptosis in cerebrocortical neurons. J Neurochem 2000; 74:134-142.

Toulmond S., Serrano A., Benavides J., and Scatton B. Prevention by eliprodil (SL 82.0715) of traumatic brain damage in the rat. Existence of a large (18 h) therapeutic window. Brain Res 1993; 620:32-41.

Tsuchida E. and Bullock R. The effect of the glycine site-specific N-methyl-D-aspartate antagonist ACEA1021 on ischemic brain damage caused by acute subdural hematoma in the rat. J Neurotrauma 1995; 12:279-288.

Tsuchida E., Rice M., and Bullock R. The neuroprotective effect of the forebrain-selective NMDA antagonist CP101,606 upon focal ischemic brain damage caused by acute subdural hematoma in the rat. J Neurotrauma 1997; 14:409-417.

Uchida K., Nakakimura K., Kuroda Y., Haranishi Y., Matsumoto M., and Sakabe T. Dizocilpine but not ketamine reduces the volume of ischaemic damage after acute subdural haematoma in the rat. Eur J Anaesthesiol 2001; 18:295-302.

van Landeghem F.K., Stover J.F., Bechmann I., Bruck W., Unterberg A., Buhrer C., and Von Deimling A. Early expression of glutamate transporter proteins in ramified microglia after controlled cortical impact injury in the rat. GLIA 2001; 35:167-179.

Vespa P., Prins M., Ronne-Engstrom E., Caron M., Shalmon E., Hovda D.A., Martin N.A., and Becker D.P. Increase in extracellular glutamate caused by reduced cerebral perfusion pressure and seizures after human traumatic brain injury: a microdialysis study. J Neurosurg 1998; 89:971-982.

Vink R., Heath D.L., and McIntosh T.K. Acute and prolonged alterations in brain free magnesium following fluid-percussion induced brain trauma in rats. J Neurochem 1996; 66:2477-2483.

Vink R., McIntosh T.K., Demediuk P., and Faden A.I. Decrease in total and free magnesium concentration following traumatic brain injury in rats. Biochem Biophys Res Commun 1987; 149:594-599.

Vink R., O'Connor C.A., Nimmo A.J., and Heath D.L. Magnesium attenuates persistent functional deficits following diffuse traumatic brain injury in rats. Neurosci Lett 2003; 336:41-44.

Vornov J.J., Tasker R.C., and Coyle J.T. Direct observation of the agonist-specific regional vulnerability to glutamate, NMDA, and kainate neurotoxicity in organotypic hippocampal cultures. Exp Neurol 1991; 114:11-22.

Wahl F., Renou E., Mary V., and Stutzmann J.-M. Riluzole reduces brain lesions and improves neurological function in rats after a traumatic brain injury. Brain Res 1997; 756:247-255.

Watanabe M., Fukaya M., Sakimura K., Manabe T., Mishina M., and Inoue Y. Selective scarcity of NMDA receptor channel subunits in the stratum lucidum (mossy fibre-recipient layer) of the mouse hippocampal CA3 subfield. Eur J Neurosci 1998; 10:478-487.

Watase K., Hashimoto K., Kano M., Yamada K., Watanabe M., Inoue Y., Okuyama S., Sakagawa T., Ogawa S. , Kawashima N., Hori S., Takimoto M., Wada K., and Tanaka K. Motor discoordination and increased susceptibility to cerebellar injury in GLAST mutant mice. Eur J Neurosci 1998; 10:976-988.

Weiss J.H., Sensi S.L., and Koh J.Y. Zn(2+): a novel ionic mediator of neural injury in brain disease. Trends Pharmacol Sci 2000; 21:395-401.

Yakovlev A.G., Knoblach S., Fan L., Fox G.B., Goodnight R., and Faden A.I. Activation of CPP32-like caspases contributes to neuronal apoptosis and neurological dysfunction after traumatic brain injury. J Neurosci 1997; 17:7415-7424.

Yeiser E.C., Vanlandingham J.W., and Levenson C.W. Moderate zinc deficiency increases cell death after brain injury in the rat. Nutri Neurosci 2002; 5:345-352.

Zwienenberg M., Gong Q.Z., Berman R.F., Muizelaar J.P., and Lyeth B.G. The effect of groups II and III metabotropic glutamate receptor activation on neuronal injury in a rodent model of traumatic brain injury. Neurosurgery 2001; 48:1119-1126.

Chapter 10
EXCITOTOXICITY IN AMYOTROPHIC LATERAL SCLEROSIS: SELECTIVE VULNERABILITY OF MOTOR NEURONS

Tiziana Mennini and Caterina Bendotti
Istituto di Ricerche Framacologiche Mario Negri, Via Eritrea 62, 20157 Milano, Italy.

Abstract: Excessive glutamate concentration is toxic to motor neurons *in vitro*, thus suggesting the possibility that glutamate excitotoxicity plays a role in motor neuron degeneration in Amyotrophic Lateral Sclerosis. Indeed high glutamate levels are found in the cerebrospinal fluid of about 40% of patients, possibly related to the decrese in glutamate uptake and EAAT2 transporter. Studies *in vitro* and in animal models of motor neuron degeneration suggest that the toxicity of glutamate to motor neurons is principally mediated by the AMPA receptor subtype, which may enhance the calcium influx due to changes in density and/or editing of their GluR2 subunit. Competitive and non-competitive AMPA receptor antagonists are effective *in vitro* on cultured motor neurons and *in vivo* in animal models of amyotrophic lateral sclerosis. Results from ongoing and future clinical trials will give definitive answer to the relevance of AMPA-mediated excitotoxicity in patients with amyotrophic lateral sclerosis.

Keywords: amyotrophic lateral sclerosis, excitotoxicity, AMPA receptors, EAAT2, calcium influx.

INDEX

1. Introduction.
2. Studies *in vitro*.
3. Studies *in vivo*.
4. Conclusion.

1. Introduction.

Glutamate excitotoxicity has been demonstrated to play a crucial role in amyotrophic lateral sclerosis (ALS). Increased levels of glutamate have been reported in the cerebrospinal fluid of about 40% of sporadic ALS patients (Spreux-Varoquaux et al., 2002) which may be the consequence of decreased reuptake of the amino acid found in synaptosomes obtained from autoptic samples (Rothstein et al., 1992). A pronounced loss of the astroglial glutamate transporter (EAAT2) protein was also found in the affected regions of ALS patients (Rothstein et al., 1995). However there is no quantitative change in mRNA for glutamate transporters EAAT1, EAAT2, or EAAT3 in ALS motor cortex, even in patients with a large loss of EAAT2 protein and decreased tissue glutamate transport (Bristol and Rothstein, 1996), suggesting that the dramatic abnormalities in EAAT2 may be due to translational or post-translational processes.

In addition, it has been demonstrated that EAAT2 immunoreactivity in anterior horns correlated with the degree of neuronal loss of anterior horn cells (Sasaki et al., 2000), thus it is difficult to understand if the alteration in glutamate level and uptake observed in ALS patients are a consequence or a cause of the disease.

Alteration of glutamate transport in ALS patients is not limited to the CNS: a 43% reduction of high-affinity glutamate uptake was reported in platelets from ALS patients, compared with normal controls and patients with chronic neurological disorders. No correlation was found between platelet glutamate uptake and sex, age, bulbar versus spinal onset, duration or severity of disease (Ferrarese et al., 2001), indicating that glutamate uptake impairment may be an early phenomenon, which can be considered a "trait", rather than a "state" marker of the disease.

Thus different research groups have attempted to provide further insight as to the possible pathogenetic role of glutamate and its receptors in ALS. Here we summarize the main evidence obtained *in vitro* on cultured cells and *in vivo*, using animal models of motor neuron degeneration.

2. Studies *in vitro*.

The cerebrospinal fluid from ALS patients is toxic to neurons in primary culture and this toxicity is blocked by the non-selective AMPA/kainate receptor antagonist, CNQX, but not by NMDA receptor antagonists (Couratier et al., 1993). Although this study did not utilize the neurons target of the disease, i.e. motor neurons, the data indirectly support the role of excitotoxicity in ALS and point to the AMPA/kainate receptors as key mediators.

The AMPA receptors of the ventral spinal cord of ALS patients show changes in GluR2 mRNA editing that may enhance the calcium influx, promoting neuronal vulnerability (Takuma et al., 1999). In line with this finding, spinal motoneuron cultures are particularly sensitive to glutamate excitotoxicity exerted through the AMPA receptor subtype *in vitro* (Carriedo et al., 1996; Comoletti et al., 2001). GYKI 53655, a non-competitive AMPA receptor antagonist, inhibited the vulnerability of cultured rat spinal motor neurons to AMPA/kainate by reducing the AMPA receptor current density (Vandenberghe et al., 2000). This study also demonstrated that glutamate triggered a 2.5 time larger AMPA receptor-mediated Ca^{2+} influx in motor neurons than in dorsal horn neurons, suggesting that a high density of functional AMPA receptors in motor neurons may account for their selective vulnerability (Vandenberghe et al., 2000).

Increased function of the AMPA receptor may also depend on their subunit composition in the motor neurons. A lack of GluR2 subunit increases the permeability of the neuron to Ca^{2+} (Pellegrini-Giampietro et al., 1997) and cultured spinal motor neurons contain both GluR2-immunopositive and GluR2-immunonegative AMPA receptor clusters, the latter begin a majority in the dendrites where their activation might greatly amplify the local Ca^{2+} signal (Vandenberghe et al., 2001). The comparison of cultured rat spinal cord motor neurons with dorsal horn neurons (which are resistant to excitotoxicity) and with motor neurons that survived an excitotoxic insult confirmed that motor neurons are relatively deficient in GluR2. The AMPA receptor properties correlated well with the selective vulnerability of motor neurons because motor neurons surviving an excitotoxic event had similar characteristics as dorsal horn neurons. These data indicate that the relative abundance of GluR2 in functional AMPA receptors may be a major determinant of the selective vulnerability of motor neurons to excitotoxicity in vitro (Van Damme et al., 2002).

It is of interest to consider that the type of motor neurons death induced by excitotoxins *in vitro* varies in function of concentration, time of exposure and cellular model. For instance, using the same experimental protocol, consisting of mixed motor neuron and glia cultures, exposed for 48 h to glutamate agonists concentration inducing about 50% cell death, two different death pathways are activated, namely the apoptotic pathway induced by kainic acid and the necrotic pathway induced by AMPA and NMDA (Comoletti et al., 2001; Mennini et al., 2003).

No clear explanation can be obtained about the reasons for which, using equipotent low concentration of the three glutamate agonists, only kainic acid induces DNA fragmentation and caspase activation typical of apoptosis (Mennini et al., 2003a). An important difference may be that AMPA (like glutamate) is a strongly desensitizing receptor agonist, while kainate is a weakly desensitizing AMPA receptor agonist (Vandenberghe et al., 2000). In addition, long exposure to kainate affected motor neurons through both Ca (2+)-permeable and Ca (2+)-impermeable AMPA receptors, while short kainate exposures selectively affected motor neurons containing Ca (2+)-permeable AMPA receptors (Vanderberghe et al., 2000).

Another point to be considered is that the sensitivity of cultured motor neurons to excitotoxins is greatly enhanced by the presence of glial cell in co-culture, by a mechanism involving nitric oxide release (Comoletti et al., 2001) (Figure 1).

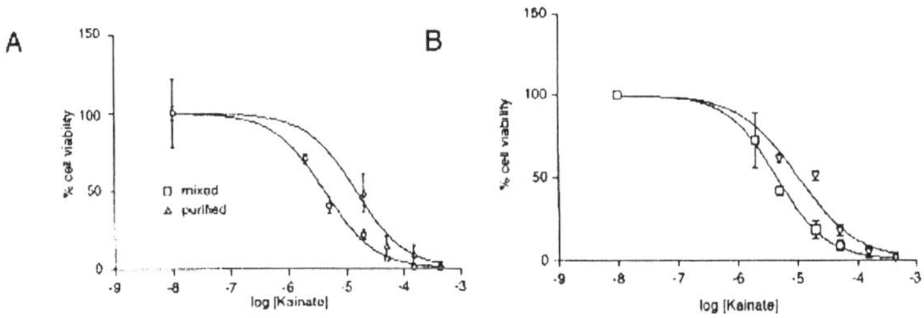

Figure 1. Kainate toxicity in primary motor neuron cultures. (A) Percent viabilitry of SMI-32 positive motor neurons is plotted against kainate concentrations in mixed neuron/glia culture (open square) or purified cultures (open triandgles). (B) Mixed neuron/glia cultures were exposed for 48 hours to different concentrations of kainate in the absence (open square) or presence (open triangle) of 40 μM l-NAME. Each point is the mean ± S.D. of three to four replicates. Non linear fitting of the dose-response curves was obtained according to the logistic equation. Adapted from Comoletti et al., 2001.

These results indicate that ALS is not a cell autonomous disease, and that glial cell could actively participate to the pathology progression. Indeed microglia, which is activated in the spinal cord of ALS patients (McGeer and McGeer, 2002) also expresses AMPA receptors whose activation induces release of tumor necrosis factor α (Noda et al., 2000), and this might contribute to the potentiation of glutamate-mediated neural death induced by this cytokine (Hermann et al., 2001).

Motoneuronal cell lines, like NSC-34, are also utilized for studies *in vitro,* however they are not a useful tool for excitotoxicity (Eggett et al., 2000), since they are relatively resistant to the toxicity induced by NMDA, kainate or glutamate, although immunostaining with NR1 and GluR2/3 antibodies indicates the expression of these NMDA and AMPA/kainate receptor subunits (Rizzardini et al., 2003).

3. Studies *in vivo.*

As discussed in the introduction, the loss of EAAT2 in patients could be consequence of neuronal death either direclty, due to the localization of the transporter on motor neurons (Sasaki et al., 2000), or indirectly, since the expression of glial glutamate transporters is regulated by neuronal activity, and the glial transporters expression in primary astrocyte-neuron cocultures undergo down regulation following glutamate induced neuronal death (Perego et al., 2000). To understand if the loss of EAAT2 could be considered a cause or a consequence of the motor neuron degeneration in ALS, time-course studies have been done in transgenic mice carrying a mutation in G93A of Cu/Zn-dependent superoxide dismutase (SOD1G93A), typical of familial ALS. This provides a reliable model of ALS showing progressive muscular weakness associated with spinal motor neuron loss and leading to early death (Gurney, 1994). A significant decrease in the level of the glial glutamate transporter (GLT1/EAAT2), but not its mRNA, has been reported in the ventral horn of spinal cord from SOD1G93A mice only at the advanced stage of the disease in presence of reactive astrocytosis (Bendotti et al., 2001) thus suggesting that this could be a consequence rather than a cause of the neuronal loss (Figure 2).

In the wobbler mice, spontaneous mutants showing early-life selective motor neuron degeneration in the cervical spinal cord (Duchen and Strich, 1968), no changes in the level of astroglial transporters in the cervical spinal cord has been reported, even in the presence of consistent motor neuron degeneration and reactive astrocytosis (Bigini et al., 2001). It is possible that in this animal adaptive decrease in GLT1 could not be seen due to the early and rapid pathology progression.

Conversely a decrease of GLT1, paralleled by increased plasma glutamate concentration was found already at the pre-symptomatic stage in the spinal cord of *mnd* mice (Mennini et al., 1998), in the absence of clear motor neurons degeneration (Mennini et al., 2002). The *mnd* mice show a late onset of progressive paralysis of the hind limb, correlated with altered innervation and reduced nerve-evoked isometric twitch tension in the extensor digitorum longus (EDL) muscle (Gorio et al., 1999). Due to the heavy accumulation of lipofuscin in almost all neurons since early age, these mice have been proposed as a model of lipofuscinosis (Bronson et al., 1998). Thus, the relation between GLT1 decrease and motor neuron degeneration needs to be further clarified.

Recently Howland et al. (2002), using transgenic rats expressing high levels of mutated SOD1, demonstrated an early decrease of GLT1 well before neuronal damage, suggesting that GLT1 expression could be regulated by factors, like oxidative stress, which has been widely demonstrated in ALS patients (Shaw, 1999). The function of GLT1 is in fact impaired by mutant SOD1, in presence of free radicals (Trotti et al., 1999). In line with this possibility reduced glutamate uptake has been reported in human neuroblastoma cell lines transfected with FALS-SOD1 mutant G93A (Ferrarese and et al., 2001), and down-regulation of GLT1 but not GLAST or the neuronal EAAC1 is observed in a cellular model of ALS expressing high levels of the G93A mutant of SOD1 (Pietrini G., personal communication).

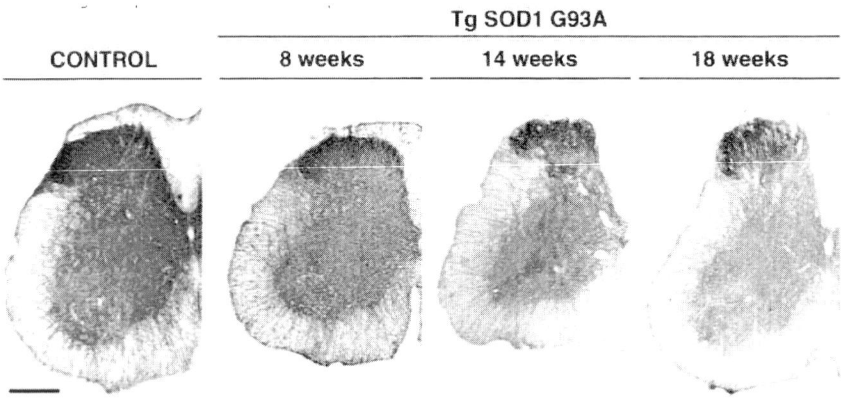

Figure 2. GLT1 immunoreactivity in representative semisections of the lumbar spinal cord of mice. In non-transgenic mice GLT1 immunostaining is mainly distributed throught the grey matter with the highest signal shown at the level of the substantia gelatinosa of dorsal horn. In TgSOD1 G93A mice a gradual decrease of GLT1 immunostaining is seen at 14 and 18 weeks of age in the ventral and intermediate region of the grey matter, while the dorsal horn maintains high immunostaining. Scale bar= 250 μm. Modified from Bendotti et al., 2001.

As consequence of the increased extracellular glutamate levels, activation of glutamate receptors play a potential role in the degeneration of motor neurons. These neurons in mice and human express different subtypes of glutamate receptors (Shaw et al., 1994; Mennini et al., 2002) although, as mentioned above, studies *in vitro* have demonstrated that they are more sensitive to AMPA receptor stimulation. This prompt different groups to test the effect of treatment with glutamate receptor antagonists, in particular AMPA antagonist, on the disease progression and motor neuron degeneration in animal models.

Competitive AMPA antagonists like NBQX and RPR119990 were shown to protect SOD1G93A mice against deterioration of motor functions and to increase their survival (Canton et al., 2001; Van Damme et al., 2003). A similar result was found in the same animal model with the non competitive AMPA inhibitor ZK187638, that has the advantage, in respect to the previous ones of a good availability in the central nervous system after oral administration. Repeated administration of this drug after the first manifestation of motor symptoms significantly improved the motor behavioral impairment, showed protection of motor neurons and increase the survival of SOD1G93A mice (Bendotti et al., 2002; Mennini et al., 2003b). This is an important result from the therapeutic point of view. In fact most of the treatments tested in SOD1 mutant mice so far have been started before the appearance of symptoms, a situation impracticable in patients with ALS. This is likely the main reason for the discrepancy in the efficacy of a drug therapy between mice and patients.

The effect of the AMPA antagonists treatments on the survival of SOD1 mutant mice is comparable to that reported for riluzole which is the only currently licensed drug that affords some clinical benefit to ALS patients. However, unlike the AMPA antagonist, riluzole was not effective on the onset of the disease (Gurney et al., 1996). This is in line with the observation in ALS patients.

Riluzole was also found to be significantly effective in ameliorating the motor dysfunction and in increasing motor neuron number and choline acetyltransferase activity in the spinal cord of wobbler mice (Mennini et al., 2003c). This was somewhat unexpected, since, as already mentionned, the glial glutamate transporters are unchanged in the spinal cord of these mice suggesting a mechanism of motor neurodegeneration independent from excitotoxicity (Bigini et al., 2001). Consistently, treatment of wobbler mice with RPR119990 failed to improve motor deficits (Fumagalli et al., 2003). Although glutamate receptor expression and sensitivity in motor neurons of wobbler mice is still under characterization, these studies open new perspective about the mechanism of riluzole protection in motor neuron diseases.

Increased levels of spinal GluR2 receptor subtypes and decreased GluR3 in motor neurons have been reported in *mnd* mice (Mennini et al., 2002), which were also responsive to antiglutamatergic drugs. Indeed treatment of *mnd* mice with the competitive AMPA agonist NBQX

(Mennini et al., 1999) or a novel non competitive AMPA inhibitor, ZK 187638 (Bendotti et al., 2002; Mennini et al., 2003b), results in amelioration of motor symptoms.

4. Conclusion.

As summarized in Table 1, treatments with competitive or non-competitive AMPA receptor antagonists improve motor behavior in SOD1G93A and in *mnd* mice, suggesting that glutamate excitotoxicity mediated by AMPA receptor subtype plays a role in motor neuron degeneration, and confirming evidence form *in vitro* studies on cultured motor neuron.

Table 1. Effect of antiglutamatergic drugs in animal models of motor neuron diseases.

Drug	Animal model	Effect	Reference
NBQX	*mnd*	protection	Mennini et al., 1999
NBQX	SOD1G93A	protection	Van Damme et al., 2003
RPR119990	SOD1G93A	protection	Canton et al., 2001
RPR119990	wobbler	not active	Fumagalli et al., 2003
ZK 187638	*mnd*	protection	Mennini et al, 2003b
ZK 187638	SOD1G93A	protection	Bendotti et al., 2002
Riluzole	SOD1G93A	protection	Gurney et al., 1996
Riluzole	wobbler	protection	Mennini et al., 2003c

It is however still not clear if excitotoxicity is a primary cause of motor neuron degeneration in ALS or represents a late event contributing to the pathology progression, but generated by other primary insults like oxidative stress. Results from ongoing and future clinical trials will give definitive answer to the relevance of AMPA-mediated excitotoxicity in patients with amyotrophic lateral sclerosis.

REFERENCES

Bendotti, C., Tortarolo, M., Suchak, S.K., Calvaresi, N., Carvelli, L., Bastone, A., Rizzi, M., Rattray, M. ,Mennini, T. Transgenic SOD1 G93A mice develop reduced GLT-1 in spinal cord without alterations in cerebrospinal fluid glutamate levels. J Neurochem 2001. 79: 737-746.

Bigini, P., Bastone, A. ,Mennini, T. Glutamate transporters in the spinal cord of the wobbler mouse. Neuroreport 2001. 12: 1815-1820.

Bristol, L.A. ,Rothstein, J.D. Glutamate transporter gene expression in amyotrophic lateral sclerosis motor cortex. Ann Neurol 1996. 39: 676-679.

Bronson, R.T., Donahue, L.R., Johnson, K.R., Tanner, A., Lane, P.W. ,Faust, J.R. Neuronal ceroid lipofuscinosis (nclf), a new disorder of the mouse linked to chromosome 9. Am J Med Gen 1998. 77: 289-297.

Canton, T., Bohme, G.A., Boireau, A., Bordier, F., Mignani, S., Jimonet, P., Jahn, G., Alavijeh, M., Stygall, J., Roberts, S., Brealey, C., Vuilhorgne, M., Debono, M.-W., Le Guern, S., Laville, M., Briet, D., Roux, M., Stutzmann, J.-M. ,Pratt, J.

RPR 119990, a Novel alpha -Amino-3-hydroxy-5-methyl-4-isoxazolepropionic Acid Antagonist: Synthesis, Pharmacological Properties, and Activity in an Animal Model of Amyotrophic Lateral Sclerosis. J Pharmacol Exp Ther 2001. 299: 314-322.

Carriedo, S.G., Yin, H.Z. ,Weiss, J.H. Motor neurons are selectively vulnerable to AMPA/kainate receptor-mediated injury in vitro. J Neurosci 1996. 16: 4069-4079.

Comoletti, D., Muzio, V., Capobianco, A., Ravizza, T. ,Mennini, T. Nitric oxide produced by non-motoneuron cells enhances rat embryonic motoneuron sensitivity to excitotoxins: comparison in mixed neuron/glia or purified cultures. J Neurol Sci 2001. 192: 61-69.

Couratier, P., Hugon, J., Sindou, P., Vallat, J.M. ,Dumas, M. Cell culture evidence for neuronal degeneration in amyotrophic lateral sclerosis being linked to glutamate AMPA/kainate receptors. Lancet 1993. 341: 265-268.

Duchen, L.W. ,Strich, S.J. An hereditary motor neurone disease with progressive denervation of muscle in the mouse: the mutant 'wobbler'. JNeurol Neurosurg Psychiatry 1968. 31: 535-542.

Eggett, C.J., Crosier, S., Manning, P., Cookson, M.R., Menzies, F.M., McNeil, C.J. ,Shaw, P.J. Development and characterisation of a glutamate-sensitive motor neurone cell line. J Neurochem 2000. 74: 1895-1902.

Ferrarese, C., Sala, G., Ceresa, C., Tremolizzo, L., Ferri, A., Carrì, M.T. Interactions between oxidative stress and excitotoxicity in a cellular model of familial amyotrophic lateral sclerosis. Amyotroph Lateral Scler Other Motor Neuron Disord 2001. 2, suppl. 2: 146-147.

Ferrarese, C., Sala, G., Riva, R., Begni, B., Zoia, C., Tremolizzo, L., Galimberti, G., Millul, A., Bastone, A., Mennini, T., Balzarini, C., Frattola, L. ,Beghi, E. Decreased platelet glutamate uptake in patients with amyotrophic lateral sclerosis. Neurol 2001. 56: 270-272.

Fumagalli, E., Bigini, P. ,Mennini, T. Effect of the competitive AMPA antagonists RPR 119990 on motor neuron degenration in wobbler mice..Proceedings of the congress on: "Molecular mechanisms of neurodegeneration: from triplet repeat expansions to protein toxicity". Milano, Italy, 2-4 May 2003 2003.

Gorio, A., Germani, E., Lesma, E., Rossoni, G., Muller, E.E. ,Di Giulio, A.M. Long-term neuroprotective effects of glycosaminoglycans-IGF-I cotreatment in the motor neuron degeneration (mnd) mutant mouse. Eur J Neurosci 1999. 11: 3395-3404.

Gurney, M.E. Transgenic-mouse model of amyotrophic lateral sclerosis. N Engl J Med 1994. 331: 1721-1722.

Gurney, M.E., Cutting, F.B., Zhai, P., Doble, A., Taylor, C.P., Andrus, P.K. ,Hall, E.D.

Benefit of vitamin E, riluzole, and gabapentin in a transgenic model of familial amyotrophic lateral sclerosis. Ann Neurol 1996. 39: 147-157.

Hermann, G.E., Rogers, R.C., Bresnahan, J.C. ,Beattie, M.S. Tumor necrosis factor-alpha induces cFOS and strongly potentiates glutamate-mediated cell death in the rat spinal cord. Neurobiol Dis 2001. 8: 590-599.

Howland, D.S., Liu, J., She, Y., Goad, B., Maragakis, N.J., Kim, B., Erickson, J., Kulik, J., DeVito, L., Psaltis, G., DeGennaro, L.J., Cleveland, D.W. ,Rothstein, J.D. Focal loss of the glutamate transporter EAAT2 in a transgenic rat model of SOD1 mutant-mediated amyotrophic lateral sclerosis (ALS).Proc Natl Acad Sci U S A 2002. 99: 1604-1609.

McGeer, P.L. ,McGeer, E.G. Inflammatory processes in amyotrophic lateral sclerosis. Muscle Nerve 2002. 26: 459-470.

Mennini, T., Bastone, A., Crespi, D., Comoletti, D. ,Manzoni, C. Spinal cord GLT-1 glutamate transporter and blood glutamic acid alterations in motor neuron degeneration (Mnd) mice. J Neurol Sci 1998. 157: 31-36.

Mennini, T, Bigini, P, Bonomi, M, Calvaresi, N, Carvelli, L, Colovic, M, Fracasso, C, Grignaschi, G, Guiso, G, Totrarolo, M, Elger, B, Seilheimer, B, Caccia, S, Bendotti, C. The non-competitive AMPA receptor antagonist ZK 187638 attenuates motor behavior impairment and increases the survival in mouse models of motor neuron disorders. Amyotroph Lateral Scler Other Motor Neuron Disord. 2003b, submitted

Mennini, T., Bigini, P., Clerici, F., Curti, D., De Paola, M., Larini, S., Pasquali, C., Brines, M., Cerami, A. ,Ghezzi, P. Erythropoietin protects primary cultured motor neurons from apoptotic but not necrotic death in vitro. Neuron 2003a, submitted.

Mennini, T., Bigini, P., Ravizza, T., Vezzani, A., Calvaresi, N., Tortarolo, M. ,Bendotti, C. Expression of glutamate receptor subtypes in the spinal cord of control and mnd mice, a model of motor neuron disorder.J Neurosci Res 2002. 70: 553-560.

Mennini, T., Cagnotto, A., Carvelli, L., Comoletti, D., Manzoni, C., Muzio, V., Rizzi, M. ,Vezzani, A. Biochemical and pharmacological evidence of a functional role of AMPA receptors in motor neuron dysfunction in mnd mice. Eur J Neurosci 1999. 11: 1705-1710.

Mennini, T., Fumagalli, E. ,Bigini, P. Riluzole protects against motor neuron degeneration in the wobbler mice. Amyotroph Lateral Scler Other Motor Neuron Disord 2003c , in press.

Noda, M., Nakanishi, H., Nabekura, J. ,Akaike, N. AMPA-kainate subtypes of glutamate receptor in rat cerebral microglia. J Neurosci 2000. 20: 251-258.

Pellegrini-Giampietro, D.E., Gorter, J.A., Bennett, M.V. ,Zukin, R.S. The GluR2 (GluR-B) hypothesis: Ca(2+)-permeable AMPA receptors in neurological disorders. Trends Neurosci 1997. 20: 464-470.

Perego, C., Vanoni, C., Bossi, M., Massari, S., Basudev, H., Longhi, R. ,Pietrini, G. The GLT-1 and GLAST glutamate transporters are expressed on morphologically distinct astrocytes and regulated by neuronal activity in primary hippocampal cocultures.J Neurochem 2000. 75: 1076-1084.

Rizzardini, M., Lupi, M., Bernasconi, S., Mangolini, A. ,Cantoni, L. Mitochondrial dysfunction and death in motor neurons exposed to the glutathione-depleting agent ethacrynic acid. JNeurol Sci 2003. 207: 51-58.

Rothstein, J.D., Martin, L.J. ,Kuncl, R.W. Decreased glutamate transport by the brain and spinal cord in amyotrophic lateral sclerosis.N Engl J Med 1992. 326: 1464-1468.

Rothstein, J.D., Van Kammen, M., Levey, A.I., Martin, L.J. ,Kuncl, R.W. Selective loss of glial glutamate transporter GLT-1 in amyotrophic lateral sclerosis. Ann Neurol 1995. 38: 73-84.

Sasaki, S., Komori, T. ,Iwata, M. Excitatory amino acid transporter 1 and 2 immunoreactivity in the spinal cord in amyotrophic lateral sclerosis. Acta Neuropathol (Berl) 2000. 100: 138-144.

Shaw, P.J. Motor neurone disease. Bmj 1999. 318: 1118-1121.

Shaw, P.J., Chinnery, R.M. ,Ince, P.G. Non-NMDA receptors in motor neuron disease (MND): a quantitative autoradiographic study in spinal cord and motor cortex using [3H]CNQX and [3H]kainate. Brain Res 1994. 655: 186-194.

Spreux-Varoquaux, O., Bensimon, G., Lacomblez, L., Salachas, F., Pradat, P.F., Le Forestier, N., Marouan, A., Dib, M. ,Meininger, V. Glutamate levels in cerebrospinal fluid in amyotrophic lateral sclerosis: a reappraisal using a new HPLC method with coulometric detection in a large cohort of patients. J Neurol Sci 2002. 193: 73-78.

Takuma, H., Kwak, S., Yoshizawa, T. ,Kanazawa, I. Reduction of GluR2 RNA editing, a molecular change that increases calcium influx through AMPA receptors, selective in the spinal ventral gray of patients with amyotrophic lateral sclerosis. Ann Neurol 1999. 46: 806-815.

Trotti, D., Rolfs, A., Danbolt, N.C., Brown, R.H., Jr. ,Hediger, M.A. SOD1 mutants linked to amyotrophic lateral sclerosis selectively inactivate a glial glutamate transporter. Nat Neurosci 1999. 2: 427-433.

Van Damme, P., Leyssen, M., Callewaert, G., Robberecht, W. ,Van Den Bosch, L. The AMPA receptor antagonist NBQX prolongs survival in a transgenic mouse model of amyotrophic lateral sclerosis. Neurosci Lett 2003. 343: 81-84.

Van Damme, P., Van Den Bosch, L., Van Houtte, E., Callewaert, G. ,Robberecht, W. A., GluR2-dependent properties of AMPA receptors determine the selective vulnerability of motor neurons to excitotoxicity. JNeurophysiol 2002. 88: 1279-1287.

Vandenberghe, W., Bindokas, V.P., Miller, R.J., Robberecht, W. ,Brorson, J.R. Subcellular localization of calcium-permeable AMPA receptors in spinal motoneurons. Eur JNeurosci 2001. 14: 305-314.

Vandenberghe, W., Ihle, E.C., Patneau, D.K., Robberecht, W. ,Brorson, J.R. AMPA receptor current density, not desensitization, predicts selective motoneuron vulnerability. JNeurosci 2000. 20: 7158-7166.

Chapter 11
GLUTAMATE TRANSMISSION IN THE PATHOGENESIS OF PARKINSON'S DISEASE

Antonio Pisani and Paolo Calabresi
Clinica Neurologica, Dipartimento di Neuroscienze, Università di Roma Tor Vergata, and Fondazione Santa Lucia IRCCS, Rome, Italy.

Abstract: Parkinson's disease (PD), a progressive neurodegenerative disorder, is a common cause of disability. The pathological hallmarks are the presence of Lewy bodies and massive loss of dopaminergic neurons in the pars compacta of the substantia nigra. The current pathophysiological concept of PD postulates a multifactorial origin, where alterations in neurotransmitter content are combined with genetic and environmental factors. With nigrostriatal dopamine depletion, a complex set of changes occurs in the functional anatomy of the basal ganglia circuitry. As a result, the firing pattern of certain glutamatergic pathways has been shown to change significantly, and to play a central role in the pathogenesis of parkinsonian symptoms. Advances in genetics have led to the discovery of gene mutations underlying some forms of PD. The mutated genes encode proteins of unknown function, such as alpha-synuclein and parkin. Moreover, compelling evidence supports the involvement of mitochondrial metabolism failure as an essential cofactor in the pathogenesis of PD. Interestingly, some environmental toxins are thought to be able to act as mitochondrial toxins. The comprehension of the pathways leading to PD requires an intense effort in order to identify and establish a plausible connection between genetic causes, altered neurotransmission and metabolic impairment.

Keywords: Parkinson's disease, glutamate, NMDA, NR2B, metabotropic, mitochondria, therapy.

INDEX

1. Introduction.
2. Animal models of Parkinson's disease (PD).
3. Evidence for the involvement of ionotropic glutamate receptors in the pathogenesis of PD and motor fluctuations associated with levodopa treatment.
4. Glutamatergic pathways and PD.
5. Possible involvement of mGlu receptors.
6. Conclusions.

1. Introduction.

Parkinson's disease (PD) is a progressive degenerative neurological disorder characterized primarily by motor symptoms that include rigidity, hypokinesia and tremor. Current knowledge on the pathogenesis of idiopathic PD include several potential factors, including neurotransmitter toxicity, mitochondrial metabolism failure, environmental toxins and genetic predisposition. The inherited movement disorders comprise a growing category of human diseases. Recent advances in genetics have led to the identification of the gene mutation in Huntington's disease, of three different gene mutations involved in PD and to the discovery of gene mutations underlying some forms of dystonia. In the late 1990s, the discovery of the gene that causes a rare autosomal-dominant form of Parkinson's disease led to renewed interest in the search for pathogenic mechanisms. The mutated gene encodes alpha-synuclein, a synaptic protein of unknown function (Polymeropoulos et al., 1997; Kruger et al., 1998). It has been suggested that it plays a role in the recycling of vesicles that have released their neurotransmitter content into the synaptic cleft. Moreover, further rare forms of PD have been identified that result from mutations in parkin and ubiquitin carboxy-terminal hydrolase L1 (Kitada et al., 1998; Leroy et al., 1998; Vaughan et al., 2001). These proteins participate in the ubiquitin-proteasome pathway, which is responsible for the degradation of unwanted proteins. The most pronounced neuropathological feature is a loss of dopaminergic neurons in the substantia nigra pars compacta, which leads to a reduction in dopamine (DA) levels in the target region, the striatum. In the substantia nigra, some neurons have intracytoplasmic inclusions known as Lewy bodies, containing alpha-synuclein.

Glutamate is the major excitatory transmitter in the basal ganglia. It has been suggested that glutamate acting both as a neurotoxin and a neurotransmitter might play a central role in the pathophysiology of PD. Intrinsic energetic defects of nigral neurons may render this area more vulnerable to otherwise innocuous glutamate doses, rendering glutamate a toxin in the presence of impaired energy metabolism (for review see Greene and Greenamyre, 1996; Greenamyre, 2001).

2. Animal models of Parkinson's disease (PD).

Different new animal models of PD and advances in older models are providing important information about the pathogenesis of PD (for review see Dawson et al., 2002). Early animal models relied on the depletion of DA by a pharmacologically-induced depletion of DA. These experimental paradigms of parkinsonism are represented by a permanent unilateral or bilateral lesion of the ascending nigrostriatal tract by intracerebral injection of 6-hydroxydopamine (6-OHDA) or by reversible depletion by reserpine. The former procedure induces a complete dopaminergic denervation of the ipsilateral striatum, and mimics parkinsonian pathology (for review, see Beal, 2001). The 6-OHDA model has been quite predictive of symptomatic and cell replacement therapies for PD and has been proven useful in investigating regenerative therapies (Ungerstedt and Arbuthnott, 1970). Several adaptive changes occur after 6-OHDA injection. These include modifications at the receptor level (Schwarting and Huston, 1996; Oh et al., 1998; Picconi et al., 2002), as well as changes in the levels of neurotransmitters (Schwarting and Huston, 1996). Reserpine reversibly depletes DA stores by binding to and thereby blocking the operation of the vesicular monoamine transporters (VMATs) (Lechardeur et al., 1993). DA level is transiently reduced by 95% bilaterally in the striatum, as measured by in vivo microdialysis (Heeringa and Abercrombie, 1995) without inducing morphologic changes in the nigral dopaminergic neurons. Behaviourally, the rats show a hypokinetic state (Betarbet et al., 2000).

Systemic methyl-.4-phenyl-1,2,3,6-tetrahydropyridine (MPTP) intoxication in a variety of different species is probably the most reliable parkinsonian model, with the only exception that it is an acute, non-progressive insult. In primates, MPTP intoxication reproduces the clinical signs observed in human PD and leads also to neuronal nonfibrillar inclusions that contain alpha-synuclein (Beal, 2001). Evidence emerged in support of an intramitochondrial localization of MPTP in dopaminergic nigral neurons, where it selectively blocks complex I of the mitochondrial electron transport chain. Use of MPTP primate models led to an improvement in the symptomatic surgical and medical treatment of PD. Recent studies suggest the involvement of environmental toxins in the pathogenesis of PD. Systemic administration of the herbicide paraquat has been shown to cause nigral neuronal loss, striatal dopaminergic denervation and formation of aggregates of alpha-synuclein in rodents (Manning-Bog et al., 2002). Similarly, rotenone, another insecticide produces parkinsonism in rats (Betarbet et al., 2000). Rotenone is a mitochondrial toxin, able to selectively inhibit complex I activity thereby leading to oxidative damage in neurons. Interestingly, rats develop progressive degeneration of the nigrostriatal pathway and neuronal inclusions similar to Lewy bodies. This model is quite remarkable not only for the slow progression in the development of the lesions, but also for the appearence of clinical signs reminiscent of parkinsonian symptoms.

Moreover, these rodents have been shown to respond to the broad spectrum dopamine agonist apomorphine. To date, these models provide an important source for testing new drugs.

3. Evidence for the involvement of ionotropic glutamate receptors in the pathogenesis of PD and motor fluctuations associated with levodopa treatment.

Glutamate activates two major classes of receptors, divided into ionotropic and metabotropic. Ionotropic glutamate receptor activation leads to the opening of ion channels, whereas metabotropic receptors are linked to G-proteins and in turn are linked to cyclic nucleotide signaling. Ionotropic receptors can be further classified into N-methyl-D-aspartate (NMDA), alpha-amino-3-hydroxy-5-methylisoxazole propionate (AMPA) and Kainate receptors. Molecular biology studies have identified the kinetic properties of each receptor subtype, demonstrating that the physiology and pharmacology of these receptors depends upon their subunit composition (Dingledine et al., 1999). Thus, the assembly of receptor subunits confers specific properties to each receptor. NMDAR1 (NR1) and NMDAR2 (NR2) subunits have been identified for the NMDA receptor subtype. NR2 comprises several splice variants, NR2A-D. As for the AMPA receptor, four subunits (GluR1-4) have been cloned and each exists in alternatively spliced forms. Five subunits have been identified for Kainate receptors.

Several lines of experimental evidence suggests that neurodegeneration of dopaminergic neurons in substantia nigra pars compacta observed in PD involves a toxic action exerted by glutamate (Klockgether and Turski, 1993; Greene and Greenamyre, 1996; Greenamyre, 2001). Excitotoxicity can be the direct result of overactivation of the NMDA receptor subtypes, as a result of either increased levels of glutamate or decreased removal from the synaptic left. Direct excitotoxicity was initially proposed as a causative factor in the pathogenesis of neurdegeneration. However, it has become evident that glutamate excitotoxicity is more likely to play a role in acute neuronal damage, rather than in chronic disorders, such as PD. An indirect toxic action by glutamate has been also postulated. Under normal condition, the NMDA receptor channel is blocked by magnesium. If a neuron is depolarized, then the magnesium block is relieved, and glutamate may induce a calcium overload. Maintenance of membrane polarity requires great energy expenditure: consequently, any bioenergetic deficit, able to impair mitochondrial function might depolarize the cell and render glutamate toxic (Beal et al., 1993; Greene and Greenamyre, 1996). It has been shown, in rats with unilateral 6-OHDA lesions of the nigrostriatal tract, a reduction in the abundance of the NR1 and NR2B subunits in the membrane fractions of lesioned striatum, whereas the presence of NR2A subunits is unchanged (Dunah et al., 2000). Similarly, the phosphorylation state of the NR1 and NR2B subunits was also modified. Interestingly, chronic levodopa treatment was found to normalize the composition and phosphorylation of striatal NMDA receptors (Dunah et al., 2000). It appears conceivable that the alterations in NMDA composition and phosphorylation state influence the functional properties of NMDA receptors (Laurie and Seeburg, 1994). Alterations of both tyrosine and serine phosphorylation of striatal NMDA receptors have been described in experimental models of parkinsonism, and recent studies suggest that dopaminergic denervation enhances the sensitivity of NMDA receptors on striatal spiny neurons (Oh et al., 1998, 1999). Such sensitization has been linked to events leading to the motor response fluctuations associated with the administration of dopaminomimetics to parkinsonian patients. Sensitization to glutamate might in turn alter striatal output in ways that would induce the appearence of the motor fluctuations observed in PD patients.

Drugs blocking NMDA receptors effectively reduce these motor fluctuations in levodopa treated parkinsonian rats (Papa et al., 1995). In MPTP-lesioned primates, that usually develop dyskinesias during a long-term levodopa treatment, NMDA receptor antagonists were able to reduce significantly these motor complications (Blanchet et al., 1998). Amantadine, and more recently dextrorphan and dextromethorphan were demonstrated to reduce dyskinetic movements and motor fluctuations (Verhagen Metman et al., 1996; Luginger et al., 2000; Rascol et al., 2000). Recent studies have shown that NR2B-selective antagonists were similarly effective (Blanchet, 1999).

It has long been known the ability of amantadine, a NMDA receptor antagonist, to mitigate parkinsonian symptoms. The efficacy of the NMDA receptor antagonist, remacemide is still controversial, as emerged from clinical trials (Parkinson Study Group, 1999). In addition, consistently with an excitotoxic pathogenetic hypothesis, glutamate receptor antagonists should provide a further therapeutic benefit by exerting a neuroprotective effect (Plannels-Cases et al., 2002). More recently, interest has been focused on an uncompetitive NMDA receptor antagonist, memantine. The combination of fast offset kinetics and strong voltage-dependency allow memantine to rapidly leave the NMDA channel upon transient physiological activation by glutamate, but to block a sustained receptor activation in pathological conditions, such as during an altered neuronal energy metabolism (Parsons et al 1999). Moreover, in a double-blind crossover randomized study, memantine has been shown to improve parkinsonian symptoms independently of dopaminomimetic drugs (Merello et al., 1999).

An involvement of AMPA receptor subtypes has been proposed in the pathogenesis of the motor dysfunctions occurring in course of levodopa therapy (Konitsiotis et al., 2000). However, the role of these ionotropic receptors is still controversial and requires further work (Silverdale et al., 2002). Taken together, these observations support the notion that an altered sensitivity of glutamate receptors located on spiny neurons contribute to the characteristic motor dysfunction observed in PD patients in course of long-term levodopa therapy.

4. Glutamatergic pathways and PD.

The loss of the nigrostriatal dopaminergic fibers has been shown to induce a complex rearrangement in the functional anatomy of the basal ganglia (Figure 1). As a result, the firing pattern of some of the glutamatergic pathways has been shown to undergo profound modifications (Bergman et al., 1990, 1994, 2000; Bevan et al., 2002; Wichmann et al 1994). A significant advance in this direction has been represented by the work conducted on subthalamic nucleus. Indeed, it is widely accepted that the activity of subthalamic projection neurons dramatically increases (Bergman et al., 1990, 1994) and that focal infusion of an NMDA receptor antagonist prevents nigral degeneration in a rodent model of PD (Blandini et al., 2001). In addition, surgical ablation of this nucleus or functional inactivation by deep brain stimulation have been proven useful in PD patients (Bergman et al., 2000). Moreover, in vitro experimental data demonstrate that activation of metabotropic glutamate receptor, type 5, produces a direct excitation of neurons in this nucleus and mediates the potentiation of NMDA responses (Awad et al., 2000). Studies of idiopathic and experimental parkinsonism suggest that changes occur both in the patterns and mean rates of activity of subthalamic nucleus projection neurons. In normal conditions, complex, irregular patterns of firing are recorded in this area, whereas, in PD, neurons display more correlated, synchronous, rhythmic patterns of activity (Bevan et al., 2002). Such a synchronized, oscillatory activity involving the subthalamic nucleus and, as a consequence, its output targets, might result in an abnormal coding of motor processes and therefore, contribute to the symptoms occurring in PD. Moreover, the subthalamic projections to the substantia nigra might promote excitotoxic damage to nigral dopaminergic neurons, as shown by the beneficial effects on neuron survival obtained by glutamatergic subthalamic activity blockade in an experimental model of PD (Blandini et al., 2001).

The activity of the thalamocortical pathway has been shown to be altered in PD, in that both a decrease in cell excitability and an alteration in the firing pattern occur (Figure 1) (Ceballos-Bauman et al., 1994; Vitek et al., 1994). In a large part, this abnormalities observed in the electrical behavior of thalamocortical neurons can be ascribed to the increase in the GABAergic activity of both the internal globus pallidus and the substantia nigra pars reticulata.

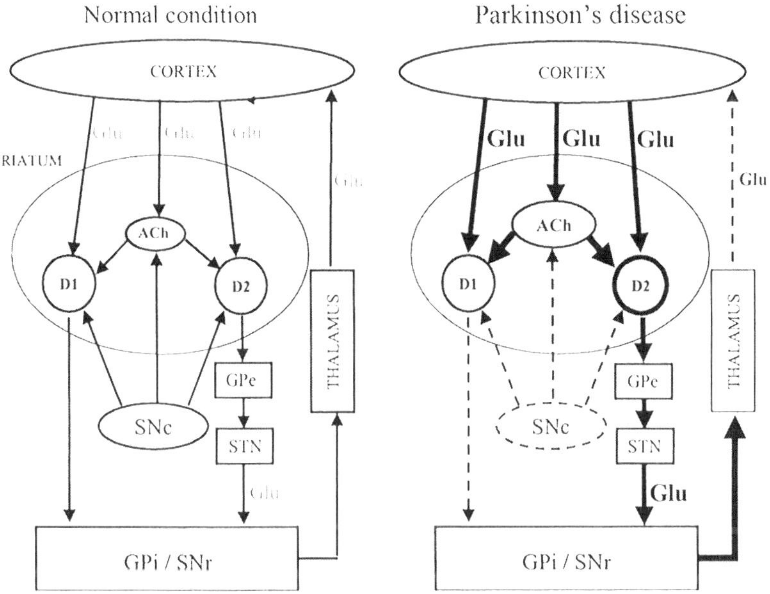

Figure 1. Simplified scheme of striatal circuitry in control condition and during parkinsonian state. Within the basal ganglia circuitry, in normal conditions, both projections neurons and ACh-containing interneurons receive glutamatergic inputs from cortex and dopaminergic influence from substantia nigra pars compacta (SNc). Striatal cholinergic interneurons exert a facilitatory effect, both on D1- and D2-containing projections neurons, that give rise, respectively, to the direct and indirect output pathways. The direct pathway projects to the internal segment of the globus pallidus (GPi) and to substantia nigra pars reticulata (SNr), whereas the indirect pathway establishes synaptic connections between the D2-containing striatal projection neurons and both the external portion of the GP (GPe) and the subthalamic nucleus (STN). In the parkinsonian state, the loss of SNc neurons leads to enhanced ACh efflux and disinhibition of DA D2-receptor-bearing striatal neurons. Both the glutamatergic input from the cortex to the striatum and from the STN to the output nuclei is significantly enhanced. The resulting imbalance of the indirect pathway leads to an increased inhibitory output to the thalamus, and to reduced thalamo-cortical feedback, resulting in parkinsonism.

Conversely, the corticostriatal glutamatergic activity is definitely increased. This notion is supported by electrophysiological recordings from 6-OHDA-denervated striatal projection neurons, showing a significant enhancement of glutamate-mediated spontaneous electrical activity and of postsynaptic potentials (Figure 2) (Calabresi et al., 1993, 2000; Picconi et al., 2002).

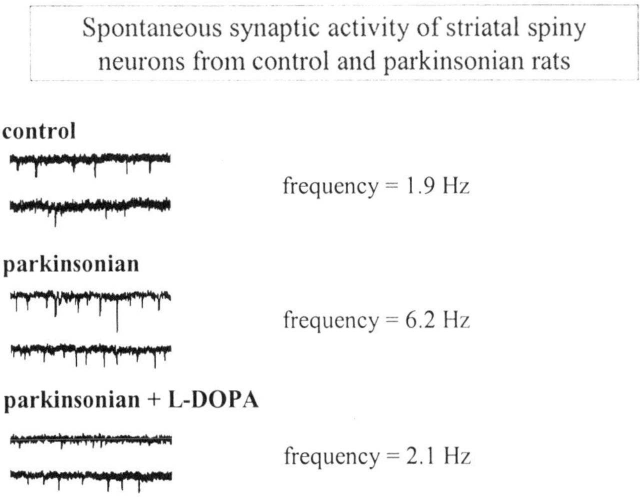

Figure 2. Dopamine denervation-induced electrophysiological changes in striatal projections neurons. Electrophysiological recordings performed in the whole-cell configuration showing spontaneous electrical activity in striatal projections neurons in normal conditions (upper trace), after 6-OHDA denervation (middle trace) and in denervated rats treated chronically with levodopa (lower trace). Note the increase in spontaneous activity in denervated rats, and the restoration of normal frequency rate by levodopa treatment.

A glutamatergic overflow from cortical terminals in the striatum plays a key role in the pathophysiology of PD. In rodent models focal administration of NMDA receptor antagonists into the striatum has antiparkinsonian effects. Similarly, it has been proposed that levodopa-induced dyskinesias are due to an overactivity of the direct pathway to the output nuclei: this increased striatal outflow might be related to the excessive corticostriatal activity.

5. Possible involvement of mGlu receptors.

Recently, in experimental models of parkinsonism it has been shown that drugs acting on mGlu receptors might prove useful in the treatment of parkinsonian symptoms. In particular, evidence accumulated in favour of an effective role of group II mGlu receptor agonists. Firstly, the selective group II agonist LY354740 administered systemically in rats blocked the parkinsonian rigidity induced by chronic haloperidol treatment (Konieczny et al., 1998). Secondly, both in the haloperidol-induced catalepsy and in the reserpine-induced akinesia models group II mGlu receptor agonists were shown to possess antiakinetic properties (Dawson et al., 2000; Murray et al., 2002). Notably, 6-OHDA denervation has been shown to induce an increase in expression and sensitivity to group II mGlu receptor agonists at corticostriatal synapses (Picconi et al., 2002). In 6-OHDA-treated rats, the potency of both group II mGlu receptor agonists LY379268 and DCG-IV in inhibiting corticostriatal EPSPs was significantly higher than in control slices. Interestingly, these changes in expression and/or regulatory properties of group II receptors determined by the lesioning procedure recovered completely after repeated levodopa treatment, testifying a regression of possible plastic changes occurring after dopaminergic treatment (Picconi et al., 2002). Besides the potential therapeutic effects produced by activation of group II mGlu receptors, very recently attention has been focused on the group I mGlu receptor subtype 5. The interest for mGlu5 receptors raised primarily because of its abundant expression within the basal ganglia. Immunohistochemical and in situ hybridization studies revealed a very high level of mGlu5 receptor expression in the striatum, and a moderate labeling in the subthalamic nucleus and its output structures, i.e. globus pallidus and substantia nigra pars reticulata (Testa et al., 1994). Systemically administered MPEP was reported to block the akinetic effects caused by haloperidol treatment (Ossowska et al., 2001). Moreover, in the 6-OHDA model of parkinsonism in rodents, obtained by lesioning the nigrostriatal tract, chronic MPEP treatment was shown to improve the motor performances and to restore reaction times in cognitive functions impaired by 6-OHDA (Breysse et al., 2002). However, what remains controversial is where mGlu5 receptor antagonists act. One possibility is represented by the subthalamic nucleus: in vitro experimental data demonstrate that mGlu5 receptor activation produces a direct excitation of neurons in this nucleus and mediates the potentiation of NMDA responses (Awad et al., 2000). In addition, surgical ablation of subthalamic nucleus has been shown useful in PD patients (Bergman et al., 2000). However, since the expression profile of mGlu5 receptors is the highest within the basal ganglia, the striatum represents a good candidate. Within this region, one possible action could be represented by the blockade of the positive interaction between NMDA and mGlu5 receptors occurring in MS neurons (Pisani et al., 1997, 2001a). Moreover, mGlu5 receptor antagonists reduce the overall excitability of cholinergic interneurons and prevent ACh release (Pisani et al., 2001b; Marti et al., 2001). Together, these observations support a beneficial pharmacological effect of mGlu5 receptor antagonists in animal models of parkinsonism.

6. Conclusions.

There is compelling evidence that the activity of glutamatergic pathways is altered in PD. Of particular interest is the evidence that the glutamatergic subthalamic nucleus becomes overactive. Indeed, both the pharmacological blockade and the surgical ablation of glutamatergic output from the subthalamus has antiparkinsonian symptomatic effects and may also exert a protective action in the survival of residual nigral neurons. This experimental evidence led several investigators to the assumption that drugs counteracting the effects of glutamate might therefore provide new protective and symptomatic strategies for therapy of PD. Levodopa and dopaminergic agonists are still the first choice drugs to treat PD patients. On a long-term basis, however, treatment with both dopamine precursors and dopamine receptor agonists is not devoid of serious side effects such as dyskinesias and motor fluctuations. These observations have prompted much effort in search of alternative strategies, and to this respect, ionotropic glutamate receptors, in particular NMDA receptors, have represented a target for pharmacological research in PD (Parsons et al., 1998; Chase and Oh, 2000). Unfortunately, NMDA receptor antagonists were active not only in reducing the occurrence of parkinsonian symptoms, but also in producing undesired side effects such as allucinations and cognitive deficits. This evidence has limited the perspectives of utilizing NMDA receptor antagonists in humans and prompted research on non-NMDA receptor-sensitive drugs.

Compared to thoroughly investigated roles of ionotropic glutamate receptors, the contribution of mGlu receptors in modulating neuronal excitability is less elucidated and remains a prime theme in most recent research on glutamate. Recent studies to assess the potential role of mGlu receptor-related compounds in in vivo models of diseases affecting the central nervous system suggest that these receptors might represent therapeutic target for different pathological conditions, including PD (Ossowska et al., 2001; Breysse et al., 2002).

These preliminary observations suggest the need for further work, in order to define the site of action for glutamate receptor antagonists, both at cell level and within the basal ganglia circuitry.

REFERENCES

Awad, H., Hubert, G.W., Smith, Y., Levey, A.I., Conn, J.P. Activation of metabotropic glutamate receptor 5 has direct excitatory effects and potentiates NMDA receptor currents in neurons of the subthalamic nucleus. J. Neurosci. 2000, 20:7871-7879.

Beal, M.F. Experimental models of Parkinson's disease. Nat. Rev. Neurosci. 2001, 2:325-334.

Beal, M.F., Hyman, B.T., and Koroshetz, W. Do detefcts in mitochondrial energy metabolism underlie the pathology of neurodegenerative diseases? Trends Neurosci. 1993, 16:125-130.

Bergman, H., Wichmann, T., DeLong, M.R. Reversal of experimental parkinsonism by lesions of the subthalamic nucleus. Science 1990, 249:1436-8.

Bergman, H, Wichmann, T., Karmon, B., and DeLong, M.R. The primate subthalamic nucleus. II. Neuronal activity in the MPTP model of parkinsonism. J. Neurophysiol. 1994, 72:507-20.

Bergman, H., Wichmann, T., Delong, M.R. Reversal of experimental parkinsonism by lesions of the subthalamic nucleus. Science 2000, 249:1346-1348.

Betarbet, R., Sherer, T.B., MacKenzie, G., Garcia-Osuna, M., Panov, A.V., and Greenamyre, J.T. Chronic systemic pesticide exposure reproduces features of Parkinson's disease. Nat. Neurosci. 2000, 3(12):1301-6.

Bevan, M.D., Magill, P.J., Terman, D., Bolam, J.P., and Wilson, C.J. Move to the rhythm: oscillations in the subthalamic nucleus-external globus pallidus network. Trends Neurosci. 2002, 25:525-31.

Blanchet, P.J., Konitsiotis, S., and Chase, T.N. Amantadine reduces levodopa-induced dyskinesias in parkinsonian monkeys. Mov. Disord. 1998, 13:798-802.

Blanchet, P.J., Konitsiotis, S., Whittemore, E.R., Zhou, Z.L., Woodward, R.M., and Chase, T.N. Differing effects of N-methyl-D-aspartate receptor subtype selective antagonists on dyskinesias in levodopa-treated 1-methyl-4-phenyl-tetrahydropyridine monkeys. J. Pharmacol. Exp. Ther. 1999, 290(3):1034-40.

Blandini, F., Nappi, G., and Greenamyre, J.T. Subthalamic infusion of an NMDA antagonist prevents basal ganglia metabolic changes and nigral degeneration in a rodent model of Parkinson's disease. Ann. Neurol. 2001, 49:525-9.

Breysse, N., Baunez, C., Spooren, W., Gasparini, F., and Amalric, M.. Chronic but not acute treatment with a metabotropic glutamate 5 receptor antagonist reverses the akinetic deficits in a rat model of parkinsonism. J. Neurosci. 2002 22, 5669-5678.

Calabresi, P., Mercuri, N.B., Sancesario, G., and Bernardi, G. Electrophysiology of dopamine-denervated striatal neurons. Brain 1993, 116:433-452.

Calabresi, P., Centonze, D., and Bernardi, G. Electrophysiology of dopamine in normal and denervated striatal neurons. Trends Neurosci. 2000, 23 Suppl. 10:S57-63.

Ceballos-Bauman, J, Obeso, J., Vitek, J., Delong, M., and Bakay, R. Restoration of thalamocortical activity after posteroventrolateral pallidotomy in Parkinson's disease. Lancet 1994, 344-814.

Chase, T.N., and Oh, J.D. Striatal dopamine and glutamate-mediated dysregulation in experimental parkinsonism. Trends Neurosci. 2000, 23 (Suppl. 10):S86-91.

Dawson, T.M., Mandir A.S, and Lee, M.K. Animal models of PD: pieces of the same puzzle? Neuron 2002, 35:219-222.

Dingledine, R., Borges, K., Bowie, D., and Traynelis, S.F. The glutamate receptor ion channels. Pharmacol. Rev. 1999, 51:7-61.

Dunah, A.W., Wang, Y., Yasuda, R.P., Kameyama, K., Huganir, R.L., Wolfe, B.B., and Standaert, D.G. Alterations in subunit expression, composition, and phosphorylation of striatal N-methyl-d-aspartate glutamate receptors in a rat 6-hydroxydopamine model of Parkinson's disease. Mol. Pharmacol. 2000, 57:342-352.

Greene, J.G., and Greenamyre, T.J. Bioenergetics and glutamate excitotoxicity. Prog. Neurobiol. 1996, 48:613-634.

Greenamyre T.J. Glutamatergic influences on the basal ganglia. Clin. Neuropharmacol. 2001, 24:65-70.

Heeringa, M.J., and Abercrombie, E.D. Biochemistry of somatodendritic dopamine release in substantia nigra: an in vivo comparison with striatal dopamine release. J. Neurochem. 1995, 65(1):192-200.

Kitada, T, Asakawa S, Hattori, N, Matsumine, H., Yamamura, Y., Minoshima, S., Yokochi, M., Mizuno, Y., and Shimizu, N. Mutations in the parkin gene cause autosomal recessive juvenile parkinsonism. Nature 1998, 392:605-608.

Klockgether, T., and Turski, L. Toward an understanding of the role of glutamate in experimental parkinsonism: agonist-sensitive sites in the basal ganglia. Ann. Neurol. 1993, 34:585-593.

Konieczny, J., Ossowska, K., Wolfarth, S., and Pilc, A. LY354740, a group II metabotropic glutamate receptor agonist with potential antiparkinsonian properties in rats. Naunyn-Schmiedeberg's Arch Pharmacol. 1998, 358:500-502.

Konitsiotis S, Blanchet PJ, Verhagen L, Lamers E, Chase TN. AMPA receptor blockade improves levodopa-induced dyskinesia in MPTP monkeys. Neurology. 2000 54(8):1589-95.

Kruger, R., Kuhn, W., Muller, T, Woitalla, D., Graeber, M., Kosel, S., Przuntek, H., Epplen, J.T., Schols, L., and Riess, O. Ala30Pro mutation in the gene encoding alpha-synuclein in Parkinson's disease. Nat. Genet. 1998, 18:106-108.

Laurie, D.J., and Seeburg, P.H. Ligand affinities at recombinant N-methyl-D-aspartate receptors depend on subunit composition. Eur. J. Pharmacol. 1994, 268:335-345.

Lechardeur, D., Castel, M.N., Reibaud, M., Scherman, D., and Laduron, P.M. Axonal transport of dopamine-containing vesicles labelled in vivo with [3H]reserpine. Eur. J. Neurosci. 1993, 5(5):449-53

Leroy, E., Boyer, R., Auburger, G., Leube, B., Ulm, G., Mezey, E., Harta, G., Brownstein, M.J., Jonnalagada, S., Chernova, T., Dehejia, A., Lavedan, C., Gasser, T., Steinbach, P.J., Wilkinson, K.D., and Polymeropoulos, M.H. The ubiquitin pathway in Parkinson's disease. Nature 1998, 395:451-452.

Luginger, E., Wenning, G.K., Bosch, S., and Poewe, W. Beneficial effects of amantadine on L-dopa-induced dyskinesias in Parkinson's disease. Mov Disord. 2000, 15:873-8.

Manning-Bog, A.B., McCormack, A.L., Li, J., Uversky, V.N., Fink, A.L, and Di Monte, D.A. The herbicide paraquat causes up-regulation and aggregation of alpha-synuclein in mice: paraquat and alpha-synuclein. J. Biol. Chem. 2002, 277(3):1641-4.

Marti, M., Paganini, F., Stocchi, S., Bianchi, C., Beani, L., Morari, M. Presynaptic group I and II metabotropic glutamate receptors oppositely modulate striatal acetylcholine release. Eur. J. Neurosci. 2001 14:1181-1184.

Merello, M., Nouzeilles, MI., Cammarota, A., and Leiguarda, R. Effect of memantine (NMDA antagonist) on Parkinson's disease: a double-blind crossover randomized study. Clin. Neuropharmacol. 1999, 22:273-276.

Murray, T.K., Messenger, M.J., Ward, M.A., Woodhouse, S., Osborne, D.J., Duty, S., O'Neill, M.J. Evaluation of the mGluR2/3 agonist LY379268 in rodent models of Parkinson's disease. Pharmacol., Biochem., and Behav. 2002, 73:455-466.

Oh, J.D., Russell, D., Vaughan, C.L., and Chase, T.N. Enhanced tyrosine phosphorylation of striatal NMDA receptor subunits: effect of dopaminergic denervation and L-DOPA administration. Brain Res. 1998, 813:150-159.

Oh, J.D., Vaughan, C.L., and Chase, T.N. Effect of dopamien denervation and dopamine agonists administration on serine phosphorylation of striatal NMDA receptor subunits. Brain Res. 1999, 821:433-442.

Ossowska, K., Konieczny, J., Wolfarth, S., Wieronska, J., and Pilc, A. Blockade of the metabotropic glutamate receptor subtype 5 (mGluR5) produces antiparkinsonian-like effects in rats. Neuropharmacology 2001, 41:413-420.

Papa, S.M., Boldry, R.C., Engber, T.M., Kask, A.M., and Chase, T.N. Reversal of levodopa-induced motor fluctuations in experimental parkinsonism by NMDA receptor blockade. Brain Res. 1995, 701:13-18.

Parsons, C.G., Danysz, W, and Quack, G. Glutamate in CNS disorders as a target for drug development: an update. Drug new Perspect. 1998, 11:523-569.

Parsons, C.G., Danysz, W, and Quack, G. Memantine is a clinically well tolerated N-methyl-D-aspartate (NMDA) receptor antagonist a review of preclinical data. Neuropharmacology 1999, 38:735-767.

Parkinson Study Group. The glutamate receptor antagonist remacemide improves motor performances in levodopa-treated Parkinson's disease. Neurology 1999, 52(2):A262.

Picconi, B., Pisani, A., Centonze, D., Battaglia, G., Storto, M., Nicoletti, F., Bernardi, G., Calabresi, P. Striatal metabotropic glutamate receptor function following experimental parkinsonism and chronic levodopa-treatment. Brain 2002, 125:1-11.

Pisani, A., Calabresi, P., Centonze, D., and Bernardi, G. Enhancement of NMDA responses by group I metabotropic glutamate receptors activation in striatal neurons. Br. J. Pharmacol. 1997, 120: 1007-1014.

Pisani, A., Bonsi, P., Calabresi, P., Centonze, D., Bernardi, G. Functional coexpression of excitatory mGluR1 and mGluR5 on striatal cholinergic interneurons. Neuropharmacology 2001a, 40:460-463.

Pisani, A., Gubellini, P., Bonsi, P., Conquet, F., Picconi, B., Centonze, D., Bernardi, G., and Calabresi, P. Metabotropic glutamate receptor 5 mediates the potentiation of NMDA responses in medium spiny striatal neurons. Neuroscience 2001, 106:579-587.

Planells-Cases, R., Montoliu, C., Humet, M., Fernandez, A.M., Garcia-Martinez, C., Valera, E., Merino, J.M., Perez-Paya, E., Messeguer, A., Felipo, V., and Ferrer-Montiel, A. A novel N-methyl-D-aspartate receptor open channel blocker with in vivo neuroprotectant activity. J. Pharmacol. Exp. Ther. 2002 Jul;302(1):163-73

Polymeropoulos, M.H., Lavedan, C, Leroy, E., Ide, S.E., Dehejia, A., Dutra, A., Pike, B., Root, H., Rubinstein, j., Boyer, R., Stenroos, E.S., Chandrasekharappa, S., Athanassiadou, A., Papapetropoulos, T, Johnson, W.G., Lazzarini, A.M., Duvoisin, R.C., Di Iorio, G., Golbe, L.I., and Nussbaum, R.L. (1997) Mutation in the alpha-synuclein geneidentified in families with Parkinson's disease. Science 276:2045-2047.

Rascol, O. The pharmacological therapeutic management of levodopa-induced dyskinesias in patients with Parkinson's disease. J Neurol. 2000, 247 Suppl. 2:II51-7.

Schwarting R.K., and Huston, J.P. Unilateral 6-hydroxydopamine lesions of meso-striatal dopamine neurons and their physiological sequelae. Prog Neurobiol. 1996, 49(3):215-66.

Silverdale, M.A., Crossman, A.R., and Brotchie, J.M. Striatal AMPA receptor binding is unaltered in the MPTP-lesioned macaque model of Parkinson's disease and dyskinesia. Exp Neurol. 2002 174:21-8.

Testa, C.M., Standaert, D.G., Young, A.B., and Penney, J.B. Jr. Metabotropic glutamate receptor mRNA expression in the basal ganglia of the rat. J. Neurosci. 1994 14:3005-3018.

Ungerstedt, U., and Arbuthnott, G.W. Quantitative recording of rotational behavior in rats after 6-hydroxydopamine lesions of the nigrostriatal dopamine system. Brain Res. 1970, 24:485-493.

Vaughan, J.R., Davis, M.B., and Wood, N.W. (2001). Genetics of parkinsonism: a review. Ann. Hum. Genet. 65:111-126.

Verhagen Metman, L., Blanchet, P.J., Mouradian, M.M., and Chase, T.N. Dextromethorphan and levodopa combination therapy in Parkinson's disease patients with motor fluctuations. Mov. Disord. 1996, 11:184.

Vitek, J., Ashe, J, and Kaneoke, Y. Spontaneous neuronal activity in the motor thalamus: alteration in pattern and rate in parkinsonism. Neuroscience 1994, 20:561.

Wichmann, T., Bergman, H., and DeLong, M.R. The primate subthalamic nucleus. I. Functional properties in intact animals. J Neurophysiol. 1994, 72(2):494-506.

Chapter 12

EXCITOTOXICITY IN HUNTINGTON'S DISEASE

M. Flint Beal
Department of Neurology and Neuroscience, Weill Medical College of Cornell University, New York Presbyterian Hospital, 525 E 68th Street, New York, NY 10021, USA.

Abstract: There is a large body of evidence implicating excitotoxicity in Huntington's Disease (HD) pathogenesis. The concept that excitotoxicity might play a role was developed after the observation that kainic acid induced striatal lesions which mimic many of the neuropathologic features of HD. We and others subsequently showed that quinolinic acid lesions, which spare NADPH-diaphorase neurons, produced an improved excitotoxic model. There however is no evidence for an increase in excitatory amino acids in HD brain tissue. Subsequent studies by ourselves and others demonstrated that mitochondrial toxins could produce striatal lesions, which are mediated by excitotoxicity. A defect in energy metabolism sensitizes cells to excitatory amino acids. We found that 3-nitropropionic acid lesions produce a model of HD in both rodents and primates, which mimics not only the neuropathologic features but also the motor and cognitive deficits, which occur in HD. The development of transgenic mouse models has furthered the evidence that excitotoxicity plays a role in HD pathogenesis. Although mice with short fragments of huntingtin show resistance to excitotoxic lesions, transgenic mice with full-length huntingtin show increased sensitivity to quinolinic acid lesions. Cultured medium spiny neurons expressing full-length huntingtin with 72 repeats, also demonstrate enhanced excitotoxicity, which is abolished by NR2B subunits specific antagonists such as, ifenprodil. There is, therefore, substantial evidence that excitotoxicity plays a critical role in HD pathogenesis.
Keywords: Huntington's disease, excitotoxicity, free radicals, nitric oxide, animal models.

INDEX

1. Excitotoxic models of Huntington's disease.
2. Mitochondrial toxin models.
3. Transgenic mouse models of Huntington's disease.
4. Excitotoxicity in transgenic mouse models of Huntington's disease.

1. Excitotoxic models of Huntington's disease.

The role of excitotoxicity in neurodegenerative diseases is based on circumstantial evidence. Some of the best evidences is for a role of excitotoxicity in Huntington's disease (HD). Initial observations showed that kainic acid striatal lesions could mimic many of the neuropathologic features of HD (Coyle and Schwarcz, 1976; McGeer and McGeer, 1976). They however do not produce sparing of striatal interneurons containing the histochemical marker NADPH-diaphorase, which are spared in HD (Dawbarn et al., 1985; Ferrante et al., 1985). Beal and co-workers and others subsequently showed that quinolinic acid and other NMDA agonists produce an improved animal model, since they result in relative sparing of NADPH-diaphorase neurons (Beal et al., 1986; Beal et al., 1989). This relative sparing is much more dramatic with chronic striatal lesions in which there is striatal shrinkage (Beal et al., 1991; Bazett et al., 1994). Parvalbumin neurons which are spared in HD are also relatively preserved by NMDA agonists, but preferentially vulnerable to kainate (Harrington and Kowall, 1991; Waldvogel et al., 1991; Bazett et al., 1994). In primates, quinolinic acid produces striking sparing of NADPH diaphorase neurons as well as an apomorphine inducible movement disorder (Ferrante et al., 1993).

Further support for an NMDA excitotoxic process comes from studies of NMDA receptors in HD postmortem tissue. If the neurons containing these receptors are preferentially vulnerable one would expect a depletion of NMDA receptors. This was shown to be the case in HD striatum (Young et al., 1988; Dure IV et al., 1991) as well as in the striatum of an asymptomatic at-risk patient, who showed a 50% depletion of NMDA receptors, suggesting that this occurs early in the disease process (Albin et al., 1990).

2. Mitochondrial toxin models.

A role for mitochondrial energy metabolism dysfunction in the pathogenesis of neuronal degeneration in HD is further supported by observations, in both humans and in experimental animals that the basal ganglia neurons are particularly vulnerable to mitochondrial toxins. These include the complex II inhibitors 3-nitropropionic acid (3-NP) and malonate, AOAA (complex I), potassium cyanide and sodium azide (complex IV) (Browne and Beal, 1994). Ingestion of 3-NP, an irreversible inhibitor of succinate dehydrogenase (complex II), produces selective basal ganglia lesions and delayed dystonia in humans (Ludolph et al., 1990).

Systemic administration of 3-NP to both rats and primates produces age-dependent striatal lesions which are strikingly similar to those seen in HD (Brouillet et al., 1993; Brouillet et al., 1995). In primates, chronic 3-NP administration produces selective striatal lesions which spare NADPH-d neurons, and induce proliferative changes in the dendrites of spiny neurons. Animals also show both spontaneous and apomorphine-inducible movement disorders resembling HD (Brouillet et al., 1995). 3-NP basal ganglia lesions in rats are associated with elevated lactate levels, similar to the increased lactate production seen in HD patients. 3-NP lesions can be prevented by prior removal of glutamatergic excitatory cortico-striatal imputs by decortication, by glutamate release inhibitors and by glutamate receptor antagonists, suggesting that 3-NP toxicity is mediated by secondary excitotoxic mechanisms (Beal, 1994; Schulz et al., 1996a).

Intrastriatal injection of either malonate or 3-NP in rats is also associated with increased oxidative damage. We found that the rate of hydroxyl free radical production is elevated in the striatum, as detected by microdialysis (Schulz et al., 1996b). Increased OH^8dG levels in striatum are also detected following systemic 3-NP administration, and elevated 3-nitrotyrosine concentrations are reported after either systemic 3-NP or intrastriatal malonate injection. Further, the finding that 3-NP-induced lesions and concomitant increase in oxidative damage markers were markedly attenuated in mice overexpressing the superoxide free radicals scavenger Cu/Zn superoxide dismutase (SOD1) imply that oxidative free radicals contribute to lesion formation (Beal et al., 1995). Furthermore, malonate and 3-NP striatal lesions were attenuated by free radical spin traps and nitric oxide synthase (NOS) inhibitors. Inhibition of nitric oxide (NO) generation in mice lacking the gene for the neuronal isoform of NOS (nNOS), also resulted in reduced volume of malonate lesions (Schulz et al., 1996c). Hence there is substantial evidence that nitric oxide-mediated oxidative damage is involved in cell death processes following energetic disruption in these models.

3. Transgenic mouse models of HD.

One of the major drawbacks of relying on human tissue for assessment of neurologic disease progression is the inability to adequately map early events in the disease etiology. Substantial evidence of a causative role in the disorder would be provided by evidence of occurrence prior to symptoms and pathology in models of HD. Over the last few years, the development of methodology to generate transgenic mouse lines expressing the physiological phenotypes associated with human gene mutations, has provided much needed *in vivo* models to circumvent many of these issues. A number of different groups have developed several different transgenic and "knock-in" mouse models of HD, which vary in terms of the transgene incorporation technique employed. As a result, mouse phenotypes vary between lines, the features manifest by the animals depending on the nature of the transgene incorporated (i.e. full length human mutant huntingtin, or a huntingtin *HD* gene fragment incorporating the mutant region in exon 1, or merely an expansion inserted into the murine HD homologue Hdh; CAG repeat length; copy number of the mutant gene incorporated; promoter used, and hence cellular specificity of expression; background strains; and expression levels of the mutant gene).

4. Excitotoxicity in transgenic mouse models of HD.

Surprisingly, it was previously demonstrated that R6/1 and R6/2 mice are resistant to excitotoxic lesions produced by quinolinic acid and malonate (Hansson et al., 1999; Hanson et al., 2001). It is possible that this may be due a marked reduction in dopamine levels in the R6/1 mice and R6/2 mice (Hickey et al., 2002; Petersen et al., 2002b). In the R6/1 mice, extracellular dopamine levels are reduced by 70 percent as compared to wild type litter mates (Petersen et al., 2002b). Intrastriatal administration of malonate in R6/1 mice resulted in a short lasting attenuated increase in local dopamine release, and the size the malonate induced striatal lesion was 80 percent smaller in these animals.

This finding, however appears to be relatively specific for the R6/1 and R6/2 mice. In other studies the excitatory post-synaptic potentials showed a more prominent component mediated by activation on N-methyl-D-aspartate receptors in striatal slices from the R6/2 mice (Cepeda et al., 2001; Klapstein et al., 2001).

There however is a progressive reduction in synaptic activity in the R6/2 mice starting at 5-7 weeks of age, which worsens in older mice (Cepeda et al., 2003). There are abnormally large amplitude synaptic events most commonly at 5-7 weeks of age. It was proposed that the reduced synaptic activity may account for the resistance to excitotoxic lesions in the R6/2 mice. The fact that riluzole, a substance with glutamate antagonist properties, prolongs survival in R6/2 mice however suggests there is an excitotoxic component (Schiefer et al., 2002). Another transgenic mouse which has a longer N-terminal fragment of huntingtin shows equal susceptibility to excitotoxic lesions mediated by quinolinic acid as compared to wild-type littermate controls (Petersen et al., 2002a).

More recent studies demonstrated that in brain slices from YAC72 transgenic HD mice, which contain human genomic DNA spanning the full-length huntingtin gene and all its regulatory elements with 72 CAG repeats there is an increase in responses of medium size striatal neurons to NMDA (Zeron et al., 2001). Subsequently, the role of excitotoxicity in YAC 72 mice aged 6 and 10 months old was compared to wild-type littermate controls (Zeron et al., 2002). It was demonstrated that quinolinic acid lesions were significantly greater in the YAC 72 transgenic mice as compared to wild-type mice. The authors also demonstrated enhanced excitotoxicity in cultured neonatal medium spiny neurons expressing full length huntingtin with 72 repeats. The increased excitotoxicity was seen after NMDA, but not AMPA exposure in culture. The NMDA induced cell death was abolished by the NR2B subunit specific antagonist, ifenprodil. The authors therefore hypothesized that regional expression of NR2B subunits may correlate with the severity of neuronal degeneration in HD. The levels of expression of NR2B were much lower in cerebellum, and there was no difference in vulnerability of wild type in the YAC 72 cerebellar granule cells to NMDA mediated excitotoxicity. There was NMDA current amplitude and caspase-3 activity in transgenic striatal neurons from YAC72 mice. These findings therefore provide further evidence for a role for excitotoxicity in HD pathogenesis.

Acknowledgements. Greta Strong is thanked for secretarial assistance. This work was supported by grants from NINDS, The Hereditary Disease Foundation and the Huntington's Disease Society of America.

REFERENCES

Albin R.L., Young A.B., Penney J.B. et al. Abnormalities of striatal projection neurons and N-methyl-D-aspartate receptors in presymptomatic Huntington's Disease. *N Engl J Med* 1990, **322**:1293-1298

Bazett T.J., Becker J.B., Falik R. C., et al. Chronic intrastriatal quinolinic acid produces reversible changes in perikaryal calbindin and parvalbumin immunoreactivity. *Neuroscience* 1994,**60:**837-841

Beal M.F. Huntington's disease, energy and excitotoxicity. *Neurobiol Aging* 1994, **15**:275-276

Beal M.F., Ferrante R.J., Swartz K.J. et al. Chronic quinolinic acid lesions in rats closely resemble Huntington's Disease. *J Neurosci* 1991, **11**:1649-1659

Beal M.F., Kowall N,W., Ellison D.W. et al. Replication of the neurochemical characteristics of Huntington's Disease with quinolinic acid. *Nature* 1986, **321:**168-171

Beal M.F., Kowall N.W., Swartz K.J., et al. Differential sparing of somatostatin-neuropeptide Y and cholinergic neurons following striatal excitotoxic lesions. *Synapse,* 1989, **3**:38-47

Brouillet E., Hantraye P., Ferrante R.J., Dolan R., Leroy-Willig A., Kowall N.W., Beal M.F. Chronic mitochondrial energy impairment produces selective striatal degeneration and abnormal choreiform movements inprimates. Proc Natl Acad Sci USA **92**:7105-7109

Brouillet E.P., Jenkins B.G., Hyman B.T., Ferrante R. J., Kowall N.W., Srivastava R., Samanta-Roy D., Rosen B.R., Beal M.F. Age-dependent vulnerability of the striatum to the mitochondrial toxin 3-nitropropionic acid. *J Neurochem* 1993, **60**:356-359

Browne S.E., Beal M.F. Oxidative damage and mitochondrial dysfunction in neurodegenerative diseases. *Biochem Soc Trans* 1994, **22**:1002-1006

Cepeda C., Ariano M.A., Calvert C.R., Flores-Hernandez. J., Chandler S.H., Leavitt B.R., Hayden M.R., Levine M.S. NMDA receptor function in mouse models of Huntington's Disease. *J Neurosci Res* 2001, **66**:525-539

Cepeda C., Hurst R.S. Calvert C.R., Hernandez-Echeagaray E., Nguyen O.K. Jocoy E., Christian L.J., Ariano M.A., Levine M.S. Transient and progressive electrophysiological alterations in the corticostriatal pathway in a mouse model of Huntington's disease. *J Neurosci* 2003,**23:**1-9

Coyle J.T. and Schwarcz R. Lesions of striatal neurons with kainic acid provides a model of Huntington's chorea. *Nature* 1976, **263**:244-246

Dawbarn D., De Quidt M.E. and Emson P.C. Survival of basal ganglia neuropeptide Y-somatostatin neurons in Huntington's Disease. *Brain Res.*, 1985, **340**:251-260

Dure IV L.S., Young A.B., Penney J.B. Excitatory amino acid binding sites in the caudate nucleus and frontal cortex of Huntington's Disease. *Ann Neurol* 1991, **30**:785-793

Ferrante R.J., Kowall N.W. Cipolloni P.B. et al. Excitotoxin lesions in primates as a model for Huntington' Disease: histopathologic and neurochemical characterization. *Exp Neurol* 1993,**119**:46-71

Ferrante R.J., Kowall N.W., Beal M.F. et al. Selective sparing of a class of striatal neurons in Huntington's Disease. *Science* 1985, **230**:561-563

Hanson O, Guatteo E., Mercuri N.B., Bernardi G., Li X.J., Castilho R.F., Brundin P. Resistance to NMDA toxicity correlates with appearance of nuclear inclusions, behavioral deficits and changes in calcium homeostasis in mice transgenic for exon 1 of the huntington gene. *Eur J Neurosci* 2001, **14**:1492-1504

Hanson O., Petersen A., Leist M., Nicotera P., Castilho R.F., Brundin P. Transgenic mice expressing a Huntington's Disease mutation are resistant to quinolinic acid-induced striatal excitotoxicity. *Proc Natl Acad Sci USA* 1999, **96**:8727-8732

Harrington K.M. and Kowall N.W. Parvalbumin immunoractive neurons resist degeneration in Huntington's Disease striatum. *J Neuropathol Exp Neurol* 1991, **50**:309

Hickey M.A., Reynolds G.P., Morton A.J. The role of dopamine in motor symptoms in the R6/2 transgenic mouse model of Huntington's Disease. *J Neurochem* 2002,**81**:46-59

Klapstein G.J., Fisher R.S., Zanjani H., Cepeda C., Jokel E.S., Chesselet M.F., Levine M.S. Electrophysiological and morphological changes in striatal spiny neurons in R6/2 Huntington's Disease. *J Neurophysiol* 2001, **86**:2667-2677

Ludolph A.C., He F., Spencer P.S. Hammerstad J, Sabri M. 3-Nitropropionic acid: exogenous animal neurotoxin and possible human striatal toxin. *Can J Neurol Sci* 1990, **18**:492-498

McGeer E.G. and McGeer P.L.. Duplication of biochemical changes of Huntington's chorea by intrastriatal injections of glutamic and kainic acids. *Nature* 1976, **263**: 517-519

Petersen A., Chase K., Puschban Z., DiFiglia M., Brundin P., Aronin N. Maintenance of susceptibility to neurodegeneration following intrastriatal injections of quinolinic acid in a new transgenic mouse model of Huntington's disease. *Exp Neurol* 2002,**175**:297-300

Petersen A., Puschban Z., Lotharius J., NicNiocaill B., Wiekop P., O'Connor W.T., Brundin P. Evidence for dysfunction of the nigrostriatal pathway in the R6/1 line of transgenic Huntington Disease mice. *Neurobiol Dis* 2002, **11**:134-146

Schiefer J., Landwehrmeyer G.B., Luesse H.G., Sprunken A., Puls C., Milkereit E., Kosinski C.M. Riluzole prolongs survival time and alters nuclear inclusion formation in a transgenic mouse model of Huntington's Disease. *Mov Disord* 2002, **17**:748-757

Schulz J.B., Henshaw D.R., MacGarvey U and Beal M.F. Involvement of oxidative stress in 3-nitropropionic acid neurotoxicity. *Neurochem Intl* **29**:167-171

Schulz J.B., Matthews R.T., Henshaw D.R. and Beal M.F. Neuroprotective strategies for treatment of lesions produced by mitochondrial toxins: implications for neurodegenerative diseases. *Neurosci* **71**:1043-1048

Waldvogel H.J., Faull R.L.M., Williams M.N., et al. Differential sensitivity of calbindin and parvalbumin immunoreactive cells in the striatum to excitotoxins. *Brain Res* 1991, **546**:329-335

Young A.B., Greenamyre J.T., Hollingsworth Z., et al. NMDA receptor losses in putamen from patients with Huntington's Disease. *Science* 1988,**241**:981-983

Zeron M.M., Chen N., Moshaver A., Lee A.T., Wellington C.L. Hayden M.R., Raymond L.A. Mutant huntingtin enhances excitotoxic cell death. *Mol Cell Neurosci* 2001, **17**:41-53

Zeron M.M., Hansson O., Chen N., Wellington C.L., Leavitt B.R., Brundin, P., Hayden M.R., Raymond L.A. Increased sensitivity to N-methyl-D-aspartate receptor-mediated excitotoxicity in mouse model of Huntington's Disease. *Neuron* 2002, **33:**849-860

Chapter 13

THE GLUTAMATERGIC SYSTEM IN ALZHEIMER'S DISEASE BRAIN: DYSFUNCTION ASSOCIATED WITH AMYLOID β-PEPTIDE AND OXIDATIVE STRESS

D. Allan Butterfield
Department of Chemistry, Center of Membrane Sciences, and Sanders-Brown Center on Aging, University of Kentucky, Lexington, KY 40506 USA

Abstract: Excitotoxic mechanisms of neuronal death can occur in the presence of excess glutamate. Hence, means to remove or convert extra-neuronal glutamate exist in brain. However, in Alzheimer's disease (AD) brain, the glutamate transporter and glutamine synthestase function with considerably reduced activity. Consistent with the observed oxidative stress in AD brain and the oxidative stress induced by amyloid β-peptide (Aβ), both the glutamate transporter (EAAT2) and glutamine synthetase are oxidatively modified in AD brain. Moreover, Aβ causes EAAT2 and glutamine synthetase to be oxidized, consistent with the notion that excess Aβ in AD brain contributes to the oxidative stress observed. This chapter reviews studies that support the notion that the dysfunctional glutamatergic system in AD brain results from oxidative modifications of key components of this system, and that Aβ plays a role in this oxidative modification. Consequent excitotoxicity could be one means to account for neuronal and synapse loss in AD brain.

Keywords: glutamate, oxidative stress, HNE, lipid peroxidation, amyloid β-peptide.

INDEX

1. Introduction.
2. The glutamatergic system is dysfunctional in AD.
3. Conclusions.

1. Introduction

Alzheimer's disease (AD), an age-related neurodegenerative disorder, is characterized clinically by progressive loss of memory and cognition and subsequently speech (Katzman and Saitoh, 1991). Pathologically, AD is characterized by three principal hallmarks: senile (neuritic) plaques, consisting mostly of fibrillar amyloid β-peptide (Aβ) surrounded by dystrophic neurites; neurofibrillary tangles, consisting mostly of hyperphosphoryated tau, a cytoskeletal protein; and loss of synapses (Katzman and Saitoh, 1991). Except for a small percentage of AD patients who inherit the disease in an autosomal dominant manner, the pathogenesis of this dementing disorder is not known.

Oxidative stress could be experimentally measured by assaying protein oxidation (assayed by protein carbonyl levels), lipid peroxidation (assayed by a number of lipid peroxidation end-products, including 4-hydroxynonenal [HNE] and acrolein [2-propen-1-al), free radical formation (assayed often by increased fluorescence of 2',7'-dichlorofluoroscein); 3-nitrotyrosine (formed by action of NO with superoxide radical and subsequent reactions of the product, peroxynitrite) and advanced glycation end products (aggregates initiated by binding of a reducing sugar with a protein amino group followed by complex Amadori chemistry).

Oxidative stress is extensive in AD brain (Markesbery, 1997; Butterfield et al., 2001; Butterfield et al., 2002; Butterfield and Lauderback, 2002). Increased protein oxidation (Hensley et al., 1995; Castegna et al., 2002a,b), increased levels of the lipid peroxidation products HNE and acrolein (Markesbery et al., 1998; Lovell et al., 2001), excess 3-nitrotyrosine (Smith et al., 1997), increased advanced glycation end products (Smith et al., 1994), and increased DNA oxidation (Gabbita et al., 1998) are reported for AD brain.

Many researchers agree that Aβ is central to the pathogenesis of AD (Selkoe, 2001). For example, subjects who inherit AD due to mutations in the genes for presenilin-1, presenilin-2, and amyloid precursor protein (APP) always have excessive amounts of the 42-mer, Aβ(1-42). In addition, patients with Down's syndrome always develop AD if they live long enough, and the gene for APP is located on chromosome 21, the locus of the trisomy of Down's syndrome. Transgenic mice that overexpress mutant human APP and mutant tau, develop plaques before developing neurofibrillar deposits. [Reviewed in Selkoe, 2001].

Our laboratory combined these two concepts: the centrality of Aβ(1-42) to the pathogenesis of AD on the one hand with the oxidative stress under which the AD brain exists on the other into a comprehensive, Aβ(1-42)-centered model for neurodegeneration in AD brain based on the oxidative stress associated with Aβ(1-42) (Butterfield et al. 2001; 2002; Butterfield and Lauderback, 2002; Varadarajan et al., 2000; Butterfield, 2002). Consistent with the model, Aβ(1-42) induces protein oxidation (Yatin et al., 1999); lipid peroxidation (Lauderback et al., 2001; Butterfield et al., 2002); reactive oxygen species production (Yatin et al., 2000) and many other markers of oxidative stress (Butterfield, 2002). These effects of Aβ(1-42) are blocked by antioxidants.

Glutamate is the principal excitatory neurotransmitter in the mammalian CNS and is essential for synaptic transmission. When glutamate levels increase dramatically, excessive neuronal stimulation causes damage to and subsequent loss of neurons in a process known as excitotoxicity (Maragakis and Rothstein, 2001). Therefore, clearance of excitotoxic glutamate from the synaptic cleft is essential for neuronal survival. Astrocyte-mediated glutamate uptake is the primary mean of maintaining synaptic glutamate concentrations (Anderson et al. 2000; Rothstein et al., 1996). If glutamate transport is inhibited, neurodegeneration and death ensue (Rothstein et al., 1993; Robinson et al., 1993; Lievens et al., 2000), consistent with a role for impaired glutamate transport in excitotoxicity.

Although evidence is scant in some cases, excitotoxicity and subsequent neuronal death have been implicated in several neurodegenerative disorders, including stroke, amyotrophic lateral sclerosis (ALS) and AD (Maragakis and Rothstein, 2001). In AD, glutamate transport is inhibited (Masliah et al., 1996), though until recently (Lauderback et al., 2001), the mechanism for inhibition of glutamate tranport in AD brain was unknown (see below).

2. The glutamatergic system is dysfunctional in AD

Glutamate is involved in cognitive processes, (Collingridge and Singer, 1990), possibly involving synaptic protein remodeling (Sheng and Kim, 2002). Synapse-resident enzymes to degrade glutamate are not known. Consequently, as mentioned above, this excitatory neurotransmitter must be cleared from the synapse by high-affinity presynaptic and glial transporters. Five different forms of glutamate transporters have been identified: GLAST (EAAT1), GLT-1 (EAAT2), EAAC-1 (EAAT3), EAAT4 and EAAT5. GLAST and GLT-1 are found exclusively in astroglial cells. GLT-1 is observed throughout maturity, (Furuta, et. al., 1997), while GLAST expression is observed during early stages of development.

Consistent with a role for Aβ(1-42) in AD, recent studies suggest that glutamate-induced NMDA receptor activation stimulates APP processing to produce Aβ peptide (Gordon-Krajcer et. al., 2002). Consequently, over-stimulation of the receptor by excess glutamate could lead to excess Aβ(1-42) production with consequent oxidative stress-induced neurotoxicity (Butterfield et al., 2001; 2002; Butterfield and Lauderback, 2002). Similarly, the reported decreased number of NMDA receptors in AD brain (Greenamyre, 1986) may lead to their overstimulation by excessive glutamate. Along with Aβ(1-42), other sources of free radicals upon NMDA receptor activation have been reported (LaFon-Cazal et al., 1993).

Neurodegeneration in AD was recently reported to be associated with expression of EAAT-2 glutamate transporter, which was localized to tangle-bearing neurons (Thal, 2002). The expression of a related transporter, EAAT-1 (GLAST), also was reported for AD brain (Scott et al., 2002).

The EAAT-2 (GLT-1) glutamate transporter system in AD brain has a significant reduction of activity (Masliah et al., 1996; Scott, et al., 1995), increasing the possibility for excitotoxicity. Table I shows that this transporter is oxidatively modified by HNE, a lipid peroxidation product, in the inferior parietal region of AD brain (Lauderback et al., 2001). HNE is known to bind
to proteins by Michael addition (Butterfield and Stadtman, 1997), and in synaptosomes HNE alters the structure of synaptosomal membrane proteins (Subramaniam et al., 1997).

Table I. Relative percent increase in HNE binding to GLT-1 in AD brain or following exposure of synaptosomes to Aβ(1-42)*

	Mean +/- SEM	N	P-Value
AD Brain	155 +/- 7	4-7	<0.002
Following exposure of synaptosomes to Aβ(1-42)	182 +/- 6	5-6	<0.002

*Lauderback et al., 2001

This alkenal also decreases the activity of numerous transmembrane enzymes and transporters (Butterfield et al., 2002), presumably because covalent modification of the proteins changes their structure and thus their function. Consequently, increased HNE binding to

GLT-1 in AD brain (Lauderback et al., 2001) is likely to be the cause of the loss of activity of this transporter in AD brain (Masliah et al., 1996). Aβ(1-42) added to synaptosomes leads to lipid peroxidation (Butterfield et al., 2002; Lauderback et al., 2001; Butterfield, 2002) and increased HNE binding to GLT-1 (Lauderback, et al., 2001), suggesting that, in AD brain, where excess Aβ(1-42) deposition occurs, Aβ-induced lipid peroxidation could account for the increased HNE binding to GLT-1. Inhibition of glutamate transport, coupled with decreased activity of glutamine synthetase in AD brain (see below), might lead to increased extra-neuronal glutamate with consequent stimulation of NMDA receptors and excitotoxic processes involving excess intraneuronal Ca^{2+} accumulation and eventual cell death (Figure 1). In astrocytes, Aβ (Harris et al., 1995) and HNE (Blanc et al., 1998) inhibit glutamate uptake. Their inhibitory effects are modulated by antioxidants or glutathione, respectively.

Recent studies report that, in contrast to the usual case and in certain cells of altered shape, GLT-1 is expressed in neurons (Thal, 2002). A different alkenal, acrolein, which also alters the structure of synaptosomal membranes (Pocernich et al., 2000; 2001), decreased uptake of glutamate in neurons (Lovell et al., 2001). In contrast to the decreased activity of glutamate transport in AD brain (Masliah et al., 1996, Scott et al., 1995), where excess Aβ is observed, and in contrast to the many studies suggesting that Aβ peptides decrease glutamate uptake (reviewed in Butterfield and Pocernich, 2003), one study recently reported that Aβ enhances glial glutamate uptake activity with subsequent decreased synaptic activity (Ikegaya et al., 2002). Impaired learning was predicted if this phenomenon occurred in AD brain.

In glia, glutamate is converted to glutamine by glutamine synthetase (GS). The glutamine is released, taken up into neurons and converted to glutamate by mitochondrial glutaminase. The activity of GS is significantly decreased in AD brain (Smith et al., 1991; Hensley et al., 1995; Butterfield et al., 1997). This loss of activity may result from specific oxidation of GS in AD (see below). Consistent with the loss of GS activity in AD, the levels of glutamate + glutamine (GLX) were substantially reduced in the cingulate cortex of AD patients, and this loss strongly correlated with both the cognitive and functional status of the patients (Antuono et al., 2001). Some reports of GLX in AD revealed no significant difference in the mid-frontal or temporoparietal GLX, (Moats et al., 1994), but an increase in GLX in the occipital lobe of AD patients was found (Ernst et al., 1997).

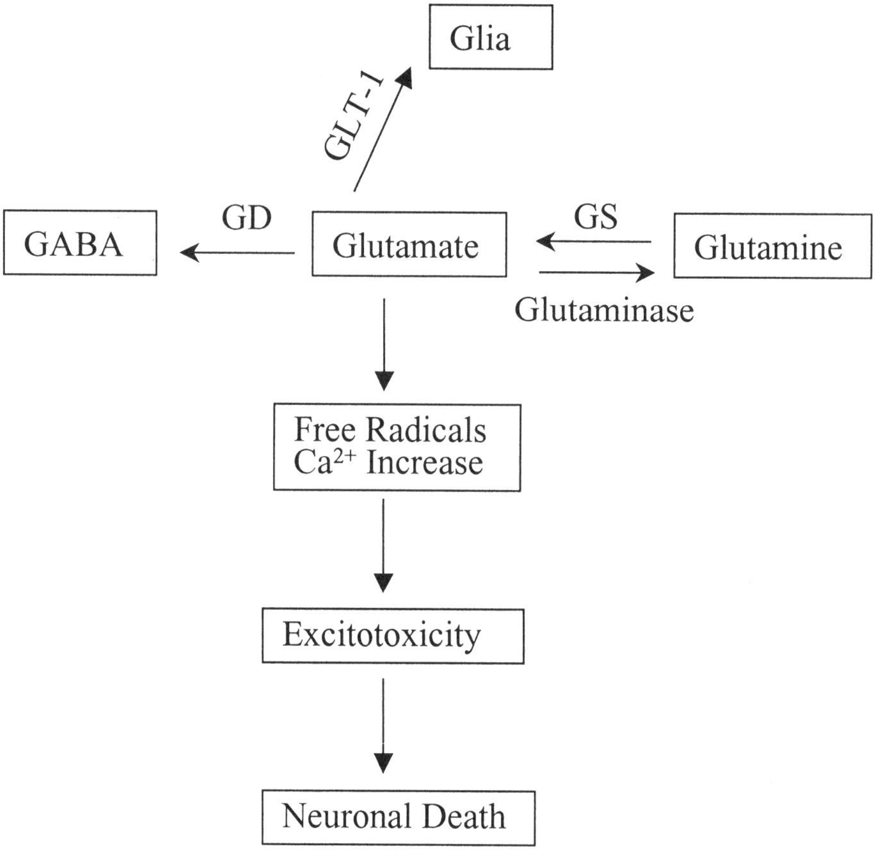

Figure 1. Glutamatergic system. In AD brain, excess glutamate, resulting from decreased glutamine synthetase (GS) activity and decreased GLT-1 transport function due to oxidative modification in both cases, could lead to excitotoxicity and neuronal death. See text.

GS is sensitive to oxidative stress (Aksenov et al., 1995; 1997; Butterfield et al., 1997). Novel electron paramagnetic resonance (EPR) experiments suggested that GS isolated and purified from AD brain was more oxidized than that isolated and purified from control brain (Butterfield et al., 1997). This suggestion was confirmed by direct determination using proteomics that GS is specifically oxidized in AD brain (Castegna et al., 2002a). The decreased enzymatic activity of GS activity in AD hippocampus and neocortex (Hensley et al., 1995) contributes to the increased possibility of excitotoxicity being an important mechanism for neurodegeneration in AD brain. Aβ(1-42) significantly decreased GS activity in cytosolic fractions of brain homogenates and in cultured hippocampal neurons and astrocytes Varadarajan et al., 2000; Butterfield 1997), suggesting the possibility for connecting the decreased activity of GS with Aβ(1-42)-induced oxidative modification of the enzyme in AD brain. In support of this suggestion, Aβ added to GS caused increased protein carbonyl on the enzyme (Aksenov et al., 1997). Fibrils are normally formed upon incubation of Aβ; however, in the presence of GS, small, non-fibillar aggregates of Aβ, that were more toxic than a solution of Aβ alone, were formed (Aksenov et al., 1995). Recent *in vivo* studies showed that small aggregates of Aβ(1-42) induced protein oxidation in a living animal (Drake et al., 2003), in accord with the notion that these small aggregates of Aβ(1-42) may be the toxic species of this peptide (Oda et al., 1995; Aksenov et al., 1995; Lambert et al., 1998; Walsh et al., 2002). Consistent with oxidative stress associated with both Aβ and excitotoxicity, treatment of non-toxic levels of Aβ potentiates excitiotoxic effects of glutamate, suggesting synergy between these two sources of oxidative stress (Mattson et al., 1993).

Diagnosis of AD is confirmed by histological examination at autopsy. Definitive diagnosis of living AD patients, though much improved in recent years, is still largely by elimination of other sources of dementia and is subject to a modest error rate (Katzman and Saitoh, 1991). Consequently, definitive diagnosis of AD requires some sort of biomarker. One such biomarker proposed for AD is the observation that GS is found in CSF and not able to be photolabeled by an azido-ATP photoprobe (Gunnersen and Haley, 1992). This lack of photolabeling may reflect the finding that GS is a specifically oxidized protein in AD brain (Castegna et al., 2002a).

Glutamate is also converted to γ-aminobutyric acid (GABA), an inhibitory neurotransmitter, by glutamate decarboxylase (GD). The AD messenger RNA for GD reportedly is increased in the caudate nucleus and putamen (Boissiere et al., 1998). In contrast, no alteration was found in the AD ventral striatum, suggesting that GABAergic neurotransmission may be increased in the dorsal striatum but not in the ventral striatum. Consistent with the increased messenger RNA in AD brain, GD expression is increased, potentially resulting in increased GABA levels and cell death (Hertz et al., 1988).

As noted above, neurons produce glutamate from glutamine by the enzyme glutaminase. In AD temporal cortex, no significant changes were seen in the levels of phosphate-activated glutaminase (Haug et al., 1996), suggesting that neurons in the cortex area of AD still have the ability to produce glutamate. In contrast, glutamate and glutaminase immunoreactive pyramidal neurons in the hippocampal dentate gyrus were decreased in number in AD brain (Kowall and Beal, 1991). These same neurons were found to contain neurofibrillary tangles. This regional difference in glutamatergic neurons conceivably could mirror the role of the hippocampus in processing memory, which is severely diminished in AD patients.

3. Conclusions

Oxidative stress occurs in AD brain and is associated with Aβ(1-42) (Butterfield et al., 2001; 2002; Butterfield and Lauderback, 2002; Butterfield, 2002). This oxidative stress leads to formation of reactive alkenals, like HNE, and protein oxidation (Butterfield and Stadtman, 1997; Lauderback et al., 2001). Oxidative stress and its sequelae provide mechanisms for the oxidative modification and subsequent loss of activity of the glutamate transporter EAAT-2 [GLT-1] and GS in AD brain (Masaliah, et al., 1996; 2000; Hensley et al., 1995; Lauderback et al., 2001; Butterfield et al., 1997; Castegna et al., 2002a). Neurons, exposed to the resulting excess glutamate, could undergo excitotoxicity (Figure 1) with subsequent cell death.

A number of pharmacological interventions to stimulate glutamate transport and prevent excitotoxicity are being pursued (reviewed in Butterfield and Pocernich, 2003). However, given that a major cause of the underlying glutamatergic dysfunction in AD brain might result from oxidative modification of important components of the glutamatergic system in AD, another therapeutic strategy to be considered is to increase the endogenous and exogenous antioxidant levels in AD patients (Butterfield et al., 2002b,c).

Excitotoxicity may be an important contributor to neuronal degeneration in AD brain. Strategies to significantly modulate, if not eliminate, excitotoxicity in AD should be beneficial to patients suffering from this devastating dementing disorder.

Acknowledgements. This work was supported in part by grants from The National Institutes of Health [AG-10836; AG-05119].

REFERENCES

Aksenov M.Y., Aksenova M.V., Harris M.E., Hensley K., Butterfield D.A., Carney J.M. Enhancement of Aβ(1-40) neurotoxicity by glutamine synthetase. J. Neurochem. 1995; 65:1899-1902.

Aksenov M.Y., Aksenova M.V., Carney J.M., Butterfield D.A. Oxidative modification of glutamine synthetase by amyloid beta peptide. Free Radical Res 1997; 27:267-281.

Anderson C.M., Swanson R.A. Astrocyte glutamate transport: review of properties, regulation, and physiological functions. Glia 2000; 32:1-14.

Antuono P.G., Jones J.L., Wang Y., Li S.J. Decreased glutamate + glutamine in Alzheimer's disease detected in vivo with ^{1}H-MRS at 0.5T. Neurology 2001; 56:737-742.

Blanc E.M., Keller J.N., Fernandez S. Mattson M.P. 4-Hydroxynonenal, a lipid peroxidation product, impairs glutamate transport in cortical astrocytes. Glia 1998; 22:149-60.

Boissiere F., Faucheux B., Duyckaerts C., Hauw J.J., Agid Y., Hirsch E.C. Striatal expression of glutamic acid decarboxylase gene in Alzheimer's disease. Neurochem 1998; 71:767-774.

Butterfield D.A. Beta-amyloid-associated free radical oxidative stress and neurotoxicity: implications for Alzheimer's disease. Chem Res Toxicol 1997; 10: 495-506.

Butterfield D.A. Amyloid β-peptide (1-42)-induced oxidative stress and neurotoxicity: implications for neurodegeneration in Alzheimer's disease brain. A Review. Free Radical Res 2002; 36:1307-1313.

Butterfield D.A., Stadtman E.R. Protein oxidation processes in aging brain. Adv. Cell Aging Gerontol 1997; 2:161-191.

Butterfield D.A., Lauderback C.M. Lipid peroxidation and protein oxidation in Alzheimer's Disease brain: potential causes and consequences involving amyloid β-peptide-associated free radical oxidative stress. Free Radical Biol. & Med 2002; 32:1050-1060.

Butterfield D.A., Pocernich C.B. The glutamatergic system and Alzheimer's disease: therapeutic implications. CNS Drugs 2003; in press.

Butterfield D.A., Hensley K., Harris M., Mattson M., Carney J. β–Amyloid peptide free radical fragments initiate synaptosomal lipoperoxidation in a sequence-specific fashion: implications to Alzheimer's disease. Biochem. Biophys. Res. Commun 1994; 200:710-715.

Butterfield D.A., Hensley K., Cole P. Subramaniam R., Aksenov M., Aksenova M., Bummer P.M., Haley B.E., Carney J.M., Oxidatively-induced structural alteration of glutamine synthetase assessed by analysis of spin labeled incorporation kinetics: relevance to Alzheimer's disease. J. Neurochem 1997; 68:2451 2457.

Butterfield D.A., Drake J., Pocernich C., Castegna A. Evidence of oxidative damage in Alzheimer's disease brain: central role of amyloid β-peptide. Trends Molecular Med 2001; 7:548-554.

Butterfield D.A., Castegna A., Lauderback C.M., Drake J. Review: Evidence that amyloid beta-peptide-induced lipid peroxidation and its sequelae in Alzheimer's disease brain contributes to neuronal death. Neurobiology of Aging 2002a; 23:655-664.

Butterfield D.A., Pocernich C.B., Drake J. Elevated gutathione as a therapeutic strategy in Alzheimer's disease. Drug Discovery Research 2002b; 56:428-437.

Butterfield D.A., Castegna A., Pocernich C.B., Drake J., Scapagnini G., Calabrese V. Nutritional approaches to combat oxidative stress in Alzheimer's disease. J. Nutritional Biochemistry 2002c: 13:444-461.

Castegna A., Aksenov M., Aksenova M. Thongboonkerd V., Klein J.B., Pierce W.M., Booze R., Markesbery W.R., Butterfield D.A., Proteomic identification of oxidatively modified proteins in Alzheimer's disease brain. Part I: creatine kinase BB, glutamine synthase, and ubiquitin carboxy-terminal hydrolase L-1. Free Radical Biol. & Med 2002a; 33:562-571.

Castegna A., Aksenov M., Aksenova M., Thongboonkerd V., Klein J.B., Pierce W.M., Booze R., Markesbery W.R., Butterfield D.A. Proteomic identification of oxidatively modified proteins in Alzheimer's disease brain. Part II: dihydropyrimidinase-related protein 2, ∀-enolase, and heat shock cognate 71. J. Neurochemistry 2002b; 82:1524-1532.

Collingridge G.L., Singer W. Excitatory amino acid receptors and synaptic plasticity. Trends Pharmacol Sci 1990; 11:290-296.

Drake J., Link C.D., Butterfield D.A. Oxidative stress precedes fibrillar deposition of Alzheimer's disease amyloid β-peptide (1-42) in a transgenic *Caenorhabditis elegans* model. Neurobiology of Aging 2003; in press.

Ernst T., Chang L., Melchor R., Mehringer C.M. Frontotemporal dementia and early Alzheimer disease: differentiation with frontal lobe H-1 MR spectroscopy. Radiology 1997; 203:829-836.

Furuta A., Rothstein J.D., Martin J.L. Glutamate transporter protein subtypes are expressed differentially during rat central nervous system development. J Neurosci 1997; 17:8363-8375.

Gabbita S.P., Lovell M.A., Markesbery W.R. Increased nuclear DNA oxidation in the brain in Alzheimer's disease. J. Neurochem. 1998; 71:2034-2040.

Gordon-Krajcer W., Salinska E., Lazarewicz J.W. N-methyl-D-aspartate receptor-mediated processing of beta-amyloid precursor protein in rat hippocampal slices: in vitro--superfusion study. Folia Neuropathol 2002; 40:13-17.

Greenamyre J.T. The role of glutamate in neurotransmission and neurologic disease. Arch Neurol 1986; 43:1058-63.

Gunnersen, D., Haley B. Detection of glutamine synthetase in the cerebrospinal fluid of Alzheimer diseased patients: a potential diagnostic biochemical marker. Proc. Nat. Acad. Sci. USA 1992; 89:11949-11953.

Harris M.E., Hensley K., Butterfield D.A., Leedle R.A., Carney J.M. Direct evidence of oxidative injury produced by the Alzheimer's amyloid beta peptide (1-40) in cultured hippocampal neurons. Exp. Neurol 1995; 131:193-202.

Haug L.S., Ostvold A.C., Cowburn R.F., Garlind A., Winblad B., Bogdanovich N., Walaas S.I. Decreased inositol (1,4,5)-trisphosphate receptor levels in Alzheimer's disease cerebral cortex: selectivity of changes and possible correlation to pathological severity. Neurodegeneration 1996; 5:169-176.

Hensley K., Hall N., Subramaniam R., ColeP., Harris M., Aksenov M., Aksenova M., Gabbita S.P., Wu J.F., Carney J.M., Lovell M., Markesbery W.R., Butterfield D.A. Brain regional correspondence between Alzheimer's disease histopathology and biomarkers of protein oxidation. J. Neurochem 1995; 65:2146-2156.

Hertz L., Drejer J., Schousboe A. Energy metabolism in glutamatergic neurons, GABAergic neurons and astrocytes in primary cultures. Neurochem Res 1988; 13:605-610.

Ikegaya Y., Matsuura S., Ueno S., Baba A., Yamada M.K., Nishiyama N., Matsuki N. Beta-amyloid enhances glial glutamate uptake activity and attenuates synaptic efficacy. J Biol Chem. 2002; 277:32180-6.

Katzman R., Saitoh T. Advances in Alzheimer's disease. FASEB J. 1991; 5:278-86.

Kowall N.W., Beal M.F. Glutamate-, glutaminase-, and taurine-immunoreactive neurons develop neurofibrillary tangles in Alzheimer's disease. Ann Neurol 1991; 29:162-167.

Lafon-Cazal M., Pietri S., Culcasi M., Bockaert J. NMDA-dependent superoxide production and neurotoxicity. Nature.1993; 364:535-537.

Lambert M.P., Barlow A.K., Chromy B.A., Edwards C., Freed R., Liosatos M., Morgan T.E., Rozovsky I., Trommer B., Viola K.L., Wals P., Zhang C., Finch C.E., Krafft G.A., Klein W.L. Diffusible, nonfibrillar ligands derived from Abeta1-42 are potent central nervous system neurotoxins. Proc Natl Acad Sci 1998; 95:6448-53.

Lauderback C.M., Hackett J.M., Huang F.F., Keller J.N., Szweda L.I., Markesbery W.R., Butterfield D.A. The glial glutamate transporter, GLT-1, is oxidatively modified by 4-hydroxy-2-nonenal in the Alzheimer's disease brain: Role of Aβ1-42. J. Neurochemistry 2001; 78:413-416.

Lievens, J.C., Bernal, F., Forni, C., Mahy, N., Kerkerian-Le Goff, L. Characterization of striatal lesions produced by glutamate uptake alteration: cell death, reactive gliosis, and changes in GLT1 and GADD45 mRNA expression. Glia 2000; 29:222-232.

Lovell M.A., Xie C., Markesbery W.R. Acrolein is increased in Alzheimer's disease brain and is toxic to primary hippocampal cultures. Neurobiol Aging 2001; 22:187-94.

Maragakis N.J., Rothstein J.D. Glutamate transporters in neurologic disease. Arch Neurol. 2001; 58:365-70. Review.

Masliah E., Alford M., DeTeresa R., Mallory M., Hansen L. Deficient glutamate transport is associated with neurodegeneration in Alzheimer's disease. Ann Neurol 1996; 40:759-66.

Masliah E., Alford M., Mallory M., Rockenstein E., Moechars D., Van Leuven F. Abnormal glutamate transport function in mutant amyloid precursor protein transgenic mice. Exp. Neurol 2000; 163:381-387.

Markesbery W.R. Oxidative stress hypothesis in Alzheimer disease. Free Radic Biol Med 1997; 23:134-47.

Markesbery W.R., Lovell M.A. 4-Hydroxynonenal, a product of lipid peroxidation, is increased in the brain in Alzheimer's disease. Neurobiol Aging 1998; 19:33-6.

Mattson M.P., Barger S.W., Cheng B., Lieberburg I., Smith-Swintosky V.L., Ryel R.E. β-amyloid precursor protein metabolites and loss of neuronal Ca^{2+} homeostasis Alzheimer's disease. Trends Neurosci. 1993; 16:409-414.

Moats R.A., Ernst T., Shonk T.K., Ross B.D. Abnormal cerebral metabolite concentrations in patients with probable Alzheimer disease. Magn Reson Med 1994; 32:110-115.

Oda T., Wals P., Osterburg H.H., Johnson S.A., Pasinetti G.M., Morgan T.E., Rozovsky I., Stine W.B., Snyder S.W., Holzman T.F., Krafft G.A., Finch C.E. Clusterin (apoJ) alters the aggregation of amyloid beta-peptide (A beta 1-42) and forms slowly sedimenting A beta complexes that cause oxidative stress. Exp Neurol 1995;136:22-31.

Pocernich C.B., La Fontaine M., Butterfield D.A. In-vivo glutathione elevation protects against hydroxyl free radical-induced protein oxidation in rat brain. Neurochem Int 2000

36:185-191.

Pocernich C.B., Cardin A.L., Racine C.L., Lauderback C.M., Butterfield D.A. Glutathione elevation and its protective role in acrolein-induced protein damage in synaptosomal membranes: relevance to brain lipid peroxidation in neurodegenerative disease. Neurochem Int 2001; 39:141-149.

Robinson, M.B., Djali, S., Buchhalter, J.R. Inhibition of glutamate uptake with L-trans-pyrrolidine-2,4-dicarboxylate potentiates glutamate toxicity in primary hippocampal cultures. J. Neurochem 1993; 61:2099-2103.

Rothstein, J.D., Jin, L., Dykes-Hoberg, M., Kuncl, R.W. Chronic inhibition of glutamate uptake produces a model of slow neurotoxicity. Proc. Natl. Acad. Sci. 1993; 90:6591-6595.

Rothstein, J.D., Dykes-Hoberg, M., Pardo, C.A., Bristol, L.A., Jin, L., Kuncl, R.W., Hediger, M.A., Wang, Y., Schielke, J.P., Welty, D.F. Knockout of glutamate transporters reveals a major role for astroglial transport in excitotoxicity and clearance of glutamate. Neuron 1996; 16:675-686.

Scott H.L., Tannenberg A., Dodd P.R. Variant forms of neuronal glutamate transporter sites in Alzheimer's disease cerebral cortex. J Neurochem 1995; 64:2193-2202.

Scott H.L., Pow D.V., Tannenberg A.E., Dodd P.R. Aberrant expression of the glutamate transporter excitatory amino acid transporter 1 (EAAT1) in Alzheimer's disease. J Neurosci. 2002; 22:RC206.

Selkoe D.J. Alzheimer's disease results from the cerebral accumulation and cytotoxicity of amyloid beta-protein. J Alzheimers Dis. 2001; 3:75-80.

Sheng M., Kim M.J. Postsynaptic signaling and plasticity mechanisms. Science 2002; 298:776-780.

Smith C.D., Carney J.M. Starke-Reed P.E., Oliver C.N., Stadtman E.R., Floyd R.A., Markesbery W.R. Excess brain protein oxidation and enzyme dysfunction in normal aging and Alzheimer disease. Proc. Natl. Acad. Sci. 1991; 88:10540-10543.

Smith M.A., Richey P.L., Taneda S., Kutty R.K., Sayre L.M., Monnier V.M., Perry G. Advanced Maillard reaction end products, free radicals, and protein oxidation in Alzheimer's disease. Proc Natl Acad Sci 1994; 91:5710-4.

Smith M.A., Richey Harris P.L., Sayre L.M., Beckman J.S., Perry G. Widespread peroxynitrite-mediated damage in Alzheimer's disease. J Neurosci 1997; 17:2653-7.

Subramaniam R., Roediger F., Jordan B., Mattson M.P., Keller J.N., Waeg G., Butterfield D.A. The lipid peroxidation product, 4-Hydroxy-2-trans-nonenal, alters the conformation of cortical synaptosomal membrane proteins. J. Neurochem 1997; 69:1161-1169.

Thal D.R. Excitatory amino acid transporter EAAT-2 in tangle-bearing neurons in Alzheimer's disease. Brain Pathol 2002; 12:405-11.

Varadarajan S., Yatin S., Aksenova M., Butterfield D.A. Review: Alzheimer's amyloid β-peptide-associated free radical oxidative stress and neurotoxicity. J. Struct Biol 2000 130:184-208.

Walsh D.M., Klyubin I., Fadeeva J.V., Cullen W.K., Anwyl R., Wolfe M.S., Rowan M.J., Selkoe D.J. Naturally secreted oligomers of amyloid beta protein potently inhibit hippocampal long-term potentiation in vivo. Nature 2002 416:535-9.

Yatin S.M., C. D. Link C.D., Butterfield D.A. In-vitro and in-vivo oxidative stress associated with Alzheimer's amyloid β-peptide (1-42). Neurobiol of Aging 1999; 20:325-330.

Yatin S.M., Varadarajan S., Butterfield D.A. Vitamin E Prevents Alzheimer's amyloid β-peptide (1-42)-induced protein oxidation and reactive oxygen species formation. Alzheimer's Disease 2000; 2:123-131.

Chapter 14
NEUROTOXICITY AND PRION DISEASE

Judyth Sassoon, Filmore Banks and David R. Brown
Dept. Biology and Biochemistry, University of Bath, BA2 7AY, U.K.

Abstract: Prion diseases are characterised by the conversion of a normal glycoprotein, the prion protein, to an abnormal protease resistant form, which is suggested to be both the infectious agent and the cause of neuronal cell death in the disease. Death of the patient results from neurodegeneration. In this review, we present data arguing that neurodegeneration in prion diseases is a complex process, involving the action of other factors in addition to the abnormal prion protein. Clearly, an understanding of the mechanisms by which prion mediated neuronal death occurs might lead to treatments that would prevent the life threatening nature of these diseases.

Keywords: astrocytes, exitotoxicity, glutamate, microglia, neurotoxicity, prion protein.

INDEX

1. Introduction
2. Prion protein and copper binding
3. Neurotoxicity studies
   3.1 Toxic peptide mimics of prp^{sc}
      3.3.1. Animal models of prion induced neurodegeneration
      3.3.2. Oxidative stress and cell death
   3.2. Proposed mechanisms of prp106-126 action
   3.3. Mechanisms of cell death
   3.4. Possible toxic domains in prp^{sc}
4. Excitotoxicity and neuronal cell death
5. Conclusions

1. Introduction

Prion diseases (transmissible spongiform encephalopathies) are a class of transmissible, fatal, neurodegenerative diseases, which include

Creutzfeldt-Jakob disease in humans as well as bovine spongiform encephalopathy (BSE) and scrapie in animals. They are characterized by the accumulation of a disease-associated isoform of the PrP protein, termed PrP^{Sc}, in the brains of affected individuals. PrP^{Sc} is the only molecule consistently found in infectious preparations and it has been proposed to be the main, if not the only component of the transmissible agent. The transmissible agent is now commonly referred to as the "prion" (Prusiner, 1982; Prusiner, 1993; Weissmann, 1996). It is generally accepted that PrP^{Sc} is derived through an ill-defined posttranslational process involving conformational changes from the normal cellular isoform of the prion protein (PrP^C). PrP^C is a 30-33 kDa cell surface glycoprotein, which is highly expressed in neurons (Kretzschmar et al., 1986), in glia (Moser et al., 1995), and to a lesser degree in other cell types (Manson et al., 1992; Moser et al., 1995). A homologue of PrP^C is encoded in the genome of all animals investigated to date. PrP^{Sc} can be distinguished from PrP^C by the high level of beta-sheet conformations in its structure and its partial resistance to protease digestion (Meyer et al., 1986; Pan et al., 1993). All prion diseases, including Creutzfeldt-Jakob disease (CJD), bovine spongiform encephalopathy (BSE), scrapie of sheep, and chronic wasting disease (CWD) of deer can be transmitted between individuals or even across species by injecting a brain extract from an infected individual into the brain of a recipient. In nature, however, prion diseases are not spread in this way and indeed the majority of prion diseases are sporadic, occurring with no evidence of transmission of any kind. Nevertheless, research on the mechanism by which prion disease can be transmitted experimentally has continued with many researchers believing that the prion hypothesis is somewhat limited and that other factors must also be involved in the spread of the diseases.

The most convincing evidence that expression of PrP^c is essential production of the infectious agent in prion disease transmission comes from work with mice in which PrP^C expression was ablated or knocked out (Büeler et al., 1992). Without PrP^C expression, mice were completely resistant to scrapie infection (Büeler et al., 1993). The reason for this was that there was no PrP^C to be converted to PrP^{Sc}. Furthermore transplants of mouse embryonic tissue, overexpressing $PrP^{C,}$ were made into a PrP-knockout mouse brain and the transplanted tissue infected with mouse scrapie. This transplanted tissue generated PrP^{Sc} and showed signs of neurodegeneration while the PrP-knockout brain tissue surrounding it remained untouched. PrP^{Sc} accumulated throughout the PrP-knockout mouse brains but caused no neurodegeneration. (Brandner et al., 1996). These results show that neurones must express PrP^C in order to be killed by the agent of neurodegeneration in these diseases. Clearly an understanding of the conversion process from PrP^C to PrP^{Sc} is a very important area of study. However, while exogenous PrP^{Sc} is an important consideration for investigations of the conversion mechanism, sporadic diseases, where there is no exogenous PrP^{Sc} introduced can not be explained by this process.

2. Prion protein and copper binding

When PrP^{C} converts to PrP^{Sc}, it undergoes a structural change and this is associated with the loss of its normal *in vivo* function. A knowledge of the exact nature of that function would provide clues as to why the loss of PrP^{C} is associated with such a devastating disease. It has been demonstrated that both recombinant and brain-derived PrP^{C} have superoxide dismutase (SOD)-like activity when bound to copper ions (Cu^{2+}) (Brown et al., 2001; Brown et al., 1999c). The copper complexation in PrP is different, however, to that in cellular Cu/Zn-SOD (van Doorslaer et al., 2001). It also appears that the protein transports copper ions, thereby increasing cellular resistance to copper toxicity (Brown et al., 1998b). When PrP^{C} converts to PrP^{Sc}, this SOD-like function is lost (Thakray et al., 2002). It is interesting to consider this property in relation to the consequent oxidative stress that would arise in the cell when this enzymatic function is lost, as an excess of reactive oxygen species may prove to be an important cause of neuronal damage.

NMR studies have shown that PrP^{C} consists of a structured C-terminal region, which is primarily α-helical, and an unstructured N-terminal region (Riek et al., 1996; Zahn et al., 2000). PrP^{C} exhibits high affinity, cooperative copper binding through a histidine-containing octapeptide repeat domain in the unstructured N-terminal region (Viles et al., 1999; Brown et al., 1997a). It may also bind copper along the more structured C-terminal domain of the protein (Cereghetti et al., 2001; van Doorslaer et al., 2001). Continuous wave electron paramagnetic resonance studies suggested that copper first binds and fills the C-terminal binding sites before occupying the octarepeats at the N-terminus (Cereghetti et al., 2001). Recombinant PrP^{C} was also found to have the capacity to bind other metal ions such as manganese (Mn^{2+}) (Brown et al., 2000) in both the octarepeats and the C-terminal sites (Collinge, 2001). *In vitro* metal ion occupancy experiments showed that when manganese replaced the copper ion in the normal prion protein, PrP^{C} altered its structure and took on a more PrP^{Sc}-like conformation (Brown et al., 2000) while also losing its SOD-like function (Brown et al., 2000). Thus metal ions might provide one potential external factor that could catalyse the conversion from normal prion to abnormal prion. The consequent loss of the prion protein's antioxidant function might then provide the basis for neurodegenerative processes.

3. Neurotoxicity studies

3.1. Toxic peptide mimics of PrP^{Sc}

The first studies on neuronal death as a consequence of prion disease were made using cell culture models (Müller et al., 1993; Forloni et al., 1993) and this has remained the preferred model system for the majority of investigations on neurodegeneration. Synthetic peptides of PrP have been generated for these studies, primarily because of the practical difficulties in isolating and purifying PrP^{Sc} from infected tissues. A peptide (PrP106-126), twenty-one amino acids in length, was identified by Forloni et al. (1993) and appeared to represent the neurotoxic core of the prion protein. Since then, virtually every group studying neuronal death in prion diseases has used this peptide or sequences equivalent to it. Any concerns that this peptide was only neurotoxic *in vitro* and not *in vivo* were shown to be unfounded following studies demonstrating that PrP106-126 was toxic to retinal cells in living rats (Ettaiche et al., 2001). Additionally, the "miniprion", a 106 amino acid residue PrP peptide, expressed in transgenic mice, retained the 106-126 region (Suppattapone et al., 1999). This miniprion was sufficient to induce prion disease and neuronal death *in vivo*. The research of the current authors' group with the PrP106-126 peptide has been extensive and a number of reviews are available (Brown, 2001b, 2001c).

Other short peptides based on the sequence of PrP^{Sc} have been studied in addition to PrP106-126. However, the toxicity of other peptides was apparently less significant (Brown, 2000a) or dependent on mutations involved in inherited prion disease (Forloni et al., 1999). In some cases different groups reported that these peptides were non-toxic (Forloni et al., 1993) while others claimed the exact opposite (Rymer and Good, 2000). In the absence of consistent findings most interest has remained with the peptide PrP106-126.

3.2. Proposed mechanisms of PrP106-126 action

PrP106-126 has many physicochemical qualities like those of PrP^{Sc} including protease resistance, high β-sheet content and the propensity to aggregate into fibrils (Forloni et al., 1993; Tagliavini et al., 1993; De Giola et al., 1994; Salmona et al., 1999). Studies have shown that in many ways PrP106-126 mimics PrP^{Sc}, a feature which is clearly important when considering the neurotoxic nature of either PrP^{Sc} or PrP106-126. The propensity to aggregate was suggested to be a factor contributing to the toxicity of PrP106-126. However, it was also shown that soluble non-aggregated peptide had toxic properties (Brown et al., 1998a). Other authors have argued that aggregation is unnecessary for toxicity (Salmona et al., 1999). Research using PrP106-126 continues to be an expanding field with many groups proposing new insights or new mechanisms for the action of this peptide (e.g. Haïk et al., 2000; Rymer and Good, 2000; Bate et al., 2001; Brown, 2001b, 2001c; Della-Bianca et al., 2001; Deli et al., 2001; Fabrizi et al., 2001; Gu et al., 2001; Kourie et al., 2001; Le et al., 2001; O'Donavan et al., 2001; Stewart et al., 2001; White et al., 2001). The different mechanisms suggest a variety of factors that might be involved. They range from proposed interactions between PrP106-126 at the membrane with proteins such as the p75 NGF receptor (Della-Bianca et al., 2001), formyl peptide receptor-like 1 (Le et al., 2001) L-type calcium channels (Florio et al., 1996), extendin (Martins et al., 1997; Lopes et al., 2001), laminin (Graner et al., 2000), vitronectin (Hajj et al., 2001), tubulin (Brown et al., 1998b), Bcl-2 (Kuschner and Morgen, 1995), PrP itself (Brown, 2000b) or other factors such as DNA (Nandi, 1997). Additionally the peptide can bind copper (Brown, 2000; Jobling et al., 2000) as does the whole protein (reviewed in Brown, 2001a). This expanding list of binding partners requires rationalisation. It is likely that many of these interactions will prove to be laboratory artefacts occurring *in vitro* and with no physiological relevance. For example, detergent conditions used during the isolation of protein extracts can frequently cause the artefactual interactions of proteins (Hsu and Youle, 1997). Unfortunately, the more binding partners that are suggested for the protein or the peptide the less credibility any such proposed partner has either for the function of the protein or for the activation of cell death pathways.

Some authors have suggested that PrP106-126 can interact directly with the membrane (Rymer and Good, 2000) and generate a transmembrane channel (Kourie and Culverson, 2000; Kourie et al., 2001) or alter membrane fluidity (Salmona et al., 1997). These interactions are not irreconcilable with other possible effects of the peptide. Indeed, given that PrP106-126 is quite hydrophobic it is likely that the peptide will readily enter the lipid bilayer. The hydrophobicity might explain why it readily binds to other proteins and may explain the multiple binding partners mentioned above. PrP^{C} is normally a GPI anchored protein tethered to the cell surface without passing through it (Stahl et al., 1990). Studies with full length proteins have shown that under abnormal conditions PrP can also become transferred into the membrane because of a potential stop transfer element (STE). The domain, which is then transmembrane, is the same region that contains the PrP106-126 sequence (Hedge et al., 1998). Mutations in the *prp* gene altered the proportion of PrP that was inserted in this transmembrane form. Unfortunately, the relevance of these observations for prion disease remains disputed with reports stating that at best only a few point mutations linked to inherited forms of prion disease would lead to transmembrane PrP (Stewart and Harris, 2000). Additional claims that transmembrane PrP accounts for neurodegeneration, independent of association with inherited mutations in the gene are still speculative (Hedge et al., 1999). For example, a recent paper argued that on prolonged incubation (4 months) the interaction between PrP106-126 and PrP^C expressed by neuronal-like tumour cells resulted in a truncated, transmembrane form of PrP (Gu et al., 2001). However, a previous study by David Brown (Brown, 2000b) clearly showed that such truncated PrP accumulated in aggregates of PrP106-126 after only one day of exposure. The aggregates were easily separated from cells and were not inserted into the membrane. This suggests that truncated PrP is most probably not a transmembrane form. Another report showed that the same region of PrP proposed by Hegde et al. (1999) to be transmembrane was also neurotoxic (Haïk et al., 2000). However, this peptide is virtually identical in sequence to PrP106-126 and there is little to suggest that the majority of peptide added by the authors to their cultures actually becomes inserted in the membrane. These peptides also aggregate rapidly when added to a salt solution and form floating clumps in the culture medium. Therefore neither of these studies could be said to provide the necessary evidence that PrP, in a transmembrane form, is neurotoxic.

3.3. Mechanisms of cell death

3.3.1. Animal models of prion induced neurodegeneration

PrP-knockout mice are resistant to prion infection (Büeler et al., 1993). Because these mice lack PrP^C, they are unable to generate PrP^{Sc}. They therefore can not be used to determine if PrP^{Sc} is toxic in the absence of neuronal expression of PrP^C *in vivo.* However, an ingenious transplantation experiment showed that PrP^C expression is necessary for the toxicity of PrP^{Sc} (Brandner et al., 1996). Transplantation of embryonic tissue from a mouse overexpressing PrP^C was made into a PrP-knockout mouse brain. The transplanted tissue was then infected with mouse scrapie. This transplanted tissue generated PrP^{Sc} and showed signs of neurodegeneration and gliosis while the PrP-knockout brain tissue surrounding it remained untouched by neurodegeneration. PrP^{Sc}

accumulated in the PrP-knockout mouce brains but surprisingly did not cause neurodegeneration. These results show that neurones must express PrP^C in order to be killed by the agent of neurodegeneration in prion disease. Assuming that this agent is solely PrP^{Sc} then PrP^{Sc} requires neuronal expression of PrP^C to be neurotoxic. PrP106-126 also requires neuronal expression of PrP^C to be neurotoxic as first shown in 1994 (Brown et al., 1994). This finding in culture has also been shown for PrP^{Sc} itself (Giese et al., 1998) and has been verified by other groups (Jobling et al., 1999). As this finding as been confirmed *in vivo*, *in vitro* and by a number of groups then it can be assumed that neuronal PrP^C expression is the first essential component of the neurotoxic mechanism of both PrP^{Sc} and PrP106-126.

3.3.2. Oxidative stress and cell death

Oxidative stress has been shown to be a hallmark of prion diseases (Guentchev et al., 2000; Wong et al., 2001) as well as many other neurodegenerative diseases. Numerous groups have shown that PrP106-126 causes oxidative stress, disturbs the expression of antioxidant proteins and that the toxicity of PrP106-126 can be inhibited by antioxidants (Brown et al., 1996; Perovic et al., 1997). Cultures infected with PrP^{Sc} are more susceptible to the toxicity of reactive oxygen species (Milhavet et al., 2000) but there is currently no evidence that antioxidants can inhibit neuronal death *in vivo*. However, it is clear that oxidative damage is involved in the mechanism by which PrP106-126 is toxic to cells in a culture system.

Programmed cell death, or apoptosis, has been described in several *in vivo* prion disease models (Lucassen, 1995; Williams, 1997) and has also been observed in cell cultures treated with neurotoxic prion fragments, such as PrP106-126 (Forloni, 1993). There is now little doubt that PrP106-126 causes cell death via an apoptotic mechanism and as the apoptosis of neurones involves such changes as caspase activation (White et al., 2001), mitochondrial depolarization (O'Donavan et al., 2001) and enhanced calcium entry through either NMDA receptors (Müller et al., 1993) or L-type voltage gated calcium channels (Brown et al., 1997b) then it is logical to assume that any or all of these changes might be induced by the peptide. However, none of these changes are specific and can be safely termed as "downstream" events that occur as a result of the triggering of apoptosis.

What triggers the apoptotic mechanism? Brown et al. (1996) proposed that the neurotoxic mechanism of PrP106-126 is rather complex involving a direct and indirect component. The direct component is the necessity for neurones to become sensitive to oxidative stress (for example when the activity of PrP^{C} is compromised) and the indirect component involves an increased production of reactive oxygen species. The source of reactive oxygen species can be microglial cells, though microglia *per se* are not essential for the mechanism to act as long as some alternative source of oxidative stress exists. Microglial cells are activated by the presence of either PrP106-126 or PrP^{Sc}. This finding has been confirmed by a number of groups (Brown et al., 1996; Giese et al., 1998; Combs et al., 1999; Fabrizi et al., 2001). Activated microglia release cytotoxic substances such as superoxide and other reactive oxygen species or tumour necrosis factor (TNF-α). Many experiments from different laboratories have shown they enhance the toxicity of PrP^{Sc} (Giese et al., 1998; Bate et al., 2001) and that they are activated before the onset of neurodegeneration in mouse scrapie (Williams et al., 1994; Giese et al., 1998). The effect is reproducible and confirms that an indirect influence of substances released by microglia contributes to the neurotoxic mechanism in prion disease.

Thus, in summary those components of the neurotoxic mechanism of PrP106-126 or PrP^{Sc} that have been confirmed are: (1) the peptide or PrP^{Sc} is not toxic in the absence of PrP^{C} expression (2) the mechanism involves both direct and indirect effects (3) the indirect effects include the generation of oxidative stress (4) the direct effects result in reduced resistance to oxidative stress.

3.4 Possible toxic domains in PrP^{Sc}

An important study based on an animal model showed that there may be two toxic domains in PrP^{Sc}. Mice expressing a truncated form of PrP starting at either amino-residue 121 or 136 in the mouse sequence on a PrP-knockout background (i.e. no wild-type PrP is expressed in these mice) showed rapid neurodegeneration soon after birth (Shmerling et al., 2000). The PrP expressed by these mice lacked the PrP106-126 region but there was nevertheless clear evidence for the neurotoxicity of the C-terminal portion of PrP. This was confirmed by Daniels et al. (2001), who showed that the prion protein contains a second toxic domain. This second domain, stretching from amino acid residue 147-220, appeared to be even more toxic to neurones than PrP106-126 (Daniels et al., 2001) lacking the expression of PrP^{C}. This fits with the observations of using transgenic mice expressing PrP121-231 (Shmerling et al., 2000). These mice showed extensive loss of neurones in the cerebellum shortly after birth. Shmerling et al. (2000) suggesting that full length PrP^{C} inhibited the toxicity of PrP121-231 when co-expressed in the mice. Experiments with full-length recombinant PrP^C in culture experiments also confirm this to be the case (Daniels et al., 2001).

A more surprising finding however was that the neurotoxic peptide PrP106-126 also inhibited the toxicity of PrP121-231 (Daniels et al., 2001). Thus PrP not only has two potentially neurotoxic domains but it is highly likely that in an intact molecule these two domains neutralise their respective neurotoxicity. Current studies of neurodegeneration in prion disease do not take these observations into account. The toxicity of PrP121-231 is not sufficient or adequate to explain the toxicity of PrP^{Sc}. As pointed out above PrP^{Sc} is not toxic to PrP-knockout neurones. Nevertheless, the cell death that occurs in prion diseases *in vivo* might not be explained by results from *in vitro* studies of PrP 106-126 or PrP^{Sc} alone. These recent studies may change the focus of research away from PrP106-126. What is yet to be determined is to what extent the new toxic domain contributes to neurodegeneration *in vivo*. It is quite possible that changes in the secondary structure of $PrP^{C,}$ when converted to PrP^{Scm} cause the two toxic domains to be unmasked. It is also possible that in the presence of functional PrP^C the C-terminal toxic domain is not involved in mediating cell death. Much further research is needed to unravel the importance of these new observations.

4. Excitotoxicity and neuronal cell death

Let us now look more closely at the current understanding of the neurodegeneration mechanisms that occur during prion infections. Apart from the build up of PrP^{Sc}, the classical pathological features of spongiform encephalopathies are vacuolation of the neuropil, gliosis and neuronal loss (Masters et al., 1978; Kretschmar et al., 1996). Gliosis is one of the clearest hallmarks of prion diseases and the possible association between the proliferation of glial cells and neuronal death has been investigated in detail. Gliosis in itself is not indicative of specific changes in the central nervous system peculiar to prion disease, as many neurodegenerative changes are accompanied by this process. Glial cells respond to damage to or death of neurones. However, in prion disease there is strong evidence that gliosis occurs even before neurodegeneration has begun (Jendroska et al., 1991; DeArmond et al., 1992; Giese et al., 1998).

Gliosis (the proliferation of microglia and astrocytes) is often believed to be a response to damaged neurones. However, this does not imply that gliosis in prion disease is a response to neuronal damage. As stated above there are indications that gliosis occurs before neurodegeneration begins. It also coincides with the deposition of PrP^{Sc}.

Therefore there is the possibility that PrP^{Sc} deposition can stimulate gliosis. It was demonstrated previously that the toxic peptide mimic of PrP^{Sc}, PrP106-126, could induce astrocyte proliferation in culture (Forloni et al., 1994). PrP106-126 could also induce microglia to proliferate (Brown et al., 1996b). The initial studies of astrocyte proliferation were on mixed glial preparations from rat cortex (Forloni et al., 1994). However, when astrocytes were purified free of contaminating cells PrP106-126 could not induce their proliferation (Brown et al., 1996a, 1998a). It was found that, like the neurotoxic effect seen in cerebellar cell cultures, the induction of astrocyte proliferation was also dependent on the presence of microglia (Brown et al., 1996b). However, microglia on their own were also not sufficient for this effect. In order for PrP106-126 to induce astrocyte proliferation, the astrocytes had to express the PrP^{C}. PrP106-126 could not induce proliferation of astrocytes in mixed glial cultures from PrP^{C}-deficient mice (Brown et al., 1996b) and addition of wild-type microglia was insufficient for PrP106-126 to induce PrP^{C}-deficient astrocyte proliferation. (Brown et al., 1998a). Furthermore changes induced directly on the astrocytes, by PrP106-126 were necessary for the astrocytes to respond to the mitogenic factors released by PrP106-126 stimulated microglia (Brown et al., 1998a).

Hafiz and Brown (2000) studied the mechanism by which PrP106-126 induces astrocyte proliferation in the presence of microglia. They identified the proteins involved in the induction of multiplication in astrocytes and also identified changes in astrocytes induced by the PrP106-126 peptide, which might make them more sensitive to signals in the environment. Interleukin-1 and interleukin-6 were identified as two factors present in co-cultures of astrocytes and microglia, that were necessary for the peptide induced proliferation of astrocytes. Previous studies showed that PrP106-126 induced the release of these cytokines from microglia (Peyrin et al., 1999). These factors stimulated the proliferation of astrocytes in the presence of the peptide though it was found that their effect was further enhanced by superoxide radicals generated in the microglia by xanthine oxidase. Thus the signal from the microglia, which induces astrocyte proliferation is multifactorial and an important role is apparently played by reactive oxygen species (Hafiz and Brown, 2000). The multifactorial signal from microglia is, in itself, insufficient to produce astrocyte proliferation and there is a requirement for the PrP106-126 peptide to promote the full effect. It seems that PrP106-126 enhances the signal from microglia to induce significant astrocyte multiplication.

Hafiz and Brown also set out to understand the effects of PrP106-126 on astrocytes, which primes them to respond more readily to the proliferation signal from microglia. They identified the MAP kinase/ERK pathway as being involved in the peptide induced changes in astrocytes.

They showed that PrP106-126 induces increased progression through the cell cycle to the late G1 phase and enhances the level of both p53 and phosphorylated ERKs in astrocytes. High levels of cyclin E were observed in peptide treated astrocytes, which suggests that they progressed through the late stages of the G1 phase of the cell cycle, preceding DNA replication in the S phase. The phosphorylation of ERKs could be related to both induction of stress (Tournier et al 1997) leading either to apoptosis or to the proliferation of astrocytes (Neary et al., 1998). The increased expression of p53 was also linked to cellular stress induced by the peptide PrP106-126. Hafiz and Brown (2000) also shown that peptide induced proliferation of astrocytes in culture could be inhibited by antibodies to cytokines or by MEK inhibitors and suggested that their work could be important for indicating possible points of intervention to block the astrogliosis observed in prion disease.

Astrocytes might actually be the cell type in which PrP^{Sc} is first replicated in the central nervous system (Diedrich et al., 1991). Indeed, animals that express PrP^{C} only in astrocytes are susceptible to prion disease suggesting that the conversion of PrP^{C} to PrP^{Sc} in astrocytes is more significant for disease than in other cell types (Raeber et al., 1997). A recent *in vitro* study also indicated that the presence of large numbers of astrocytes could accelerate the rate at which neurones were killed by peptide PrP106-126 (Brown, 1999a). The toxicity of PrP106-126 was, however, not enhanced by the astrocytes (Brown et al., 1996a) but instead the toxic effects of other factors to neurones, such as the neurotransmitter glutamate, could be enhanced by the effect of PrP106-126 on PrP^{C}-expressing astrocytes. A cell culture system is by design a simplification of the natural situation in the brain. As described above, the basic mechanism of PrP106-126 toxicity to cerebellar cells involves microglia. However, when increased numbers of astrocytes are added into this system to mimic the astrogliosis seen in prion disease, increased toxicity is observed, which can be directly attributed to effects of PrP106-126 on astrocytes.

Microglia activated by a prion protein peptide release both cytokines and toxic substances such as superoxide. The cytokines released by the microglia stimulate astroglial proliferation and alterations in the interactions between astrocytes and neurones have been shown to expose the neurones to the toxicity of the neurotransmitter, glutamate (Brown, 1999a; 1999b). Glutamate is the main fast excitatory neurotransmitter in the mammalian central nervous system. However, under certain circumstances it may cause neurotoxicity via over-excitation - a situation called "excitotoxicity". Processes that increase the sensitivity of glutamate receptors or affect glutamate homeostasis often induce cell death, which is frequently connected with calcium overload. Under normal circumstances, the clearance of glutamate released by neurones is carried out primarily by astrocytes. (McLennan, 1976) and this controls glutamate build up. Glutamate taken up by astrocytes is metabolised to glutamine, which can then be recycled back to the neurones. Thus neurones and astrocytes are metabolically coupled (van den Berg and Garfinkle, 1971). It has also been shown that signalling from neurones influences glutamate uptake by astrocytes (Hansson, 1986; Mennerick et al., 1996) and in the absence of astrocytes, glutamate becomes highly toxic to neurones (Rosenberg et al., 1992). Clearly, an imbalance in the regulation of glutamate uptake by astrocytes can lead to the excess accumulation of glutamate in the vicinity of neruones. Cell death or damage occurs because of the overstimulation by excess glutamate and the intracellular accumulation of factors such as reactive oxygen species (Bondy and Lee, 1993)

Use of radiolaballed glutamate has facilitated investigations on the regulation of glutamate uptake into astrocytes (Hansson, 1985; Flott and Seifert, 1991). The major component of glutamate uptake by astrocytes is sodium dependent (Henn et al., 1974; Schousboe et al., 1977; Flott and Seifert, 1991), although a minor chloride dependent system has also been identified (Pin et al., 1984; Bridges et al., 1987; Flott and Seifert, 1991). Sodium dependent uptake by astrocytes is through high affinity glutamate specific transporter proteins (Garlin et al., 1995). Three sodium dependent glutamate uptake proteins have been identified in the rat forebrain: EAAC1, GLAST and GLT-1. EAAC1 is expressed only on neurones (Kanai and Hediger, 1992) while GLAST and GLT-1 are expressed on glial cells (Storck et al., 1992; Pines et al., 1992). It is possible to synthesize antisense oligonucleotides to block the synthesis of GLAST and GLT-1, and this demonstrated that decreased uptake through GLAST and GLT-1 transporters enhanced the neurotoxicity of glutamate *in vivo* and *in vitro* (Rothstein et al., 1996). Thus factors influencing glutamate uptake by these transporters can influence neuronal survival.

Thus there is now some experimental evidence showing a possible relationship between the expression of the cellular prion protein, PrP^C, and glutamate uptake in astrocytes. Experiments using cell culture models have been carried out to investigate the influence of astrocytic PrP^C expression on glutamate uptake (Brown and Mohn, 1999b). Type 1 astrocytes, expressing PrP^C were shown to have a higher rate of sodium dependent glutamate uptake than PrP^C deficient type 1 astrocytes. This effect was stronger when serum free medium was used to culture the astrocytes. It was suggested that the decrease in glutamate affinity was responsible for the sensitivity of PrP^C-deficient astrocytic glutamate uptake to culture conditions. As discussed previously, PrP^C has been shown to bind copper and it is possible that a greater sensitivity to copper concentrations may be responsible for the decreased glutamate affinity observed. These findings that PrP106-126 inhibits glutamate uptake by astrocytes implies that in a situation where neurones are dependent on astrocytes for protection the PrP106-126 would possibly cause local increases in glutamate concentrations which might be sufficient to initiate excitotoxic damage.

Astroglial proliferation occurs late in the course of prion disease at a similar time to the onset of neurodegeneration (Jendroska et al., 1991). It has also been suggested that PrP^{Sc} is generated in astrocytes and may be a major source of the extracellular deposits seen in the disease (Diedrich et al., 1991). As discussed above, studies with the neurotoxic peptide PrP106-126 have shown that it can stimulate astrocyte proliferation. This proliferation is dependent on both the expression of PrP^C and factors released by microglia stimulated by the peptide. In the absence of inhibitors of mitosis such as cytosine arabinoside, PrP106-126 was shown to stimulate glial proliferation in cell cultures (Brown 1997b). Brown et al (1999a) suggested that PrP^C-expressing astrocytes actually reduce the survival of peptide treated neuronal cultures by a failure to clear glutamate and this result cannot be explained by the direct toxicity of PrP106-126 to the cerebellar cells. Normally the peptide is not toxic to cultures containing neurones deficient in PrP^C. However, it was toxic to cerebellar cells when they were co-cultured with astrocytes. This result cannot be explained by direct toxicity of PrP106-126 to the cerebellar cells as those cells are resistant to the toxicity of PrP106-126 by virtue of not expressing PrP^C. This toxicity was only found to occur in the presence of astrocytes, which were PrP^C positive and the process was mediated by glutamate. Furthermore, Brown (1999a) showed that PrP^C positive astrocytes could protect cerebellar cells from glutamate toxicity, an effect that was inhibited on addition of PrP106-126. Brown (1999b) proposed that the peptide did not enhace the toxicity of glutamate to neurones directly. When cerebellar cells were co-cultured with astrocytes, the neurones became dependent on astrocytes for protection from glutamate toxicity and expressed an increased sensitivity to glutamate. In such a system, the protective effects of astrocytes against glutamate toxicity to neurones were inhibited by PrP106-126, resulting in a greater reduction in neuronal suvival than would have been caused by the peptide when astrocytes were not present. With this model, Brown (1999b) proposed a possible mechanism by which gliosis in prion disease may accelerate the neurodegeneration observed in the later stages of the disease.

Experiments performed by the current authors (Sassoon et al., in preparation) using a co-culture system demonstrated that PrP106-126 caused excitotoxic neuronal death with the mediation of glutamate. The neuronal death resulted from the modification if the expression of NMDA receptor subtypes, stimulated by the exposure of neurones to the combination of astrocytic factors, elevated copper and the peptide. These results showed that glutamate uptake from astrocytes could be dependent on PrP^C expression and, in turn, may be related to copper metabolism. Thus a more complex picture of neuronal death in prion disease is emerging, which aside from the loss of function of PrP^C and the build up of PrP^{Sc} also involves microglia and astrocytes. The suggestion is that they may mediate neuronal death in the presence of the abnormal PrP^{Sc} isoform.

5 .Conclusion

It is clear from a variety of experiments that neurotoxic mechanisms using PrP106-126 in culture provide an acceptable model for PrP^{Sc} toxicity. PrP106-126 has basically two requirements for its toxicity to manifest, namely that the target neurones must express PrP^C and that there must also be a source of stress present in the surroundings, such as oxygen radicals. Cell death results from the impairment of protective mechanisms within the cells as a result of the interaction of PrP106-126 with the cells. The loss of PrP^C function as a direct result of the interaction between PrP106-126 and PrP^C has been demonstrated. It has also been shown that the fragment PrP106-126 can increase the production of oxygen radicals or other stress agents by microglia or astrocytes in the vicinity of the neurones. But the main reason that PrP106-126 is toxic is because it changes the response of neurones to the substances in their surrounding environment that are then able to initiate apoptosis.

The cellular environment plays a major role in the response of neurones to any stimulation. In particular, exposure to toxic substances such as glutamate or copper is regulated by astrocytes. Astrogliosis is a major pathological feature of prion diseases and PrP106-126 can induce neuronal cell death by inhibiting the ability of astrocytes to clear glutamate. Cell death induced by this mechanism involves glutamate alone and therefore does not require the neuronal expression of PrP^C. The mechanism is dependent on signalling between astrocytes and neurones when they are grown together in culture. Such signalling was experimentally disrupted by the peptide PrP106-126 and resulted in glutamate induced neuronal apoptosis. It was also shown that PrP106-126 could directly inhibit sodium dependent glutamate transport by astrocytes and the peptide was able to cause the release of both copper and glutamate by astrocytes. Thus astrocytes clearly participate actively in the induction of cell death by an independent mechanism, in this case the conversion of PrP^C to PrP^{Sc}.

These discussions suggest that although many new findings have emerged about the nature of neurodegeneration in prion disease there are still many questions for which answers have not been provided. However, from among the myriad of new ideas and new proposed mechanisms certain clear paths stand out as being worthy of further and more intensive research. If researchers focus on these then it is likely that important advances will be made. One of the most important questions to be tackled is how can neurodegeneration in prion disease be prevented? The future of research into the field has several obvious targets. These include: (1) diagnosing prion diseases before clinical signs resulting from neuronal loss occur; (2) inhibiting neuronal loss in prion disease; (3) reversing clinical signs in prion disease. Clearly, these targets are all dependent on a better understanding of changes in neurones caused by abnormal forms of PrP. The *in vitro* models that have been used for some years to study pathogenesis of prion disease have been shown to have relevance *in vivo* at a basic level. It is now time to apply the information gathered from *in vitro* experiments and to see if it will bring us closer to understanding possible targets for therapy in addition to the abnormal prion .

REFERENCES

Bate C., Reid S. Williams A. (2001) Killing of prion-damaged neurones by microglia. Neuroreport, 12, 1-6.

Bondy S.L.C., Lee D:K. (1993) Oxidative stress induced by glutamate receptor agonists. Brain Res. 610, 229-233.

Brandner S., Isenmann S., Raeber A., Fischer M., Sailer A., Kobayashi Y., Marino S., Weissmann C., Aguzzi A. (1996) Normal host prion protein necessary for scrapie-induced neurotoxicity. Nature, 379, 339-343.

Bridges R.J., Nieto-Sampedo M., Kari, M., Cotman C.W. (1987) A novel chlorine-dependent L-[^{3}H]glutamate binding site in astrocyte membrane. J. Neurochem. 48, 1709-1715.

Brown D. R., Herms J., Kretzschmar H. A. (1994). Mouse cortical cells lacking cellular PrP survive in culture with a neurotoxic PrP fragment. Neuroreport 5, 2057-2060.

Brown D. R., Schmidt B., Kretzschmar, H. A. (1996). Role of microglia and host prion protein in neurotoxicity of a prion protein fragment. Nature 380, 345-347.

Brown D. R., Schmidt B., Kretzschmar H. A. (1997a) Effects of oxidative stress on prion protein expression in PC12 cells. Int J Dev Neurosci 15: 961-972.

Brown D. R., Herms, J. W., Schmidt B., Kretzschmar H. A (1997b) Different requirements for the neurotoxicity of fragments of PrP and ß-amyloid. Euro. J. Neurosci. 9, 1162-1169.

Brown, D. R., Pitschke, M., Riesner, D. and Kretzschmar, H. A. (1998a) Cellular effects of a neurotoxic prion protein peptide are related to its β-sheet content. Neurosci. Res. Comm. 23, 119-128.

Brown D. R., Schmidt B., Kretzschmar H. A. (1998b) Effects of copper on survival of prion protein knockout neurons and glia. J. Neurochem. 70: 1686-1693.

Brown D.R. (1999a) Prion protein peptide neurotoxicity can be mediated by astrocytes. J. Neurochem. 73, 1105-1113.

Brown D.R., Mohn C.M. (1999b) Astrocytic glutamate uptake and prion protein expression. Glia 25, 282-292.

Brown D.R., Wong B.-S., Hafiz F., Clive C., Haswell S., Jones I. M. (1999c) Normal prion protein has superoxide dismutase-like activity. Biochem. J. 344, 1-5.

Brown D.R., Hafiz F., Glasssmith L.L., Wong B.-S., Jones I.M., Clive C., Haswell S.J. (2000)

Consequences of manganese replacement of copper for prion protein function and proteinase resistance. EMBO J. 19,1180-1186.

Brown D. R. (2000a) Prion protein peptides: Optimal toxicity and peptide blockade of toxicity. Mol. Cell Neurosci. 15, 66-78.

Brown D.R. (2000b) PrP^{Sc}-like prion protein peptide inhibits the function of cellular prion protein. Biochem. J. 352, 511-518.

Brown D. R., Clive C., Haswell S. J. (2001) Anti-oxidant activity related to copper binding of native prion protein. J. Neurochem. 76, 69-76

Brown D. R. (2001a) Prion and prejudice: normal protein at the synapse. Trends Neurosci. 24, 85-90.

Brown D. R. (2001b) Microglia and Prion Disease. Microscop. Res. Tech. 54, 71-80.

Brown D. R. (2001c) Prion protein peptides: agents of death for neurons. in Molecular Pathology of Prion Diseases. Ed. Baker, H. Humana Press Totowa, New Jersey, USA pp.51-70.

Büeler H., Fischer M., Lang Y., Bluethmann H., Lipp H.-P., DeArmond S.J., Prusiner S.B., Aguet M., Weissmann C. (1992) Normal development and behaviour of mice lacking the neuronal cell-surface PrP protein. Nature, 356, 577-582.

Büeler H., Aguzzi A., Sailer A., Greiner R. A., Autenried P., Aguet M., Weissmann C. (1993). Mice devoid of PrP are resistant to scrapie. Cell 73, 1339-1347.

Cereghetti G.M., Schweiger A., Glockshuber R., Van Doorslaer S. (2001) EPR evidence for binding of Cu^{2+} to the C-terminal domain of the murine prion protein. Biophys. J. 81, 516-521.

Collinge, J. (2001) Prion diseases of humans and animals: their causes and molecular basis. Annu. Rev. Neurosci. 24, 519-50.

Combs C. K., Johnson D. E., Cannady S. B., Lehman T. M., Landreth, G. E. (1999) Identification of microglial signal transduction pathways mediating a neurotoxic response to amyloidogenic fragments of α–amyloid and prion protein. J. Neurosci. 19, 928-939.

Daniels M., Cereghetti G.M. and Brown D.R. (2001) Toxicity of novel C-terminal prion protein fragments and peptides harbouring disease-related C-terminal mutations. Eur. J. Biochem. 268, 6155-6164 .

Deli M. A., Sakaguchi S., Nakaoke R., Abraham C. S., Takahata H., Kopacek J., Shigematsu K., Katamine S., Niwa M. (2001) PrP fragment 106-126 is toxic to cerebral endothelial cells expressing PrP(c). Neuroreport, 2000, 11, 3931-3936.

Della-Bianca V., Rossi F., Armato U., Dal-Pra I., Costantini C., Perini G., Politi V.,Vale G. D. (2001) Neurotrophin p75 receptor is involved in neuronal damage by prion peptide-(106-126). J. Biol. Chem. 276, 38929-38933.

De Gioia L., Selvaggini C., Ghibaudi E., Diomede L., Bugiani O., Forloni G., Tagliavini F., Salmona M. (1994) Conformational polymorphism of the amyloidogenic and neurotoxic peptide homologous to residues 106-126 of the prion protein. J. Biol. Chem. 269, 7859-7862.

Diedrich J.F., Minnigan H., Carp R.I., Whitaker J.N., Race R., Frey W., Haase A.T. (1991) Neuropathological changes in scrapie and Alzheimer's disease are associated with increased expression of apolipoprotein E and cathepsin D in astrocytes. J. Virol. 65, 4759-4768.

Ettaiche M., Pichot R., Vincent J.-P., Chabry J. (2000) In vivo cytotoxicity of prion protein fragment PrP106-126. J. Biol. Chem. 275, 36487-36490.

Fabrizi C., Silei, V., Menegazzi M., Salmona M., Bugiani O., Tagliavini F., Suzuki H., Lauro G. M. (2001) The stimulation of inducible nitric-oxide synthase by the prion protein fragment 106-126 in human microglia is tumor necrosis factor-alpha-dependent and involves p38 mitogen-activated protein kinase. J Biol Chem. 276, 25692-25696.

Flott B., Seifert W. (1991) Characterization of glutamate uptake systems in astrocyte primary cultures from rat brain. Glia 4, 293-304.

Florio T., Grimaldi M., Scorziello A., Salmona M., Bugiani O., Tagliavini F., Forloni G., Schettini, G. (1996) Intracellular rise through L-type calcium channels, as molecular mechanism for prion protein fragment 106-126-induced astroglial proliferation. Biochm. Biophys. Res. Comm. 228, 397-405.

Forloni, G., Angeretti, N., Chiesa, R., Monzani, E., Salmona, M., Bugiani, O. and Tagliavini, F. (1993) Neurotoxicity of a prion protein fragment. Nature 362, 543-546.

Forloni G., Angeretti N., Malesani P., Peressini E., Rodriguez Martin T., Della Torre P., Salmona M. (1999) Influence of mutations associated with familial prion-related encephalopathies on biological activity of prion protein peptides. Ann. Neurol. 45, 489-494.

Garlin, A.B., Sinor, A.D., Sinor, J.D., Jee, S.H., Grinspan, J.B., Robinson, M.B. (1995) Pharmacology of sodium-dependent high-affinity Lö[2H]glutamate transport in glial cultures.J. Neurochem. 64, 2572-2580.

Giese A., Brown D. R., Groschup M. H., Feldmann C., Haist, I., Kretzschmar, H. A. (1998) Role of microglia in neuronal cell death in prion disease. Brain Pathol. 8, 449-457.

Graner E., Mercadante A. F., Zanata S. M., Forlenza O. V., Cabral A. L. B., Veiga S. S., Juliano M. A., Roesler R., Walz R., Minetti A., Izauierdo I., Martins V. R., Brentani, R. R. (2000) Cellular prion protein binds laminin and mediates neurogenesis. Mol. Brain Res. 76, 85-92.

Gu Y., Fujioka H. Mishra R. S., Li R., Singh, N. (2001) Prion peptide 106-126 modulates the aggregation of cellular prion protein and induces the synthesis of potentially neurotoxic transmembrane PrP. J. Biol. Chem. (IN PRESS).

Guentchev M., Voigtlander T., Haberler C., Groschup M. H., Budka H. (2000) Evidence for oxidative stress in experimental prion disease. Neurobiol Dis. 7, 270-273.

Hafiz F.B., Brown, D.R. (2000) A model for the mechanism of astrogliosis in prion disease. Molecular and cellular neuroscience, 16, 221-232.

Haïk S., Peyrin J. M., Lins L., Rosseneu M. Y., Brasseur R., Langeveld J. P., Tagliavini F., Deslys J. P., Lasmézas C., Dormont, D. (2000). Neurotoxicity of the putative transmembrane domain of the prion protein. Neurobiol. Dis. 7, 644-656.

Hajj,G. N., Mercadante A. F., Veiga S. S. Brentani R. R., Martins V. R. (2001) Evidence of specific interaction between cellular prion protein and vitronectin. J. Neurochem. 78 (Suppl. 1) 133.

Hansson E. (1985) Transport of monamine and amino acid neurotransmitters by primary astroglial cultures. Neurochem. Res. 10, 667-675.

Hansson E. (1986) Co-cultivation of astroglial enriched cultures from striatum and neuronal containing cultures from substantia nigra. Life. Sci. 39269-277.

Hegde R.S., Mastrianni J.A., Scott M.R., Defea K.D., Tremblay P., Torchia M., DeArmond S.J., Prusiner S.B.. Lingappa V.R. (1998) A transmembrane form of the prion protein in neurodegenerative disease. Science 279, 827-834.

Hegde R.S., Tremblay P., Groth D., DeArmond S.J., Prusiner S.B., Lingappa V.R. (1999) Transmissible and genetic prion diseases share a common pathway of neurodegeneration. Nature 402, 822-826.

Henn F.A., Goldstein M.N., Hamberger A. (1974) Uptake by neurotransmitter candidate by glia. Nature 249, 633-634.

Hsu Y. T., Youle R. J. (1997) Nonionic detergents induce dimerization among members of the Bcl-2 family. J. Biol. Chem. 272, 13829-13834.

Jendroska K., Heinzel F. P., Torchia M., Stowring L., Kretzschmar H. A., Kon A., Stern A., Prusiner S. B., DeArmond, S. J. (1991) Proteinase-resistant prion protein accumulation in Syrian hamster brain correlates with regional pathology and scrapie infectivity. Neurology 41, 1482-1490.

Jobling M. F., Stewart L. R., White A. R., McLean C., Friedhuber A., Maher F., Beyreuther K., Masters C. L., Barrow C. J., Collins S. J., Cappai R. (1999) The hydrophobic core sequence modulates the neurotoxic and secondary structure properties of the prion peptide 106-126. J Neurochem. 73, 1557-1565.

Jobling M. F., Huang X., Stewart L. R., Barnham K. J., Curtain C., Volitakis I., Perugini M., White A. R., Cherny R. A., Colin L., Masters C. L., Barrow C. J., Collins S. J.,Bush A. I., Cappai R. (2001) Copper and zinc binding modulates the aggregation and neurotoxic properties of the prion peptide PrP106-126. Biochemistry. 40, 8073-8084.

Kanai Y., Hediger M.A. (1992) Primary structure and functional characterization of a high-affinity glutamate transporter. Nature 360, 467-471.

Kourie J.I. and Culverson A. (2000) Prion peptide fragment PrP[106-126] forms distinct cation channel types. J. Neurosci. Res. 62, 120-133

Kourie J. I., Farrelly P. V. Henry, C. L. (2001) Channel activity of deamidated isoforms of prion protein fragment 106-126 in plnar lipid bilayers. J. Neurosci. Res. 66, 214-220.

Kretzschmar H.A., Prusiner S.B., Stowring L.E., DeArmond S.J. (1986) Scrapie prion proteins are synthesized in neurons. Am. J. Pathol.122: 1-5

Kretzschmar H.A., Ironside J.W., DeArmond S.J., Tateishi J. (1996) Diagnostic criteria for sporadic Creutzfeldt-Jakob disease. Arch. Neurol.53: 913-920

Kurschner C. Morgan J.I. (1995) The cellular prion protein (PrP) selectively binds to Bcl-2 in the yeast two-hybrid system. Mol. Brain Res. 30, 165-168.

Le Y., Yazawa H., Gong W., Yu Z., Ferrans V. J., Murphey P. M., Wang, J. M. (2001) The neurotoxic prion peptide PrP106-126 is a chemotactic agonist for the G-protein-coupled receptor formyl peptide receptor-like 1, J. Immunol. 166, 1448-1451.

Lopes M. H., Zanata S. M., Mercadante A. F., Graner E., Huang L., Bulingame A., Martins V. R., Brentani, R. R. (2001) Biochemical interaction between cellular prion protein and extendin. J. Neurochem. 78 (Suppl.1) 133.

Lucassen P.J., Williams A, Chung W.C.J., Fraser H. (1995) Detection of apoptosis in murine scrapie. Neurosci. Lett. 198: 185-188

Manson J., West J.D., Thomson V., McBride P., Kaufman M.H., Hope J. (1992) The prion protein gene: a role in mouse embryogenesis? Development 115: 117-122

Martins V. R., Graner E., Garcia-Abreu J., DeSouza S. J., Neto V. M., Brentani R. R. (1997) Complementary hydropathy identifies a cellular prion protein receptor. Nature Med. 3, 1376-1381.

Masters C.L., Richardson E.P., Jr. (1978) Subacute spongiform encephalopathy (Creutzfeldt-Jakob disease). The nature and progression of spongiform change. Brain 101: 333-344

McLennan H. (1976) the autoradiographic localization of L-[^{3}H]glutamate in rat brain tissue. Brain Res. 115, 139-144.

Mennerick S., Benz A., Zorumski C.F: (1996) Components of glial responses to exogenous and synaptic glutamate in rat hippocampal microcultures. J. Neurosci. 16, 55-64.

Meyer R.K., McKinley M.P., Bowman K.A., Braunfeld M.B., Barry R.A., Prusiner S.B. (1986) Separation and properties of cellular and scrapie prion protein. Proc. Natl. Acad. Sci. USA83: 2310-2314

Milhavet O, McMahon HE, Rachidi W, Nishida N, Katamine S, Mange A, Arlotto M, Casanova D, Riondel J, Favier A, Lehmann S. (2000) Prion infection impairs the cellular response to oxidative stress. Proc Natl Acad Sci U S A. 97, 13937-13942.

Moser M., Colello R.J., Pott U., Oesch B. (1995) Developmental expression of the prion protein gene in glial cells. Neuron, 14, 509-517.

Müller W.E.G., Ushijima H., Schroder H.C., Forrest J.M.S., Schatton W.F.H., Rytik P.G., Heffner-Lauc M. (1993) Cytoprotective effect of NMDA receptor antagonists on prion protein (Prion^{Sc})-induced toxicity in rat cortical cell cultures. Eur. J. Pharmacol. 246, 261-267.

Nandi P. K. (1997) Interaction of prion peptide HuPrP106-126 with nucleic acid. Arch. Virol. 142, 2537-2545.

Neary J.T., McCarthy M., Kang Y., Zuniga S. (1998) Mitogenic signals from P1 and P2 purinergic receptors to mitogen activated protein kinase in human fetal astrocyte cultures. Neurosci, Lett. 242, 159-162.

O'Donovan C.N., Tobin D., Cotter T.G. (2001) Prion protein fragment PrP106-126 induces apoptosis via mitochondrial disruption in human neuronal SH-SY5Y cells. J. Biol. Chem. (IN PRESS).

Pan K.M., Baldwin M., Nguyen J., Gasset M., Serban A., Groth D., Mehlhorn I., Huang Z., Fletterick R.J., Cohen F.E. (1993) Conversion of α-helices into β-sheets features in the formation of the scrapie prion proteins. Proc. Natl. Acad. Sci. USA 90, 10962-10966

Perovic S., Schröder H.C., Pergande G., Ushijima H., Müller W. E.G. (1997) Effect of flupirtine on Bcl-2 and glutathione level in neuronal cells treated in vitro with the prion protein fragment (PrP106-126). Exp. Neurol. 147, 518-524.

Peyrin J.M., Lasmézas C.I., Haik S., Tagliavini F., Salmona M., Williams A., Richie D., Deslys J.-P., Dormont D. (1999) Microglial cells respond to amyloidogenic PrP peptide by the production of inflammatory cytokines. Neuroreport, 10, 723-729.

Pin J.-P., Bockaert J., Recasesn M. (1984) The Ca^{2+}/Cl^- dependent L[^{3}H]glutamate binding: a new receptor or a particular transport process? FEBS Lett. 175, 31-36.

Pines G., Danbolt N.C., Bjoraş M., Zang Y., Bendahan A., Eide L., Koepsell H., Storm-Mathisen J., Seeberg E., Kanner B.I. (1992) Cloning and expression of rat brain L-glutamate transporter. Nature 360, 464-467.

Prusiner S.B. (1982) Novel proteinaceous infectious particles cause scrapie. Science 216: 136-144

Prusiner S.B. (1993) Genetic and infectious prion diseases. Arch. Neurol.50: 1129-1153.

Riek R., Hornemann S., Wider G., Billeter M., Glockshuber R., Wüthrich K. (1996) Nature 382, 180-183.

Rothstein J.D., Bristol L.A., Hosler B., Brown R.H. (1994) Chronic inhibition of superoxide dismutase produces apoptotic cell death in spinal neurons. Proc. Natl. Acad. Sci. USA 91, 4155-4159.

Rosenberg P.A., Amin S., Leitner M. (1992) Glutamate uptake disguises neurotoxic potency of glutamate agonists in cerebral cortex in dissociated mouse brain hemispheres. J. Neurosci. 12, 56-61.

Rymer D.L., Good T.A. (2000) The role of prion protein structure and aggregation in toxicity and membrane binding. J. Neurochem. 75, 2536-2545.

Salmona M., Forloni G., Diomede L., Algeri M., DeGiola L., Angeretti N., Giaccone G., Tagliavini F., Bugiani O. (1997) A neurotoxic and gliotrophic fragment of the prion protein increases plasma membrane microviscosity. Neurobiol. Dis. 4, 47-57.

Schousboe A., Svenneby G., Hertz L. (1971) Uptake and metabolism of glutamate in astrocytes cultured from dissociated mouse brain hemispheres. J. Neurochem. 29, 999-1005.

Shmerling D., Hegy I., Fischer M., Blättler T., Brandner S., Götz J., Rülicke T., Flechsig E., Cozzio A., von Mering C., Hangartner C., Aguzzi A., Weissmann, C. (1998) Expression of amino-terminally truncated PrP in the mouse leading to ataxia and specific cerebellar lesions. Cell 93, 203-214.

Stahl N., Borchelt D.R., Prusiner S.B. (1990). Differential release of cellular and scrapie prion protein from cellular membranes of phosphatidylinositol specific phospholipase C. Biochemistry 29, 5405-5412.

Stewart R. S.. Harris, D. A. (2000) Most pathogenic mutations do not alter the membrane topology of the prion protein. J. Biol. Chem. 276, 2212-2220.

Storck T., Schulte S., Hofman K., Stoffel W. (1992) Structure expression and functional analysis of a Na+-dependent glutamate/aspartate transporter from rat brain. Proc. Natl. Acad. Sci. USA 89, 10955-10959.

Supattapone S., Bosque P., Muramoto T., Wille H., Aagaard C., Peretz D., Nguyen H.-O. B., Heinrich C., Torchia M., Safar J., Cohen F. E., DeARmond S. J., Prusiner S. B., Scott, M. (1999) Prion protein of 106 residues creates an artificial transmission barrier for prion replication in transgenic mice. Cell, 96, 869-878.

Tagliavini F., Prelli F., Verga L., Giaccone G., Sarma R., Gorevic P., Ghetti B., Passerini F., Forloni G., Salmona M., Bugiani O., Frangione B. (1993). Synthetic peptides homologous to prion protein residues 106-147 form amyloid-like fibrils in vitro. Proc. Natl. Acad. Sci. USA 90, 9678-9682.

Thackray A.M., Knight R., Haswell S.J., Bujdoso R., Brown, D.R. (2002) Metal imbalance and compromised antioxidant function are early changes in prion disease. Biochem J. 362, 253-258.

Tournier C., Thomas G., Pierre J., Jacquemin C., Asunier B. (1997) Mediation by arachidonic acid metabolites of the H_2O_2-induced stimulation of mitogen-activated protein kinase (extracellular-signal-regulated kinases and c-Jun NH_2-terminal kinase) Eur. J. Biochem. 244, 587-595.

Van den Berg C.J., Garfinkel D. (1971) A stimulation study of brain compartments: metabolism of glutamate and related substances in mouse brain. Biochem. J. 123, 211-218.

van Doorslaer S., Cereghetti G.M., Glockshuber R., Schweiger A. (2001) Unraveling the Cu^{2+} binding sites in the C-terminal domain of the murine prion protein: a pulse EPR and ENDOR study. J Phys Chem B. 105: 1631-1639.

Viles J. H., Cohen F. E., Prusiner S. B., Goodin D. B., Wright P. E., Dyson H. J. (1999) Copper binding to the prion protein: structural implications of four identical cooperative binding sites. Proc. Natl. Acad. Sci. USA, 96: 2042-2047.

Weissmann C. (1996) Molecular biology of transmissible spongiform encephalopathies. FEBS Lett.389: 3-11

White A.R., Guirguis R., Brazier M.W., Jobling M.F., Hill A.F., Beyreuther K., Barrow C.J., Masters C.L., Collins S.J., Cappai R. (2001). Sublethal concentrations of prion peptide PrP106-126 or the amyloid beta peptide of Alzheimer's disease activates expression of proapoptotic markers in primary cortical neurons. Neurobiol. Dis. 8, 299-316.

Williams, A. E., Lawson, L. J., Perry, V. H., Faser, H. (1994) Characterization of microglial response in murine scrapie. Neuropathol. Appl. Neurobiol. 20, 47-55.

Williams A, Lucassen PJ, Ritchie D, Bruce M. (1997) PrP deposition, microglial activation, and neuronal apoptosis in murine scrapie. Exp. Neurol. 144: 433-438

Wong B.-S., Brown D. R., Pan T., Whiteman M., Liu T., Bu X., Li R., Gambetti P., Olesik J., Rubinstein R., Sy, M.-S. (2001) Oxidative impairment in scrapie-infected mice is associated with brain metal perturbations and altered ani-oxidantion activities. J. Neurochem. 79, 689-698.

Zahn R., Liu A., Luhrs T., Riek R., von Schroetter C., Lopez Garcia F., Billeter M., Calzolai L., Wider G., Wüthrich K. (2000) NMR solution structure of the human prion protein. Proc. Natl. Acad. Sci. U.S.A. 97, 145-150.

Chapter 15
GLUTAMATE EXCITOTOXICITY IN MULTIPLE SCLEROSIS

David Pitt and Cedric S. Raine
Albert Einstein College of Medicine, New York, NY 10461, USA.

Abstract Multiple sclerosis (MS) is a disease of central white matter with prominent demyelination and oligodendrocyte loss. Recent data on MS pathogenesis has implicated disruption of glutamate homeostasis as a disease mechanism, at the center of which, molecules of the glutamate clearance pathway are selectively downregulated on oligodendrocytes. Thus, oligodendrocytes are unable to process extracellular glutamate appropriately, rendering these cells targets for glutamate excitotoxicity. This concept has been supported by data from animal models of inflammatory demyelination and from oligodendroglial cell culture studies. Since oligodendrocyte damage by glutamate excitotoxicity has been described in a number of unrelated white matter conditions, glutamate excitotoxicity may be a common pathway in white matter injury.

Keywords: multiple sclerosis, white matter excitotoxic injury, oligodendrocyes, glutamate metabolism, non-synaptic glutamate release.

INDEX

1. Introduction
   1.1 Pathology of MS lesions
   1.2 Tools in MS research
2. Glutamate excitotoxicity, an underlying cause of oligodendrocyte death in MS
3. Altered glutamate homeostasis in MS and EAE
   3.1 Evidence for reduced glutamate catabolism in MS and EAE
   3.2 Evidence for increased glutamate secretion in MS
4 The concept of white matter excitotoxicity
   4.1 Excitotoxic oligodendrocyte damage in other neurological disorders
   4.2 Glutamate uptake by astrocytes versus oligodendrocytes
   4.3 Non-synaptic glutamate release
5. Conclusions

1. Introduction

Multiple sclerosis (MS) is the most common chronic and disabling disease of the central nervous system (CNS) in young adults. It is generally accepted that this disease begins with an inflammatory assault against components of the myelin sheath. The assault becomes clinically manifest in symptoms ranging from visual impairment to sensory disturbances, ataxia and limb paralysis which may last between one day and several weeks. Patients typically recover completely or almost completely. Subsequent attacks (relapses) occur at an average rate of one to two per year and with each relapse, transient or permanent neurologic damage is added. Clinical relapses are a defining feature of MS and its diagnosis requires consequently the demonstration of at least two separate attacks, clinically or radiographically, together with disseminated CNS lesions. In about one third of patients, MS proceeds to a secondary progressive state in which the neurologic status worsens continuously rather than by distinct relapses. At this stage, patients typically are markedly paretic and enter severe disability.

In the USA, the prevalence of MS ranges between about 30 to 70 cases per 100,000. Because of the chronic and disabling character, the personal and social burden of MS is enormous and a recent estimate of the US national annual health care costs (1994) were estimated to be in excess of 6.8 billion dollars (Whetten-Goldstein et al., 1998).

MS tends to follow a north/south distribution gradient in North America and, with some exceptions, worldwide i.e. prevalence increases with increasing geographical latitude. Therefore, epidemiologically, MS appears to be a place-related disorder. It has been pointed out however that the north/south gradient also matches patterns of Northern European migration e.g. to North America, New Zealand, Israel, Australia and South Africa and thus may reflect a geographical distribution of susceptibility genes (Kurtzke, 1997). The latter would argue in favor of MS as a "person-related" disease. However, another epidemiological dimension is added by data suggesting that there is a critical age (around puberty), wherein geography is most relevant for the risk of contracting MS. People migrating from areas of high to low MS risk at an age younger than 15 acquire the lower risk intrinsic to their new residence. Those migrating after age 15 retain the higher MS risk of their birthplace (Kurtzke, 2000), suggesting, in fact, that MS is an environmental disease acquired in early adolescence.

As with other putative autoimmune diseases, it is generally believed that MS is the result of interplay between genes and immunologic stimuli from the environment. For example, it is now well established that susceptibility to MS is linked to genes of the major histocompatibility complex on chromosome 6 (Olerup et al., 1991). Beyond MHC alleles, however, no major susceptibility genes have yet been identified. Similarly, no specific environmental agents could be singled out for MS pathogenesis so far. One possible exception is that of immunological cross recognition of myelin antigens with antigens from human herpes simplex virus 6 (Wucherpfennig et al., 1995; Cermelli et al., 1997), Epstein-Barr virus (Levin et al., 2003), as well as Chlamydia trachomatis (Sriram et al., 2002), yet direct proof of causal correlation is lacking. Thus, the causes of multiple sclerosis and the factors determining its varied outcomes remain unclear.

1.1 Pathology of MS lesions

As mentioned above, MS is a demyelinating disease of CNS white matter. MS lesions have several characteristic features which help define the disease (Raine, 1997a). The most distinctive is that of primary demyelination, i.e. loss of myelin with relative preservation of axons in affected areas. It is well established that oligodendrocytes, the cells which synthesize and maintain myelin, are severely depleted or absent in chronic plaques. The fate of oligodendrocytes in acute lesions is less clear and discrepant findings of both reduced or unchanged numbers of oligodendrocytes have led to the suggestion of separate pathologic phenotypes in MS (Scolding, 1999). Another aspect of MS lesions is the discrete, well-defined demarcation in which inflammation and demyelination occur. Lesions are fueled by waves of infiltrating lymphocytes, monocytes and reactive tissue macrophages and tend to occur around blood vessels, a feature known as perivascular inflammatory cuffing. It is important to emphasize that the inflammatory process in acute MS causes widespread and selective destruction of myelin sheaths, oligodendrocytes and of some axons, while astrocytes on the whole are spared.

Chronic active lesions eventually evolve into chronic silent plaques with significant axonal loss and pronounced fibrillary astrogliosis and little or no inflammatory activity. The many sclerotic appearing lesions in longstanding MS give the disease its name.

1.2 Tools in MS research

One of the main tools in MS research is the animal model, experimental autoimmune encephalomyelitis (EAE) which was established some 70 years ago (Rivers et al., 1933). EAE is a demyelinating condition that mimics several important features of the clinical presentation and the pathology of MS. This model disease is inducible in a variety of animal species by immunization with myelin antigens. With some variations within different species, EAE manifests itself usually in the lumbosacral spinal cord from where the inflammatory demyelinating lesions progresses rostrally. Extensive involvement of the brain occurs generally only in severe cases. In particular, EAE induced in mice is widely used to test different treatment modalities and, due to the availability of murine transgenic and gene deletion technology, to study select molecular pathways.

EAE is also inducible by transfer of myelin-antigen stimulated lymphocytes from sensitized to naive animals (adoptive transfer), which formally proved the autoimmune nature of this condition (Paterson, 1960). It was shown subsequently that adoptive transfer EAE is conveyed by $CD4^+$ T_H1 cells rather than other types of myelin-specific T cells, generating the notion, that EAE is a prototypic T_H1-mediated autoimmune disorder (Zamvil et al., 1990). This has been the basis for many approaches directed at MS therapy in the last decade. Surprisingly, strategies that used this paradigm e.g. by depleting $CD4^+$ cells in MS patients or by targeting myelin specific $CD4^+$ T_H1 responses, failed to improve or even exacerbated MS (van Oosten et al., 1997; Hohlfeld, 2001). Because of these and other inconsistencies, the validity of the classic T_H1 concept has been called into question and a more complicated picture in which various immune cells play important roles is now emerging (Hemmer et al., 2002).

Regardless of the T cell immunology, most concepts of the pathogenesis in MS stem from and have been tested in EAE, a number of which have been translated into new therapeutic approaches (The IFNB Multiple Sclerosis Study Group, 1993; Miller et al., 2003). In addition, cutting edge technology applied to EAE and MS, such as transcriptional profiling via cDNA microarrays has accelerated the discovery of new players in MS pathogenesis (Steinman et al., 2002). In recent years, a spectrum of potential therapeutic targets have been revealed with this approach, such as osteopontin (Chabas et al., 2001), granulocyte colony stimulating factor (Lock et al., 2002) and the Jagged-Notch pathway (John et al., 2002), amongst others.

2. Glutamate excitotoxicity, an underlying cause of oligodendrocyte death in MS

The cause of the oligodendrocyte loss in MS lesions remains unclear. The consensus of opinion is that in MS, oligodendrocytes are lost from the inflammatory lesion as an event secondary to the progression of the inflammatory attack. Attempts to single out more distinct mechanisms responsible for oligodendrocyte loss have focused on soluble factors such as lytic autoantibodies or the pro-inflammatory cytokine TNF-α (Selmaj et al., 1988), and on cell-fate mechanisms (Bonetti and Raine, 1997). The ultimate contribution however of these mechanisms to depletion of oligodendrocytes is unknown (Raine, 1997b).

Recently, the problem of oligodendrocyte pathology in MS was linked to another distinct mechanism, glutamate excitotoxicity. Initial studies by Oka, Yoshioka and others (Oka et al., 1993; Yoshioaka et al., 1996; Matute et al., 1997) found that like neurons, oligodendrocytes are exquisitely sensitive to glutamate excitotoxicity. Excitotoxicity was demonstrated to be a primary cause for oligodendroglial damage *in vitro* in oligodendroglial cell lines and primary oligodendrocyte cultures and *in vivo* in isolated spinal dorsal columns (Li et al., 2000), optic nerve (Matute, 1997 & 1998), and subcortical white matter (McDonald et al., 1998). In contrast to neurons, excitotoxicity in oligodendrocytes is mediated by AMPA/Kainate receptors, the only subgroup of ionotropic glutamate receptors expressed by these cells (Platneau et al., 1994).

Therefore, it is conceivable that the pathway of glutamate excitotoxicity, a known cause of neuronal death in a number of CNS gray matter diseases, might have a detrimental effect on oligodendrocytes in white matter, e.g. in multiple sclerosis, in which oligodendrocyte loss is prominent.

This hypothesis was tested on mice immunized to develop EAE and then treated with the AMPA/Kainate receptor antagonist, NBQX (Pitt et al., 2000). Starting with the onset of disease, the group of NBQX-treated animals displayed substantially less clinical signs than their placebo-treated counterparts. The clinical outcome was matched at the histopathological level by minimal loss of oligodendrocytes within the lesion and the lesion vicinity, and by evidence of reduced axonal damage. Importantly, these improvements were present despite the absence of any discernable effect on inflammation, i.e. inflammatory activity was unchanged within the lesions and lesion count was similar in both groups throughout the neuraxis. The prevention of oligodendrocyte death in EAE by AMPA/Kainate antagonists suggests that AMPA/Kainate-mediated glutamate excitotoxicity was responsible for oligodendrocyte loss in white matter lesions in EAE.

In a similar experiment with rat EAE, Smith and colleagues also used AMPA/Kainate receptor antagonists but focused on neuronal death in the ventral horns of the spinal cord, a feature less typical of EAE or MS (Smith et al., 2000). The authors observed improved clinical score in addition to reduced neuronal damage in the ventral horns. In other studies addressing the effects of glutamate receptor antagonists in EAE, rats was treated with Memantine and MK 801, respectively which also resulted in improved clinical outcomes (Wallstroem et al., 1996; Bolton et al. 1997).

Although the above studies did not address oligodendroglial damage, the results imply excessive activation of glutamate receptors in EAE, possibly due to raised extracellular glutamate levels. These results therefore suggest that glutamate homeostasis is disrupted in EAE and, by extrapolation, in MS.

3. Altered glutamate homeostasis in MS and EAE

In maintaining glutamate homeostasis in the CNS, a gradient of one to several thousand is upheld between the extra- and intracellular space. This is achieved not by degradation of extracellular glutamate but by rapid sequestration into surrounding glia and neurons by glutamate transporters. Five different isoforms of glutamate transporters are known, termed excitatory amino acid transporters 1 to 5 (EAAT-1 to EAAT-5) in humans (Danbolt, 2001). The transporters rapidly take up glutamate from the synaptic cleft and thus are crucial in terminating glutamatergic synaptic transmission and limiting glutamate spillover to other synapses. Accordingly, large populations of transporters localize in the immediate vicinity of the synaptic cleft. In particular, the glial carriers EAAT-1 and EAAT-2, or GLAST and GLT-1, respectively, in rodents, have the greatest impact on glutamate clearance (Rothstein, 1996a).

Once glutamate is moved into the cell, it is catabolized by the two key enzymes, glutamine synthetase (GS) and glutamate dehydrogenase (GDH), to glutamine and α-ketoglutarate, respectively (Schousboe et al., 1993). The latter pathway supplies the metabolic precursors for tricarboxylic acids linking glutamate to the Krebs cycle. Both Krebs cycle constituents and glutamine are then supplied to neurons as energy metabolites and glutamate precursors (Westergaard et al., 1995). Several lines of evidence suggest that glutamate transport is directly coupled with intracellular glutamate metabolism and that GDH and GS contribute to maintaining non-toxic levels of extracellular glutamate (Belin et al., 1997;

Behrens et al., 2002). Derangement of either glutamate transporters or intracellular glutamate metabolism would therefore lead to a shift in glutamate homeostasis with build-up of excitotoxic levels of extracellular glutamate. The measurement of the above mentioned transporters and enzymes in demyelinating diseases was addressed in the following studies.

3.1 Evidence for reduced glutamate catabolism in MS and EAE

In an immunohistological study on MS tissue from this laboratory, it was demonstrated that GDH and GS were virtually absent from active lesions and adjacent white matter (Werner et al., 2001). This depletion of enzymes persisted in chronic silent lesions where inflammation has subsided and was replaced by astrogliosis. More intriguingly, the glutamate transporters, EAAT-1, -2 and –3, were also found to be markedly reduced. While EAAT-1 and EAAT-3 downregulation was restricted to the lesion perimeter, downregulation of the main transporter, EAAT-2, extended far into normal adjacent white matter (NAWM) suggestive of widespread impairement in glutamate uptake. The latter was demonstrated both at protein and mRNA levels (Pitt et al., 2003).

Downregulation of GLT-1 (EAAT-2, in humans), in white matter lesions and NAWM was also reported in rat EAE (Ohgoh et al, 2000). In contrast to the human study, the neuronal/glial glutamate transporter, EAAC-1, was upregulated. Interestingly, downregulation of GLT-1 as well as upregulation of EAAC-1, was suppressed by AMPA/Kainate antagonist treatment.

In mouse EAE, the two enzymes for intracellular glutamate breakdown, GDH and GS, were quantitated in spinal cords, the main site for inflammatory lesions and were found to be dramatically downregulated in comparison to normal animals (Hardin-Pouzet et al., 1997).

3.2 Evidence for increased glutamate secretion in MS

In addition to dysregulation of glutamate clearance, glutamate production is likely to be increased in MS lesions.

A well documented feature of macrophages and microglia stimulated *in vitro* is their increased glutamate production and secretion (Piani et al., 1991; Newsholme et al., 1989). It was demonstrated e.g. in microglia/neuronal co-cultures, that stimulated microglia incited excitotoxicity in neurons, damage which was abolished by NMDA receptor blockers (Piani et al., 1992).

Consistent with these findings, in MS, an intense upregulation of the main enzyme for glutamate production, glutaminase, within macrophages/microglial cells in the lesion center (Pitt et al., 2001). This feature was not specific for MS since it occurred also in other inflammatory CNS diseases. However, in diseases such as tropical spastic paraparesis and subacute sclerosing panencephalitis, upregulation of glutaminase was less intense and less focused, probably owing to the different lesion topography.

Thus, circumstantial evidence of glutamate release in inflammatory lesions suggests an increased extracellular glutamate load in MS. Together with the widespread downregulation of the two components of glutamate clearance, glutamate transport and intracellular glutamate metabolism, it implicates extracellular glutamate levels of higher than physiological concentrations in inflammatory lesions in MS. The prevention of oligodendroglial damage in mouse EAE by AMPA/Kainate antagonists corroborates excitotoxic levels of glutamate in inflammatory demyelinating lesions.

Direct quantitative measurement of glutamate in lesions e.g. with microdialysis, has not yet been undertaken. Glutamate measurement in cerebrospinal fluid (CSF), has been used as an approximate approach to determine extracellular glutamate concentration. In CSF from MS patients, different studies have yielded conflicting results of either significantly increased or normal glutamate levels (Stover et al., 1997; Klivenyi et al., 1997; Garseth et al., 2001). In any case, CSF levels probably do not reflect glutamate concentrations at the center of areas of inflammation because of dilution effect and because of absorption of glutamate by non-affected tissue next to the CSF space. More widespread and superficial inflammation, e.g. bacterial meningitis and encephalitis, has been shown unequivocally to lead to elevated glutamate in CSF and its levels tend to correlate with clinical outcome (Spranger et al., 1996).

4. The concept of white matter excitotoxicity

4.1 Excitotoxic oligodendrocyte damage in other neurological disorders

As outlined in other chapters in this book, a variety of neurological diseases have been linked by circumstantial or direct evidence to glutamate excitotoxicity. However, the majority of these disorders such as ALS, Alzheimer's disease, epilepsy and Huntington's disease (Rothstein, 1996b;

Greenamyre, 1991; Meldrum, 1993; Grunewald et al., 1999), are gray matter diseases and thus have been studied in terms of neuronal excitotoxicity.

Excitotoxicity has also long been recognized as an intrinsic mechanism for CNS injury in ischemic stroke (Lee et al., 1999). Although stroke involves both gray and white matter, research has focused mainly on cortical i.e. gray matter, pathology. Only recently, white matter damage and injury to oligodendrocytes have been examined more closely.

Data *in vitro* and *in vivo* suggest that oligodendrocytes in white matter are highly vulnerable to ischemia/hypoxia (Lyons et al., 1998), comparable to the susceptibility of neurons (Pantoni et al., 1996). Amongst several proposed mechanisms, excitotoxic pathways have been demonstrated to be of major relevance, e.g. AMPA/Kainate receptor blockade prevents oligodendroglial cell death from hypoxic injury in cell culture (Fern et al., 2000). Similarly, in brain slice preparations, oligodendrocytes can be protected from hypoxia-induced injury by AMPA/Kainate receptor antagonists or by removal of Ca^{2+} (Tekkok et al., 2001). Despite the compelling effectiveness of AMPA/Kainate receptor blockage *in vitro*, AMPA/Kainate antagonist treatment in mature rats subjected to transient focal ischemia did not prove to prevent white matter oligodendrocyte damage (Irving et al., 1997). Interestingly, in similar studies with rat pups, hypoxia produced selective damage to developing white matter and immature oligodendrocytes and was preventable by treatment with AMPA antagonists (Follett et al., 2000).

In summary, the above results suggest that oligodendrocyte injury in acute ischemia is mediated by glutamate excitotoxicity thus evoking a specific mechanism in selective damage to oligodendrocytes. These findings parallel findings in multiple sclerosis, in trauma (Li et al., 1999) and in perinatal injury (du Plessi et al., 2002). Therefore, it appears that glutamate excitotoxicity may be a common pathway for selective oligodendrocyte injury in white matter in pathologically unrelated CNS diseases.

4.2 Glutamate uptake by astrocytes versus oligodendrocytes

Based on localization studies, it is held that GLT-1 and GLAST are exclusively expressed by astrocytes, predominantly in cell processes enveloping synaptic complexes. Moreover, GLT-1 and GLAST immunoreactivity coincides with areas of dense glutamatergic innervation, i.e. in cerebral cortex, cortical projections such as hippocampus, corticofugal pathways and cerebellum (Lehre et al., 1995). This is in accord with the presumed role of glutamate transporters which involves the disposal of glutamate overflow from synaptic regions (Rothstein et al., 1994; Danbolt et al., 1998). These studies focused primarily on gray matter and glutamate transporters were reported initially to be largely absent from white matter tracts. However, this pattern of distribution has recently been revised.

A study on rat optic nerve found that GLAST was localized to interfascicular oligodendrocytes (Domercq et al., 1999). The same group reported that EAAC1 was present on oligodendroglial progenitor cells n subcortical white matter of rat brain (Gottlieb et al., 2000). In oligodendrocyte cultures from optic nerve, expression of GLT-1, GLAST and EAAC1 was observed at all stages (Domercq et al., 1999). This was confirmed by a report of robust mRNA expression of GLT-1 and GLAST in oligodendrocyte cultures which was higher in the case of GLT-1, or equal in the case of GLAST, than that in cultured astrocytes (Kondo et al., 1995). In our own studies on human white matter, we found that the glutamate transporters EAAT-1 and EAAT-2, were expressed predominantly on oligodendrocytes and to a much lesser extent, on astrocytes (Werner et al., 2001, Pitt et al., 2003). The latter was confirmed at the mRNA level by *in situ* hybridization of human white matter tissue. Similarly, we found EAAT-1 and EAAT-2 to be expressed in cultured human fetal oligodendrocytes (Pitt et al., 2003). The functionality of glutamate transporters was demonstrated by several studies with uptake of radioactive glutamate studies in cultured rat, mouse and human fetal oligodendrocytes (Reynolds et al., 1984; Domercq et al., 1999; Pitt et al., 2003).

The kinetic profile of glutamate uptake in oligodendrocytes showed glutamate transport to be mediated primarily by GLT-1 (Domercq et al., 1999). Dihydrokainate, a selective inhibitor of this type of transporter, suppressed glutamate uptake by about 80%, suggesting minor participation of the glutamate transporters, EAAC1 and GLAST, also expressed by oligodendrocytes (Pitt et al., 2003).

Further evidence for active glutamate metabolism in oligodendrocytes is provided by demonstrating in these cells glutamine synthetase (Cammer, 1990; Domercq et al., 1999; Werner et al., 2001), and glutamate dehydrogenase (Werner et al., 2001; Pitt et al., 2003). Similar to glutamate transporters, glutamine synthetase has been viewed previously as an astroglial "marker", absent from oligodendrocytes. As with glutamate transporters, the first studies focused on cerebral cortex, hippocampus and other gray matter regions where GS is localized to astrocytes but not oligodendrocytes. However, detailed examination of white matter has since revealed intense and predominant immunoreactivity for this enzyme in oligodendrocytes (Tansey et al., 1991).

Therefore, judging from their expression of glutamate transporters and enzymes, oligodendrocytes appear to be well-equipped to contribute to glutamate homeostasis in white matter. In our study (Werner et al., 2001), astrocytes in white matter expressed less glutamate transporters and metabolizing enzymes than oligodendrocytes or their counterparts in gray matter and we suggested that oligodendrocytes in human white matter were significant players in glutamate clearance. However, the relative contributions of oligodendroglial vs. astroglial glutamate uptake in white matter are not clear and warrant further investigation.

4.3 Non-synaptic glutamate release

In contrast to gray matter, glutamate release in white matter is non-synaptic and presumably takes place at a much lower level, in turn requiring less glutamate uptake capacity. Thus, transporter expression, although functionally significant in its environment, is comparatively low and might have been previously overlooked. In fact, low level and non-focal release of glutamate in white matter might be a prerequisite for oligodendrocytes to function as retrievers of extracellular glutamate, because of their exquisite vulnerability to glutamate excitotoxicity.

Under physiological conditions, glutamate can be released from astrocytes through a controlled, Ca^{2+}-dependent process (Parpura et al., 1994; Nedergaard et al., 2002), and from axons in a non-vesicular fashion following electrical activity (Martin, 1992; Kriegler et al., 1993). This non-synaptic glutamate release is thought to provide communication between glial cells or glial cells and neurons.

Under pathological circumstances, a number of mechanisms for non-synaptic glutamate are possible, depending on the nature of the insult. Astrocyte swelling, an early event after injury e.g. in brain trauma and hypoxia, has been associated with glutamate leakage through volume-sensitive anion channels (Kimelberg et al., 1990; Nedergaard et al., 2002). Furthermore, in metabolically compromised glial cells in which ion gradients are extinguished, intracellular glutamate is released by reversed electrogenic glutamate uptake, i.e. through glutamate transporters (Szatkowski et al., 1990).Reverse glutamate uptake constitutes a major source of extracellular glutamate in conditions such as ischemia. A third mechanism entails activation of the chemokine receptor, CXCR4, on astroglial cells. In the ensuing signaling cascade, glutamate and additional factors such as tumor necrosis factor-α, are released (Bezzi et al., 2001). This process gains pathological significance because activated microglia amplify CXCR4-induced glutamate release and thus convert this chemokine-activated cell–cell signaling system into an excitotoxic, pro-apoptotic cascade. For example, amongst other functions, CXCR4 is a co-receptor for the HIV-1 envelope protein, gp120. Therefore, in AIDS, gp120 binding to CXCR4 induces neuronal glutamate excitotoxicity via astroglial glutamate release, an event potentiated by activated microglia.

Activated microglia have also been linked to excitotoxic oligodendrocyte damage through other pathways. The microglial cytokines, interleukin-1β and TNF-α, both caused apoptosis of oligodendrocytes in cocultures of astrocytes and microglia and this could be averted by the AMPA/Kainate receptor antagonists, NBQX and CNQX (Takahashi et al., 2003). The presumed mechanism is that of inhibition of astroglial glutamate uptake through IL-1β and TNF-α and subsequent glutamate overload. In fact, in earlier studies, both cytokines were demonstrated to reduce glutamate uptake by human astrocytes (Chao et al., 1994; Fine et al., 1996; Hu et al., 2000), and evidence suggests that this effect was mediated by nitric oxide (Hu et al., 2000). Glutamate uptake in human oligodendrocyte cultures was similarly reduced by TNF-α, associated with a direct downregulation of the main glutamate transporter, EAAT-2 (Pitt et al., 2003).

Thus, under pathological conditions, numerous mechanisms exist which lead to the accumulation of toxic levels of glutamate. These mechanisms have been studied in large part in mixed cell cultures. Since they affect non-neuronal glutamate release and glutamate clearance, they can take place during both gray and white matter pathology. Impaired glutamate clearance has been demonstrated in astrocytes and oligodendrocytes. In contrast, pathological glutamate release through reverse transport or the CXCR4 pathway was observed only in astrocytes. It remains to be seen whether oligodendrocytes also release glutamate through similar or other mechanisms.

In CNS inflammatory disease, it appears that inflammatory mediators, i.e. cytokines and chemokines, play a specific role in the disruption of glutamate homeostasis either by inducing astroglial glutamate release or by inhibiting glutamate uptake. Reactive microglial cells are crucial in this process since they release glutamate directly and secrete cytokines which in turn stimulate glutamate release from astrocytes.

A scenario for glutamate excitotoxic damage to oligodendrocytes is proposed in the figure below.

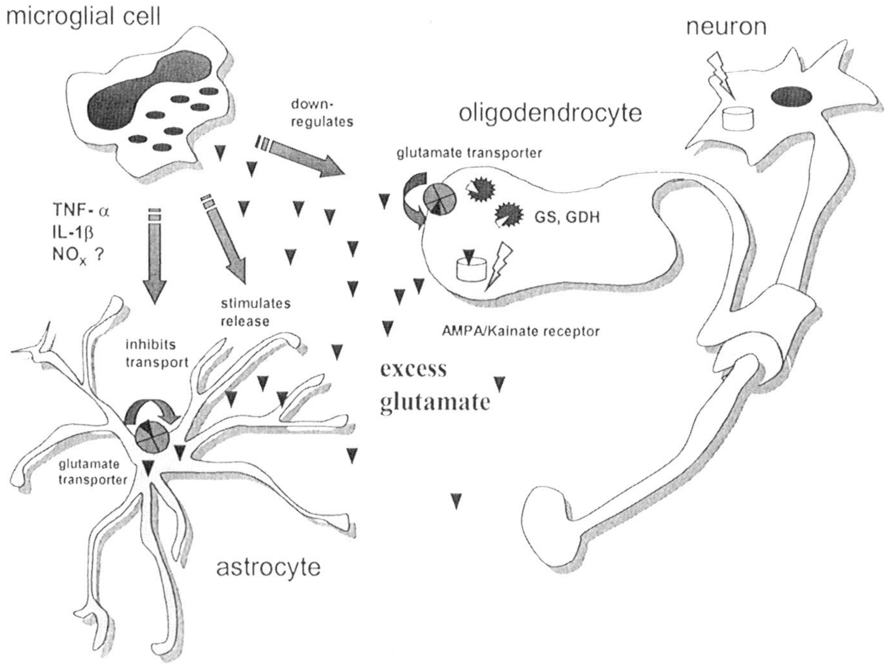

From the above diagrammatic representation, the following mechanisms may act in concert to produce glutamate imbalance and subsequent excitotoxicity in MS lesions. Activation of microglia and macrophages leads to release of excess glutamate. TNF-α and other inflammatory mediators induce astrocytes to release additional glutamate; astroglial glutamate uptake is inhibited. In oligodendrocytes, TNF-α reduces expression of glutamate transporters and glutamate metabolizing enzymes. Within the active lesion, GDH and GS, as well as transporters, are widely absent. The resultant glutamate overload leads to excitotoxic death in oligodendrocytes (modified from Steinman, 2000).

5. Conclusions

Over the last few years, it has been established that oligodendrocytes sustain injury through glutamate excitotoxicity. It has also become increasingly clear that oligodendrocyte damage coincides with deranged glutamate homeostasis in several different white matter diseases, e.g. in white matter ischemia, trauma, progressive multifocal leukoencephalopathy, and multiple sclerosis.

The authors therefore propose that excitotoxicity is a common pathway in many white matter pathologies. This implies that glutamate has a physiological role in white matter with features different from that in gray matter. Glutamate release in white matter is non-synaptic and total glutamate levels, as well as intra/extracellular glutamate turnover, are comparatively low. Furthermore, the novel concept is floated that the glutamate environment in white matter is regulated, at least in part, by oligodendrocytes.

In MS, where selective oligodendrocyte loss from lesions is a prominent hallmark, it is tempting to speculate that excitotoxicity is a major cause of oligodendrocyte demise. Several lines of evidence from animal models and from MS tissue studies support this view. The recently established synergism between reactive microglia and astrocytes causing glutamatergic excitotoxicity, further suggests that inflammation in the CNS is associated with excitotoxic damage.

These insights have major implications for novel pharmacological strategies and may pave the way for new treatments in MS and other inflammatory demyelinating diseases.

Acknowledgements. We thank Iris Nagelmeyer for providing some of the data. Supported by NMSS grant RG-1001-G-10; and HHS NS 11920 and NS 08952.

REFERENCES

Behrens PF, Franz P, Woodman B, Lindenberg KS, Landwehrmeyer GB. Impaired glutamate transport and glutamate-glutamine cycling: downstream effects of the Huntington mutation. *Brain.* 2002; 125(Pt 8):1908-22.

Belin MF, Didier-Bazes M, Akaoka H, Hardin-Pouzet H, Bernard A, Giraudon P. Changes in astrocytic glutamate catabolism enzymes following neuronal degeneration or viral infection. *Glia.* 1997; 21(1):154-61.

Bezzi P, Domercq M, Brambilla L, Galli R, Schols D, De Clercq E, Vescovi A, Bagetta G, Kollias G, Meldolesi J, Volterra A. CXCR4-activated astrocyte glutamate release via

TNFalpha: amplification by microglia triggers neurotoxicity. *Nat Neurosci.* 2001; 4(7):702-10.

Bolton C and Paul C. MK 801 limits neurovascular dysfunction during experimental autoimmune encephalomyelitis. *J Pharmacol Exp Ther.* 1997; 282:397-402.

Bonetti B, Raine CS. Multiple sclerosis: oligodendrocytes display cell death-related molecules in situ but do not undergo apoptosis. *Ann Neurol.* 1997; 42(1):74-84.

Cammer W. Glutamine synthetase in the central nervous system is not confined to astrocytes. *J Neuroimmunol.* 1990; 26(2):173-8.

Cermelli C, Berti R, Soldan SS, Mayne M, D'ambrosia JM, Ludwin SK, Jacobson S. High frequency of human herpesvirus 6 DNA in multiple sclerosis plaques isolated by laser microdissection. *J Infect Dis.* 2003; 187(9):1377-87.

Chabas D, Baranzini SE, Mitchell D, Bernard CC, Rittling SR, Denhardt DT, Sobel RA, Lock C, Karpuj M, Pedotti R, Heller R, Oksenberg JR, Steinman L. The influence of the proinflammatory cytokine, osteopontin, on autoimmune demyelinating disease. *Science.* 2001; 294(5547):1731-5.

Chao CC, Hu S. Tumor necrosis factor-alpha potentiates glutamate neurotoxicity in human fetal brain cell cultures. *Dev Neurosci.* 1994; 16:172-179.

Danbolt NC, Chaudhry FA, Dehnes Y, Lehre KP, Levy LM, Ullensvang K, Storm-Mathisen J. Properties and localization of glutamate transporters. *Prog Brain Res.* 1998; 116:23-43.

Danbolt NC. Glutamate uptake. *Prog Neurobiol.* 2001; 65(1):1-105.

Domercq M, Sanchez-Gomez MV, Areso P, Matute C. Expression of glutamate transporters in rat optic nerve oligodendrocytes. *Eur J Neurosci.* 1999; 11(7):2226-36.

du Plessis AJ, Volpe JJ. Perinatal brain injury in the preterm and term newborn. *Curr Opin Neurol.* 2002; 15(2):151-7.

Fern R, Moller T. Rapid ischemic cell death in immature oligodendrocytes: a fatal glutamate release feedback loop. *J Neurosci.* 2000; 20(1):34-42.

Fine SM, Angel RA, Perry SW, Epstein LG, Rothstein JD, Dewhurst S, Gelbard HA. Tumor necrosis factor alpha inhibits glutamate uptake by primary human astrocytes. *J Biol Chem.* 1996; 271:15303-15306.

Follett PL, Rosenberg PA, Volpe JJ, Jensen FE. NBQX attenuates excitotoxic injury in developing white matter. *J Neurosci.* 2000; 20(24):9235-41.

Garseth M, White LR, Aasly J. Little change in cerebrospinal fluid amino acids in subtypes of multiple sclerosis compared with acute polyradiculoneuropathy. *Neurochem Int.* 2001; 39(2):111-5.

Gottlieb M, Domercq M, Matute C. Altered expression of the glutamate transporter EAAC1 in neurons and immature oligodendrocytes after transient forebrain ischemia. *J Cereb Blood Flow Metab.* 2000; 20(4):678-87.

Greenamyre JT. Neuronal bioenergetic defects, excitotoxicity and Alzheimer's disease: "use it and lose it". *Neurobiol Aging.* 1991; 12(4):334-6.

Grunewald T, Beal MF. Bioenergetics in Huntington's disease. *Ann N Y Acad Sci.* 1999; 893:203-13.

Hardin-Pouzet H, Krakowski M, Bourbonniere L, Didier-Bazes M, Tran E, Owens T. Glutamate metabolism is down-regulated in astrocytes during experimental autoimmune encephalomyelitis. *Glia.* 1997; 20:79-85.

Hemmer B, Archelos JJ, Hartung HP. New concepts in the immunopathogenesis of multiple sclerosis. *Nat Rev Neurosci.* 2002; 3(4):291-301.

Hohlfeld R, Wiendl H. The ups and downs of multiple sclerosis therapeutics. *Ann Neurol.* 2001; 49(3):281-4.

Hu S, Sheng WS, Ehrlich LC, Peterson PK, Chao CC. Cytokine effects on glutamate uptake by human astrocytes. *Neuroimmunomodulation.* 2000; 7:153-9.

Irving EA, Yatsushiro K, McCulloch J, Dewar D. Rapid alteration of tau in oligodendrocytes after focal ischemic injury in the rat: involvement of free radicals. *J Cereb Blood Flow Metab.* 1997; 17(6):612-22.

John GR, Shankar SL, Shafit-Zagardo B, Massimi A, Lee SC, Raine CS, Brosnan CF. Multiple sclerosis: re-expression of a developmental pathway that restricts oligodendrocyte maturation. *Nat Med.* 2002; 8(10):1115-21.

Kimelberg HK, Goderie SK, Higman S, Pang S, Waniewski RA. Swelling-induced release of glutamate, aspartate and taurine from astrocyte cultures. *J Neurosci.* 1990; 10:1583-1591.

Klivenyi P, Kekesi K, Juhasz G, Vecsei L. Amino acid concentrations in cerebrospinal fluid of patients with multiple sclerosis. *Acta Neurol Scand.* 1997: 95(2):96-8.

Kondo K, Hashimoto H, Kitanaka J, Sawada M, Suzumura A, Marunouchi T, Baba A. Expression of glutamate transporters in cultured glial cells. *Neurosci Lett.* 1995; 188(2):140-2.

Kriegler S, Chiu SY. Calcium signaling of glial cells along mammalian axons. *J Neurosci.* 1993; 13:4229-4245.

Kurtzke JF. "The epidemiology of multiple sclerosis" in *Multiple sclerosis*, Raine CS, McFarland HF and Tourtellotte WW, eds: Chapman and Hall Medical, 1997.

Kurtzke JF. Multiple sclerosis in time and space--geographic clues to cause. *J Neurovirol.* 2000; 6:S134-40.

Lee JM, Zipfel GJ, Choi DW. The changing landscape of ischemic brain injury mechanisms. *Nature.* 1999; 399(6738 Suppl):A7-14.

Lehre KP, Levy LM, Ottersen OP, Storm-Mathisen J, Danbolt NC. Differential expression of two glial glutamate transporters in the rat brain: quantitative and immunocytochemical observations. *J Neurosci.* 1995; 15:1835-53.

Levin LI, Munger KL, Rubertone MV, Peck CA, Lennette ET, Spiegelman D, Ascherio A. Multiple sclerosis and Epstein-Barr virus. *JAMA.* 2003; 289(12):1533-6.

Li S, Mealing GA, Morley P, Stys PK. Novel injury mechanism in anoxia and trauma of spinal cord white matter: glutamate release via reverse Na+-dependent glutamate transport. *J Neurosci.* 1999; 19(14):RC16.

Lock C, Hermans G, Pedotti R, Brendolan A, Schadt E, Garren H, Langer-Gould A, Strober S, Cannella B, Allard J, Klonowski P, Austin A, Lad N, Kaminski N, Galli SJ, Oksenberg JR, Raine CS, Heller R, Steinman L. Gene-microarray analysis of multiple sclerosis lesions yields new targets validated in autoimmune encephalomyelitis. *Nat Med.* 2002; 8(5):500-8.

Lyons SA, Kettenmann H. Oligodendrocytes and microglia are selectively vulnerable to combined hypoxia and hypoglycemia injury in vitro. *J Cereb Blood Flow Metab.* 1998; 18(5):521-30.

Martin DL. Synthesis and release of neuroactive substances by glial cells. *Glia.* 1992; 5(2):81-94.

Matute C, Sanchez-Gomez MV, Martinez-Millan L, Miledi R. Glutamate receptor-mediated toxicity in optic nerve oligodendrocytes. *Proc Natl Acad Sci U S A.* 1997; 94(16):8830-5.

Matute C. Properties of acute and chronic kainate excitotoxic damage to the optic nerve. *Proc Natl Acad Sci U S A.* 1998; **95(16):**10229-34.

McDonald JW, Althomsons SP, Hyrc KL, Choi D, Goldberg MP. Oligodendrocytes from forebrain are highly vulnerable to AMPA/Kainate receptor-mediated excitotoxicity. *Nat Med.* 1998; 4(3):291-7.

Meldrum BS. Excitotoxicity and selective neuronal loss in epilepsy. *Brain Pathol.* 1993; 3(4):405-12.

Miller DH, Khan OA, Sheremata WA, Blumhardt LD, Rice GP, Libonati MA, Willmer-Hulme AJ, Dalton CM, Miszkiel KA, O'Connor PW. International Natalizumab Multiple Sclerosis Trial Group. A controlled trial of natalizumab for relapsing multiple sclerosis. *N Engl J Med.* 2003; 348(1):15-23.

Nedergaard M, Takano T, Hansen AJ. Beyond the role of glutamate as a neurotransmitter. *Nat Rev Neurosci.* 2002; 3(9):748-55.

Newsholme EA, Newsholme P, Curi R, Crabtree B, Ardawi MSM. "Glutamine metabolism in different tissues: its physiological importance" in: *Perspectives in Clinical Nutrition.* Kinney JM and Borum PG, eds: Urban and Schwarzenberg, 1989.

Olerup O, Hillert J. HLA class II-associated genetic susceptibility in multiple sclerosis: a critical evaluation. *Tissue Antigens.* 1991; 38(1):1-15.

Ohgoh M, Hanada T, Smith T, Hashimoto T, Ueno M, Yamanishi Y, Watanabe M, Nishizawa Y. Altered expression of glutamate transporters in experimental autoimmune encephalomyelitis. *J Neuroimmunol.* 2002; 125(1-2):170-8.

Oka A, Belliveau MJ, Rosenberg PA, Volpe JJ. Vulnerability of oligodendroglia to glutamate: pharmacology, mechanisms, and prevention. *J Neurosci.* 1993; 13(4):1441-53.

Pantoni L, Garcia JH, Gutierrez JA. Cerebral white matter is highly vulnerable to ischemia. *Stroke.* 1996; 27(9):1641-6.

Parpura V, Basarsky TA, Liu F, Jeftinija K, Jeftinija S, Haydon PG. Non-vesicular release of glutamate from glial cells by reversed electrogenic glutamate uptake. *Nature* 1994; 369(6483):744-7.

Paterson P.Y. Transfer of allergic encephalomyelitis in rats by means of lymphe node cells. *J Exp Med.* 1960; 111:119-135.

Piani D, Frei K, Do KQ, Cuenod M, Fontana A. Murine brain macrophages induced NMDA receptor mediated neurotoxicity in vitro by secreting glutamate. *Neurosci Lett.* 1991; 133(2):159-62.

Piani D, Spranger M, Frei K, Schaffner A, Fontana A. Macrophage-induced cytotoxicity of N-methyl-D-aspartate receptor positive neurons involves excitatory amino acids rather than reactive oxygen intermediates and cytokines. *Eur J Immunol.* 1992; 22(9):2429-36.

Pitt D, Werner P, Raine CS. Glutamate excitotoxicity in a model of multiple sclerosis. *Nat Med.* 2000; 6(1):67-70.

Pitt D, Nagelmeyer IE, Wilson H, Raine CS. Glutamate uptake by oligodendrocytes: implications for excitotoxicity in multiple sclerosis. *Neurology.* 2003. In press

Patneau DK, Wright PW., Winters C, Mayer ML, Gallo V. Glial cells of the oligodendrocyte lineage express both kainate- and AMPA-preferring subtypes of glutamate receptor. *Neuron*; 12:357-71.

Raine C.S. "The neuropathology of multiple sclerosis" in *Multiple sclerosis*, Raine C.S., McFarland H.F. and Tourtellotte W.W., eds: Chapman and Hall Medical, 1997a.

Raine CS. The Norton Lecture: a review of the oligodendrocyte in the multiple sclerosis lesion. *J Neuroimmunol.* 1997b; 77(2):135-52.

Reynolds R, Herschkowitz N. Uptake of [3H]GABA by oligodendrocytes in dissociated brain cell culture: a combined autoradiographic and immunocytochemical study. *Brain Res.* 1984; 322(1):17-31.

Rivers TM, Sprunt DH, Gerry BP. Observations on attempts to produce acute disseminated encephalomyelitis in monkeys. *J Exp Med.* 1933; 58:39-53.

Rothstein JD, Martin L, Levey AI, Dykes-Hoberg M, Jin L, Wu D, Nash N, Kuncl RW. Localization of neuronal and glial glutamate transporters. *Neuron.* 1994; 13(3):713-25.

Rothstein JD, Dykes-Hoberg M, Pardo CA, Bristol LA, Jin L, Kuncl RW, Kanai Y, Hediger MA, Wang Y, Schielke JP, Welty DF. Knockout of glutamate transporters reveals a major role for astroglial transport in excitotoxicity and clearance of glutamate. *Neuron.* 1996a; 16(3):675-86.

Rothstein JD. Excitotoxicity hypothesis. *Neurology.* 1996b; 47(4):S19-25.

Schousboe A, Westergaard N, Sonnewald U, Petersen SB, Huang R, Peng L, Hertz L. Glutamate and glutamine metabolism and compartmentation in astrocytes. *Dev Neurosci.* 1993;15(3-5):359-66.

Scolding N. "Multiple sclerosis" in *Immunological and inflammatory disorders of the central nervous system*, Scolding N., ed: Butterworth-Heinemann, 1999.

Selmaj KW, Raine CS. Tumor necrosis factor mediates myelin and oligodendrocyte damage in vitro. *Ann Neurol.* 1988; 23(4):339-46.

Smith T, Groom A, Zhu B, Turski L. Autoimmune encephalomyelitis ameliorated by AMPA antagonists. *Nat Med.* 2000; 6(1):62-6.

Spranger M, Schwab S, Krempien S, Winterholler M, Steiner T, Hacke W. Excess glutamate levels in the cerebrospinal fluid predict clinical outcome of bacterial meningitis. *Arch Neurol.* 1996; 53(10):992-6.

Sriram S, Yao SY, Stratton C, Calabresi P, Mitchell W, Ikejima H, Yamamoto Y. Comparative study of the presence of Chlamydia pneumoniae in cerebrospinal fluid of Patients with clinically definite and monosymptomatic multiple sclerosis. *Clin Diagn Lab Immunol.* 2002; 9(6):1332-7.

Steinman L. Multiple approaches to multiple sclerosis. *Nat Med.* 2000; 6(1):15-6.

Steinman L, Martin R, Bernard C, Conlon P, Oksenberg JR. Multiple sclerosis: deeper understanding of its pathogenesis reveals new targets for therapy. *Annu Rev Neurosci.* 2002; 25:491-505.

Stover JF, Pleines UE, Morganti-Kossmann MC, Kossmann T, Lowitzsch K, Kempski OS. Neurotransmitters in cerebrospinal fluid reflect pathological activity. *Eur J Clin Invest*. 1997; 27(12):1038-43.

Szatkowski M, Barbour B, Attwell D. Nonvesicular release of glutamate from glial cells by reversed electrogenic glutamate uptake. *Nature* 1990; 348:443-446.

Takahashi JL, Giuliani F, Power C, Imai Y, Yong VW. Interleukin-1beta promotes oligodendrocyte death through glutamate excitotoxicity. *Ann Neurol*. 2003; 53(5):588-95.

Tansey FA, Farooq M, Cammer W. Glutamine synthetase in oligodendrocytes and astrocytes: new biochemical and immunocytochemical evidence. *J Neurochem*. 1991; 56(1):266-72.

Tekkok SB, Goldberg MP. Ampa/kainate receptor activation mediates hypoxic oligodendrocyte death and axonal injury in cerebral white matter. *J Neurosci*. 2001; 21(12):4237-48.

The IFNB Multiple Sclerosis Study Group. Interferon beta-1b is effective in relapsing-remitting multiple sclerosis. Clinical results of a multi-center, randomized, double-blind, placebo-controlled trial. *Neurology*. 1993; 43:655-61.

van Oosten BW, Lai M, Hodgkinson S, Barkhof F, Miller DH, Moseley IF, Thompson AJ, Rudge P, McDougall A, McLeod JG, Ader HJ, Polman CH. Treatment of multiple sclerosis with the monoclonal anti-CD4 antibody cM-T412: results of a randomized, double-blind, placebo-controlled, MR-monitored phase II trial. *Neurology*. 1997; 49(2):351-7.

Wallstroem E, Diener P, Ljundahl A, Khademi M, Nilsson CG, Olsson T. Memantine abrogates neurological deficits, but not CNS inflammation, in Lewis rat experimental autoimmune encephalomyelitis. *J Neurol Sci*. 1996; 137: 89-96.

Werner P, Pitt D, Raine CS. Multiple sclerosis: altered glutamate homeostasis in lesions correlates with oligodendrocyte and axonal damage. *Ann Neurol*. 2001; 50(2):169-80.

Westergaard N, Sonnewald U, Schousboe A. Metabolic trafficking between neurons and astrocytes: the glutamate/glutamine cycle revisited. *Dev Neurosci*. 1995; 17(4):203-11.

Whetten-Goldstein K, Sloan FA, Goldstein LB, Kulas ED. A comprehensive assessment of the cost of multiple sclerosis in the United States. *Mult Scler*. 1998; 4(5):419-25.

Wucherpfennig KW, Strominger JL. Molecular mimicry in T cell-mediated autoimmunity: viral peptides activate human T cell clones specific for myelin basic protein. *Cell*. 1995; 80(5):695-705.

Yoshioka A, Ikegaki N, Williams M, Pleasure D. Expression of *N*-methyl-D-aspartate (NMDA) and non-NMDA glutamate receptor genes in neuroblastoma, medulloblastoma, and other cell lines. *J Neurosci Res*. 1996; 46:164-178

Zamvil SS, Steinman L. The T lymphocyte in experimental allergic encephalomyelitis. *Annu Rev Immunol*. 1990; 8:579-621.

Chapter 16
HIV-1 ASSOCIATED DEMENTIA

Marcus Kaul and Stuart A. Lipton
The Burnham Institute, Center for Neuroscience and Aging Research, 10901 North Torrey Pines Road, La Jolla, CA 92037, USA.

Abstract: Dementia constitutes one devastating consequence of infection with human immunodeficiency virus-1 (HIV-1). While the mechanism of the HIV-1 induced neuropathology remains incompletely understood, several lines of evidence suggest the involvement of chemokine receptors, inflammatory factors and NMDA receptor-mediated excitotoxicity. All of these can in turn trigger several downstream mechanisms, including excessive influx of Ca^{2+} ions, activation of stress-associated protein kinases, caspases, and transcription factors, and production of cytotoxic free radicals. The chapter on HIV-1 associated dementia will review recently identified pathways to neuronal injury and death triggered by the virus or parts of it, and potential approaches to development of applicable therapeutic intervention.

Keywords: Dementia, human immunodeficiency virus-1 (HIV-1), infection, chemokines, inflammation factor, excitotoxicity, glutamate receptor, *N*-methyl-*D*-aspartate (NMDA) receptor.

INDEX

1. Introduction.
2. The neuropathology of HIV-1 infection.
3. Possible roles of chemokine receptors in HAD.
4. HIV-1 infection of the brain and excitotoxicity.
5. Pathways downstream from NMDA receptor stimulation.
6. Potential approaches to therapeutic intervention in HIV-associated dementia.

1. Introduction.

Infection with human immunodeficiency virus-1 (HIV-1) can induce severely debilitating neurological problems, in particular a syndrome of cognitive and motor dysfunction that has been designated HIV-associated dementia (HAD). With the advent of highly active anti-retroviral therapy (HAART) a milder form of neurologic dysfunction, termed minor cognitive/motor disorder (MCMD), appears to become more prevalent than frank dementia. Nevertheless, HAD and MCMD remain significant independent risk factors for AIDS mortality (Kaul et al., 2001; Power et al., 2002). This situation strongly suggests that despite improvements in control of peripheral viral replication and the treatment of opportunistic infections HAART confers only insufficient protection from the development of HAD. One possible explanation for this provides the fact that HIV protease inhibitors and several of the nucleoside analogues poorly penetrate into the central nervous system (CNS) (Kaul et al., 2001; Power et al., 2002). However, while the mechanisms underlying the development of HAD are poorly understood, evidence is accumulating that the neuropathology of HIV-1 infection involves the activation of chemokine receptors, inflammatory factors and NMDA receptor-mediated excitotoxicity (Kaul et al., 2001; Power et al., 2002). This chapter will discuss recently identified pathways to neuronal damage initiated by HIV-1 with emphasis on the excitotoxic component, which HAD seems to share with other neurodegenerative diseases, such as stroke. Finally, we will briefly mention recent efforts aimed at devising applicable clinical intervention in conjunction with anti-retroviral therapies.

2. The neuropathology of HIV-1 infection.

HAD is often accompanied by a number of neuropathological findings, such as astrocytosis, infiltration of macrophages, increased number of microglia, multinucleated giant cells, myelin pallor, dendritic and synaptic damage, and apoptosis. The latter most likely causes the frank loss of neurons (Kaul et al., 2001; Power et al., 2002; Budka, 1991). The mechanisms that initiate HIV-associated dementia are only in part understood, though the accumulation of macrophages/microglia has been reported to correlate with the severity of HIV-associated dementia (Glass et al., 1995). Furthermore, we and our colleagues have shown that HIV-1 infected or immune-stimulated macrophages and microglia produce neurotoxins (Giulian et al., 1990; Gartner, 2000; Kaul et al., 2001). Most strikingly, neurons are not productively infected by HIV-1 and astrocytes only rarely, primarily in pediatric cases (Gartner, 2000; Kaul et al., 2001; Power et al., 2002).

Interestingly, even without an intact virus present, the HIV proteins gp120, gp41, gp160, Tat, Nef, Rev, and Vpr have been observed to initiate neuronal damage, at least *in vitro*, and in some cases *in vivo* in animal models (Brenneman et al., 1988;Dreyer et al., 1990;Toggas et al., 1994;Kaul et al., 2001). In this regard, the viral envelope protein gp120 has been of particular interest since it is responsible for selective binding and signaling of HIV-1 to its target cell, and for initiating viral infection (Brenneman et al., 1988; Kaul and Lipton, 1999; Meucci et al., 1998; Chen et al., 2002; Kaul et al., 2001; Power et al., 2002). Moreover, intracerebroventricular injection of gp120 causes brain injury *in vivo* in rodents (Bezzi et al., 2001), and a transgenic mouse model expressing gp120 in the CNS displays many neuropathological features that are also observed in *post-mortem* brain specimens from HAD patients (Toggas et al., 1994).

3. Possible roles of chemokine receptors in HAD.

In order to infect macrophages and lymphocytes HIV-1 binds via its viral envelope protein gp120 to one of several possible chemokine receptors in conjunction with CD4. Macrophages and microglia are primarily infected via the β-chemokine receptors CCR5 and CCR3, but the α-chemokine receptor CXCR4 may also be involved (Ohagen et al., 1999; Chen et al., 2002). The so called HIV co-receptors, CCR5 and CXCR4, among other chemokine receptors, are also expressed on neurons and astrocytes (Meucci et al., 1998; Kaul et al., 2001). Therefore it is not surprising when several *in vitro* studies strongly suggest that CXCR4 is directly involved in HIV-associated neuronal damage, while CCR5 may additionally serve a protective role (Hesselgesser et al., 1998; Meucci et al., 1998; Kaul and Lipton, 1999).

In cerebrocortical neurons and neuronal cell lines, picomolar concentrations of HIV-1 gp120, as well as intact virus, can trigger neuronal death via CXCR4 receptors (Hesselgesser et al., 1998; Ohagen et al., 1999; Kaul and Lipton, 1999; Zheng et al., 1999; Chen et al., 2002). In mixed neuronal/glial cerebrocortical cultures that mimic the cellular composition of the intact brain, this apoptotic death of neurons seems to result predominantly from the release of microglial toxins rather than direct neuronal damage (Kaul and Lipton, 1999; Chen et al., 2002). However, nanomolar concentrations of SDF-1α/β interacting with CXCR4 can induce apoptotic death of neurons, even in the absence of microglia. This suggests a possible direct interaction with neurons while interaction with astrocytes might also occur and contribute indirectly to neuronal demise (Kaul and Lipton, 1999). In contrast to these findings, somewhat higher concentrations of SDF-1α have been reported to provide neuroprotection from X4-preferring gp120-induced damage of isolated hippocampal neurons (Meucci et al., 1998). However, the results obtained on isolated neurons may be different from those observed in mixed neuronal/glial cultures since glial cells are known to modify the involved death pathways (Kaul and Lipton, 1999; Bezzi et al., 2001; Kaul et al., 2001; Power et al., 2002; Chen et al., 2002).

Using mixed neuronal/glial cerebrocortical cultures from rat and mouse, we recently investigated the role of chemokine receptors in the neurotoxicity of gp120. We found that gp120 from CXCR4 (X4)-preferring as well as CCR5 (R5)-preferring and dualtropic HIV-1 strains all were able to trigger neuronal death. However, we also observed that the β-chemokines MIP-1β and RANTES prevent gp120 neurotoxicity (Kaul and Lipton, 1999). Interesting with regard to the diversity inherent to different HIV-1 strains, gp120 of one X4-preferring virus completely lacked neurotoxicity in CXCR4-deficient cerebrocortical cultures, while gp120 of another X4-classified strain retained some residual ability to induce neuronal death (Kaul, 2002). The gp120 of the dualtropic HIV-1_{SF2} also maintained some ability to trigger neuronal death in the absence of CXCR4, but surprisingly showed even greater neurotoxicity in CCR5 knockout cultures, compared to wildtype or CXCR4-deficient cultures (Kaul, 2002). These observations are in line with a primarily neurotoxic consequence of CXCR4 activation by gp120. In contrast, stimulation of CCR5 is at least in part neuroprotective. Nonetheless, gp120 from R5-preferring HIV-1 can also induce neuronal death, suggesting the possibility that gp120s from various HIV-1 strains differ with regard to stimulation of CCR5. However, it is important to bear in mind that the classification of HIV-1 as R5 or X4-preferring is based on the virus' ability to use certain receptors for infection of a target cell rather than activation or induction of a death signaling pathway. Consequently, this classification is only valid for an interaction of the HIV envelope that leads to infection. In order to understand the mechanism(s) of HAD, gp120s from HIV-1s classified as preferring a certain chemokine receptor for infection may have to be re-assessed for their capability to also interact with other receptors that trigger activation and downstream intracellular signaling. In particular, gp120s of macrophage-tropic strains may need to be tested for their ability to initiate cellular signaling via CXCR4.

Since inhibition of microglial activation is sufficient to prevent neuronal death after gp120 exposure, at least *in vitro*, it seems likely that stimulation of CXCR4 in macrophages / microglia is one sufficient prerequisite for the neurotoxicity of gp120 (Ohagen et al., 1999; Kaul and Lipton, 1999). In contrast, SDF-1 might directly activate CXCR4 in astrocytes and neurons to trigger neuronal death, for example, by reversing glutamate uptake in astrocytes (Hesselgesser et al., 1998; Kaul and Lipton, 1999; Bezzi et al., 2001; Kaul et al., 2001).

4. HIV-1 infection of the brain and excitotoxicity.

Analysis of specimens from AIDS patients (Sardar et al., 1999) as well as *in vivo* and *in vitro* experiments indicate that HIV-1 infection creates excitotoxic conditions in the CNS (Kaul et al., 2001; Kaul and Lipton, 2002; Chen et al., 2002). In fact, Ca^{2+} channel blockers can, at least *in vitro*, completely prevent HIV-1 or gp120-induced neuronal death (Dreyer et al., 1990; Chen et al., 2002). Macrophages and microglia play a crucial role because they are the predominant cells productively infected with HIV-1 in the brain (Kaul et al., 2001; Power et al., 2002; Bezzi et al., 2001; Chen et al., 2002). Moreover, HIV-1-infected or gp120-stimulated mononuclear phagocytes have been shown to release neurotoxins that directly stimulate the NMDAR, including quinolinic acid, cysteine, platelet-activating factor (PAF), and a low-molecular weight compound designated NTox (Giulian et al., 1996; Kaul et al., 2001).

In addition, HIV-infected or activated macrophages, microglia and possibly astrocytes generate and release inflammatory mediators, such as TNF-α, arachidonic acid metabolites, free radicals (ROS and NO), and extracellular matrix degrading enzymes, including MMPs(Johnston et al., 2000; Kaul et al., 2001; Bezzi et al., 2001; Chen et al., 2002). All those factors may indirectly contribute to excitotoxic neuronal injury and eventual death.

Along these lines, gp120 has been found to aggravate excitotoxic conditions by compromising astrocyte uptake of glutamate via arachidonic acid that is released from activated macrophages and microglia (Kaul et al., 2001). SDF-1, TNF-α, and prostaglandins also can stimulate a Ca^{2+}-dependent release of glutamate by astrocytes (Bezzi et al., 2001).

Altogether, in our view, these observations strongly suggest that HIV-1 infection and its associated neuropathology involve both chemokine receptors and NMDAR-mediated excitotoxicity. Chemokine receptors are involved at different levels, first in HIV infection, and further, in the response to chemokines, which may be produced as a consequence of viral infection. For example, MCP-1, MIP-1α, MIP-1β, RANTES, and SDF-1 are all likely to interfere, directly and/or indirectly, with the physiological functions of neurons, astrocytes and microglia (Kaul et al., 2001; Bezzi et al., 2001; Langford et al., 2002; Zheng et al., 1999; Letendre et al., 1999). However, the role(s) of these chemokines and their receptors in the brain under *in vivo* conditions, with and without HIV-1 infection, are only recently beginning to emerge and are the subject of ongoing research. On the other hand, NMDARs respond to excitatory agents, such as neurotransmitters and neurotoxins, but G protein-coupled chemokine receptors might also modulate their activity, and vice versa. Indeed, a β–chemokine, RANTES, can diminish neuronal damage induced by excessive NMDAR stimulation (Bruno et al., 2000), (Paessens, L., Kaul, M. and Lipton, S.A., unpublished results). In turn, excitotoxic stimulation can enhance expression of CCR5 (Galasso et al., 1998). Whether these findings reflect mutual regulatory feedback mechanisms of chemokine signaling pathways and consequences of NMDAR stimulation remains to be elucidated.

5. Pathways downstream from NMDA receptor stimulation.

In case overstimulation of the NMDAR occurs and the excitotoxic insult is fulminant, the cells die after short time from a loss of ionic homeostasis, leading to acute swelling and lysis (necrosis). If the insult is more mild, as it appears to be the case with HAD, neurons enter a delayed, programmed death pathway known as apoptosis (Garden et al., 2002; Kaul and Lipton, 2002). Neuronal apoptosis after an excitotoxic episode involves Ca^{2+} overload, activation of stress-related protein kinases, such as p38 mitogen-activated protein kinase (MAPK), release of cytochrome c from mitochondria, caspase activation, free radical formation, lipid peroxidation, and chromatin condensation (Garden et al., 2002; Kaul et al., 2001). The p38 MAP kinase phosphorylates and activates several transcription factors, including myocyte enhancer factor 2 (MEF2). Interestingly, cleavage of MEF2 by caspases can also contribute to neuronal apoptosis (Okamoto et al., 2002). Caspase-3 and -7 generate truncated MEF2 molecules that lack transcriptional activity, but still bind to DNA. Thus, the MEF2 fragments apparently compete with intact MEF2 and consequently interfere with perhaps survival-promoting gene transcription in neurons (Okamoto et al., 2002). Interestingly, however, blockade of p38 MAPK activity abrogates *in vitro* the induction of neuronal death by HIV-1 gp120 and SDF-1β (Kaul and Lipton, 1999).

Antibody-mediated neutralization of TNF-α or inhibition of its downstream effector caspase-8 also prevents the neurotoxicity of HIVgp120 in cultured cerebrocortical neurons (Garden et al., 2002). Caspase-8 activation can initiate caspase-3 activity, leading eventually to apoptosis. At least *in vitro*, proteins of the Bcl-2 family possess the potential to abrogate neuronal damage subsequent to HIV-1 infection (Chen et al., 2002). An apparent inability of neurons to upregulate Bcl-2 or Bcl-xL in response to the excitotoxicity generated by HIV infection might increase neuronal vulnerability (Chen et al., 2002).

The NMDAR-operated ion channel is connected with neuronal nitric oxide synthase (nNOS), a Ca^{2+}-activated enzyme, via the scaffolding protein PSD-95 (postsynaptic density-95). Thus, nNOS resides in close proximity to the NMDAR-mediated Ca^{2+} influx (Sattler et al., 1999). Excessive intracellular Ca^{2+} over-stimulates nNOS and protein kinases, causing the generation of cytotoxic free radicals, including reactive oxygen species (ROS), nitric oxide (NO) and peroxynitrite ($ONOO^-$). ROS and NO can form the highly cytotoxic $ONOO^-$ (Kaul and Lipton, 2002). Interestingly, a study in *post mortem* brain specimens from HIV patients indicated that the expression of immunologic NOS (iNOS) and gp41 correlate with the occurence and severity of HAD (Adamson et al., 1996). Furthermore, the neurotoxic effect of gp120 is, at least *in vitro*, also dependent on NOS (possibly both nNOS and iNOS) (Dawson et al., 1993;Dreyer et al., 1999).

In addition to the intracellular effects of NO, we have recently identified a potential extracellular proteolytic pathway to neuronal injury that is mediated by nitrosylation and subsequent activation of MMP-9 (Gu et al., 2002). Proteolytically active MMP-9 incites neuronal death, presumably by disrupting the interaction of cellular adhesion factors and extracellular matrix. It is important to note that the expression and activation of MMPs, including MMP-2 and MMP-9, is increased in HIV infected macrophages and also in *post mortem* brain specimens from AIDS patients when compared with uninfected controls (Johnston et al., 2000).

Taken together the different lines of evidence suggest that activation of MMPs may go hand-in-hand with excitotoxic brain injury and NOS activity. Indeed, we found that nitrosylation of MMP-9 resulted in its activation (Gu et al., 2002). As a next step, we investigated the *in vitro* effects of NO-activated MMP-9 on neurons in cerebrocortical cultures (Gu et al., 2002). For these experiments, recombinant (R)-proMMP-9 was pre-activated with the endogenous NO donor S-Nitroso cysteine (SNOC) before addition to the cultures. Neuronal apoptosis was assessed 18 hours later. Since NO had already been released from SNOC by the time the cultures were incubated with the activated MMP, direct release of NO from

SNOC or the formation of peroxynitrite due to the release of NO from SNOC and subsequent reaction with superoxide anion (O_2^-) could not trigger neuronal death (Gu et al., 2002). However, NO-activated MMP-9 significantly increased apoptosis of neurons, whereas treatment with inactive proMMP-9 or the MMP inhibitor GM6001 blocked neuronal cell death (Gu et al., 2002). Furthermore, many dead neurons detached from the dish after exposure to NO-activated MMP-9. These results strongly suggest that inactive (pro)MMP-9 protein does not have a deleterious effect on neurons. However, NO-triggered activation converts MMP-9 into a neurotoxin.

6. Potential approaches to therapeutic intervention in HIV-associated dementia.

An effective pharmacotherapy for HAD has yet to be developed. Accumulating evidence regarding the pathogenesis of HAD suggests several potentially neuroprotective therapeutic strategies in addition to anti-retroviral drugs (ARVs). The clinically tolerated NMDAR antagonist Memantine is among the agents under consideration (Kaul and Lipton, 2002). Others include β-chemokines; chemokine and cytokine receptor antagonists, e.g. against CXCR4- and CCR5 activation; inhibitors of MMPs, p38 MAPK, or caspases; and antioxidants (Kaul et al., 2001; Power et al., 2002; Kaul and Lipton, 2002).

Chemokine receptors mediate HIV-1 infection and specific chemokines may even block infection *in vivo* (Cocchi et al., 1995). Additionally, elevated concentrations of β-chemokines in the CSF of HIV patients correlate with relatively better neuropsychological performance (Letendre et al., 1999; Kaul et al., 2001). Thus, the β-chemokine-CCR5 interaction is of great potential therapeutic interest.

Concerning NMDAR antagonists, our group found that Memantine is an uncompetitive, open-channel blocker of the NMDAR-associated ion channel. Memantine blocks the NMDAR-operated channel primarily when it is open for pathological periods of time. Conversely, Memantine has little effect during normal neurotransmission, when there is less NMDAR-dependent channel activity (Chen et al., 1992; Kaul and Lipton, 2002). In fact, Memantine was shown to hold promise for HAD in a recent phase II clinical trial (Le and Lipton, 2001), and demonstrated a clear positive effect in a phase III trial of Alzheimer's disease patients with moderate-to-severe dementia (Reisberg et al., 2003).

Our group has also shown that nitroglycerin, which produces a nitric oxide-related nitrosonium ion (NO^+), acts, at least in part, at redox-modulatory sites of the NMDAR/channel complex and reduces receptor activity and consequent neuronal damage due to excessive Ca^{2+} influx (reviewed in Kaul and Lipton, 2002).

In order to take advantage of two well-established, clinically safe drugs, which avoid the serious adverse side effects of many NMDAR antagonists, and for the purpose of combining their therapeutic potential for neuroprotection, we are currently studying combinatorial nitro-memantine compounds.

Acknowledgments. This work was supported by amfAR (M.K.) and NIH grants R01 NS41207, R01 NS43242, R01 EY09024, and P01 HD29587 (S.A.L.).

REFERENCES

Adamson D.C., Wildemann B., Sasaki M., Glass J.D., McArthur J.C., Christov V.I., Dawson T.M., Dawson V.L. Immunologic NO synthase: elevation in severe AIDS dementia and induction by HIV-1 gp41. Science 1996; 274:1917-1921.

Bezzi P., Domercq M., Brambilla L., Galli R., Schols D., De Clercq E., Vescovi A., Bagetta G., Kollias G., Meldolesi J., Volterra A. CXCR4-activated astrocyte glutamate release via TNFalpha: amplification by microglia triggers neurotoxicity. Nat.Neurosci. 2001; 4:702-710.

Brenneman D.E., Westbrook G.L., Fitzgerald S.P., Ennist D.L., Elkins K.L., Ruff M.R., Pert C.B. Neuronal cell killing by the envelope protein of HIV and its prevention by vasoactive intestinal peptide. Nature 1988; 335:639-642.

Bruno V., Copani A., Besong G., Scoto G., Nicoletti F. Neuroprotective activity of chemokines against N-methyl-D-aspartate or beta-amyloid-induced toxicity in culture. Eur.J.Pharmacol. 2000; 399:117-121.

Budka H. Neuropathology of human immunodeficiency virus infection. Brain Pathol 1991; 1:163-175.

Chen H.S., Pellegrini J.W., Aggarwal S.K., Lei S.Z., Warach S., Jensen F.E., Lipton S.A. Open-channel block of N-methyl-D-aspartate (NMDA) responses by memantine: therapeutic advantage against NMDA receptor-mediated neurotoxicity. J Neurosci 1992; 12:4427-4436.

Chen W., Sulcove J., Frank I., Jaffer S., Ozdener H., Kolson D.L. Development of a human neuronal cell model for human immunodeficiency virus (HIV)-infected macrophage-induced neurotoxicity: apoptosis induced by HIV type 1 primary isolates and evidence for involvement of the Bcl-2/Bcl-xL-sensitive intrinsic apoptosis pathway. J Virol. 2002; 76:9407-9419.

Cocchi F., Devico A.L., Garzino-Demo A., Arya S.K., Gallo R.C., Lusso P. Identification of RANTES, MIP-1 alpha, and MIP-1 beta as the major HIV- suppressive factors produced by CD8+ T cells. Science 1995; 270:1811-1815.

Dawson V.L., Dawson T.M., Uhl G.R., Snyder S.H. Human immunodeficiency virus type 1 coat protein neurotoxicity mediated by nitric oxide in primary cortical cultures. Proc Natl Acad Sci U S A 1993; 90:3256-9.

Dreyer E.B., Kaiser P.K., Offermann J.T., Lipton S.A. HIV-1 coat protein neurotoxicity prevented by calcium channel antagonists. Science 1990; 248:364-367.

Dreyer E.B., Zurakowski D., Gorla M., Vorwerk C.K., Lipton S.A. The contribution of various NOS gene products to HIV-1 coat protein (gp120)-mediated retinal ganglion cell injury. Invest Ophthalmol.Vis.Sci 1999; 40:983-989.

Galasso J.M., Harrison J.K., Silverstein F.S. Excitotoxic brain injury stimulates expression of the chemokine receptor CCR5 in neonatal rats. Am.J Pathol. 1998; 153:1631-1640.

Garden G.A., Budd S.L., Tsai E., Hanson L., Kaul M., D'Emilia D.M., Friedlander R.M., Yuan J., Masliah E., Lipton S.A. Caspase cascades in human immunodeficiency virus-associated neurodegeneration. J Neurosci 2002; 22:4015-4024.

Gartner S. HIV infection and dementia. Science 2000; 287:602-604.

Giulian D., Vaca K., Noonan C.A. Secretion of neurotoxins by mononuclear phagocytes infected with HIV-1. Science 1990; 250:1593-1596.

Giulian D., Yu J., Li X., Tom D., Li J., Wendt E., Lin S.N., Schwarcz R., Noonan C. Study of receptor-mediated neurotoxins released by HIV-1 infected mononuclear phagocytes found in human brain. J Neurosci 1996; 16:3139-3153.

Glass J.D., Fedor H., Wesselingh S.L., McArthur J.C. Immunocytochemical quantitation of human immunodeficiency virus in the brain: correlations with dementia. Ann Neurol 1995; 38:755-762.

Gu Z., Kaul M., Yan B., Kridel S.J., Cui J., Strongin A., Smith J.W., Liddington R.C., Lipton S.A. S-nitrosylation of matrix metalloproteinases: signaling pathway to neuronal cell death. Science 2002; 297:1186-1190.

Hesselgesser J., Taub D., Baskar P., Greenberg M., Hoxie J., Kolson D.L., Horuk R. Neuronal apoptosis induced by HIV-1 gp120 and the chemokine SDF-1 alpha is mediated by the chemokine receptor CXCR4. Curr.Biol. 1998; 8:595-598.

Johnston J.B., Jiang Y., van Marle G., Mayne M.B., Ni W., Holden J., McArthur J.C., Power C. Lentivirus infection in the brain induces matrix metalloproteinase expression: role of envelope diversity. J Virol. 2000; 74:7211-7220.

Kaul M. Chemokines and their receptors in HIV-associated dementia. J Neurovirol 8[suppl.1], 41-42. 2002.

Kaul M., Garden G.A., Lipton S.A. Pathways to neuronal injury and apoptosis in HIV-associated dementia. Nature 2001; 410:988-994.

Kaul M., Lipton S.A. Chemokines and activated macrophages in gp120-induced neuronal apoptosis. Proceedings of the National Academy of Sciences,USA 1999; 96:8212-8216.

Kaul M, Lipton SA. AIDS. In: Lodge D, Danysz W, Parsons CG, editors. Ionotropic glutamate receptors as therapeutic targets. Johnson City, TN: FP Graham Publishing Co, 2002. p. 1701-1729.

Langford D., Sanders V.J., Mallory M., Kaul M., Masliah E. Expression of stromal cell-derived factor 1alpha protein in HIV encephalitis. J Neuroimmunol. 2002; 127:115-126.

Le D.A., Lipton S.A. Potential and current use of N-methyl-D-aspartate (NMDA) receptor antagonists in diseases of aging. Drugs Aging 2001; 18:717-724.

Letendre S.L., Lanier E.R., McCutchan J.A. Cerebrospinal fluid beta chemokine concentrations in neurocognitively impaired individuals infected with human immunodeficiency virus type 1. J.Infect.Dis. 1999; 180:310-319.

Meucci O., Fatatis A., Simen A.A., Bushell T.J., Gray P.W., Miller R.J. Chemokines regulate hippocampal neuronal signaling and gp120 neurotoxicity. Proc Natl Acad Sci U S A 1998; 95:14500-14505.

Ohagen A., Ghosh S., He J., Huang K., Chen Y., Yuan M., Osathanondh R., Gartner S., Shi B., Shaw G., Gabuzda D. Apoptosis induced by infection of primary brain cultures with diverse human immunodeficiency virus type 1 isolates: evidence for a role of the envelope. J Virol 1999; 73:897-906.

Okamoto S., Li Z., Ju C., Scholzke M.N., Mathews E., Cui J., Salvesen G.S. , Bossy-Wetzel E., Lipton S.A. Dominant-interfering forms of MEF2 generated by caspase cleavage contribute to NMDA-induced neuronal apoptosis. Proc Natl Acad Sci U S A 2002; 99:3974-3979.

Power C., Gill M.J., Johnson R.T. Progress in clinical neurosciences: the neuropathogenesis of HIV infection: host-virus interaction and the impact of therapy. Can.J Neurol Sci 2002; 29:19-32.

Reisberg B., Doody R., Stoffler A., Schmitt F., Ferris S., Mobius H.J. Memantine in moderate-to-severe Alzheimer's disease. N.Engl.J Med. 2003; 348:1333-1341.

Sardar A.M., Hutson P.H., Reynolds G.P. Deficits of NMDA receptors and glutamate uptake sites in the frontal cortex in AIDS. Neuroreport 1999; 10:3513-5.

Sattler R., Xiong Z., Lu W.Y., Hafner M., MacDonald J.F., Tymianski M. Specific coupling of NMDA receptor activation to nitric oxide neurotoxicity by PSD-95 protein. Science 1999; 284:1845-1848.

Toggas S.M., Masliah E., Rockenstein E.M., Rall G.F., Abraham C.R., Mucke L. Central nervous system damage produced by expression of the HIV-1 coat protein gp120 in transgenic mice. Nature 1994; 367:188-193.

Zheng J., Ghorpade A., Niemann D., Cotter R.L., Thylin M.R., Epstein L., Swartz J.M., Shepard R.B., Liu X., Nukuna A., Gendelman H.E. Lymphotropic virions affect chemokine receptor-mediated neural signaling and apoptosis: implications for human immunodeficiency virus type 1-associated dementia. J.Virol. 1999; 73:8256-8267

Chapter 17
MARKERS OF EXCITOTOXICITY IN PATIENTS: TOOL FOR DIAGNOSIS OR THERAPY?

Simone Beretta, Angelo Aliprandi and Carlo Ferrarese
Department of Neurology, University of Milano-Bicocca, Ospedale S. Gerardo, Via Donizetti 106, 20052 Monza (MI), Italy.

Abstract: The role of excitotoxicity in single patients affected by neurological disorders is difficult to assess, but it may provide useful indications as far as prognosis and response to therapy. Glutamate levels could be measured in biological fluids, such as plasma and cerebrospinal fluid. Increased levels of the amino acid have been demonstrated in cerebrovascular diseases, amyotrophic lateral sclerosis (ALS) and AIDS dementia complex (ADC). In the case of stroke and ADC, glutamate levels correlated with disease severity. Molecular and biochemical markers of glutamatergic transmission can be assessed in peripheral tissues, such as platelets and fibroblasts, which physiologically express the three major glutamate transporters. Expression of EAAT1(GLAST) glutamate transporter was reported to be reduced in platelets and fibroblasts from Alzheimer's disease (AD) patients. Functional glutamate uptake studies showed a reduced V_{max} in platelets from patients with AD, Down syndrome (DS), Parkinson's disease (PD), ALS and stroke. Fibroblasts from AD showed both reduced V_{max} and altered regulatory responses. A number of systemic factors, such as genetics, oxidative stress and inflammatory mediators, may be responsible for the observed glutamate uptake impairment in neurological patients.

Keywords: platelets, fibroblasts, neurological patients, excitotoxicity, peripheral markers.

INDEX

1. How can we assess glutamatergic function in patients?
2. Glutamate levels in cerebrospinal fluid and plasma.
3. Human platelets and fibroblasts as peripheral models for studying glutamate transport.
   3.1 Expression and function of glutamate transporters.
   3.2 Endogenous and exogenous regulation of glutamate uptake.

4. Expression of glutamate transporters in platelets and fibroblasts from neurological patients.
5. Functional studies of glutamate uptake in platelets and fibroblasts from neurological patients.
6. The clinical significance of ex vivo markers for diagnosis and therapy.

1. How can we assess glutamatergic function in patients?

Directly measuring neuronal excitatory activity in the functional human brain is an extremely complex and difficult task for both neuroscientists and clinical neurologists.

Despite the recent optimism about being able to map brain function, derived from positron emission tomography (PET) and functional magnetic resonance imaging (fMRI), fundamental problems still remain unsolved. The imaging signal obtained by the forementioned techniques is actually a measure of neuroenergetics, in terms of blood flow and energy consumption. Nevertheless, recent results from *in vivo* magnetic resonance spectroscopy (MRS) have established quantitative relations between cerebral metabolic rates and glutamate neurotransmitter cycling, the latter measured as the isotope label flow from glutamate to glutamine (Shulman, 2001). A number of studies have been performed with carbon or proton MRS in neurological patients. Slow-rate of glutamine-glutamate cycling, decreased glutamine content and increase in glutamate content were shown in epileptic human hippocampus (Petroff et al., 2002). Excess of glutamate and glutamine was reported in the medulla of 10 amyotrophic lateral sclerosis patients compared to 7 control subject tested with MRS (Pioro et al., 1999). On the contrary, MRS showed no increased striatal glutamate in 12 Parkinson's disease patients compared to 12 healthy controls (Taylor-Robinson et al., 1999). One major limit of these neuroimaging techniques applied to excitatory neurotransmission is that their signal indicates solely the total glutamate level in the studied area, without discriminating between its intra- or extra-cellular localisation at the synaptic level.

Other means exist to indirectly assess glutamatergic function. *In vivo* microdialysis is a quantitative measure of glutamate levels in the extracellular fluid in brain tissue. This technique has allowed to confirm the pathophysiologic role of glutamate toxicity in cerebral ischemia (Berger et al., 1999; Berger et al., 2002), subarachnoid haemorrhage (Shulz et al., 2000; Hutchinson et al., 2002) and traumatic brain injury (Vespa et al., 1998; Stahl et al., 2001).

More feasible methods, which have been extensively applied in many neurological disorders, are glutamate level measurements in the cerebrospinal fluid (CSF) and plasma by high-performance liquid chromatography (HPLC). Although very informative, these analyses are not devoid of methodological problems and their use must take in consideration some fundamental aspects, such as the integrity of the blood-brain barrier and diet. Interestingly, molecular and biochemical markers of glutamatergic transmission may be assessed in the periphery, where specific glutamate receptors and transporters are found, identical to those expressed in the CNS. Peripheral tissues, such as platelets and fibroblasts, have been investigated as far as excitatory amino acid transporters (EAATs) activity, and platelets have been studied for NMDA receptor activity in psychiatric disorders. Scientific evidences coming from the latter techniques will be discussed in details in the present chapter.

2. Glutamate levels in cerebrospinal fluid and plasma

The measurement of glutamate levels in the cerebrospinal fluid and in plasma of patients has been the object of extensive investigation.

Before discussing the results obtained by these efforts, some preliminary issues concerning the reliability and physiological significance of this analysis need to be addressed.

A considerable (up to 100 fold) variability in CSF glutamate levels has been observed in different studies (Ferrarese et al, 1993), raising the issue of the consistency and reproducibility of glutamate detection in biological fluids. Our opinion is that these discrepancies might be the result of the different methods adopted for sample processing and storing. In facts, the dynamics of glutamate in the CSF are complex, as they depend not only on the release of the aminoacid from the brain parenchyma, but also on ongoing enzymatic activity in biological fluids, as both CSF and plasma contain sizable amounts of glutamic oxalacetic transaminase (Fleisher et al., 1957). As a result of the activity of this enzyme, spurious reduction of glutamate levels in vitro may occur.

Several methods have been proposed to inactivate enzymatic activity in CSF or plasma, but some of these, such as sample acidification without neutralization, may actually be the source of further bias. Our group has developed and validated a method of sample processing for aminoacid detection, which has been shown to yield stable glutamate levels (Ferrarese et al, 1993) and is thus currently used by several groups. Other inactivation techniques, including sample boiling or ultrafiltration may also be used, but their reliability has not been extensively tested. Great care to the inactivation methods used and to between runs and within assays variability of results should than be taken in the critical evaluation of the literature on this topic.

Despite these methodological issues, however, the analysis of glutamate levels in biological fluids has certainly contributed to the understanding of several neurologic disease, as will be described in the following sections:

Cerebrovascular diseases. Although the involvement of excitotoxic mechanisms in this disorders is well established and has a sound biological basis (Dirgnagl et al., 1997), data in stroke patients are still scanty. Few reports with invasive microdyalitic techniques in patients treated with decompressive craniotomy, had showed increased levels of the aminoacid in the tissue surrounding the necrotic core (Kanthan et al., 1995). CSF and plasma glutamate levels were found to be increased in ischaemic stroke patients 24h after symptom onset, and they were higher in patients with large cortical lesions or with a more severe clinical picture (Castillo et al., 1996). Furthermore, a significant difference between patients with stable and progressive stroke was evident (Castillo et al., 1997). These data, beside supporting the key role of excitotoxicity in stroke, suggest that the detection of glutamate in CSF or plasma might be a useful prognostic tool in predicting the clinical outcome of patients. More importantly, a biological marker associated with stroke in evolution may be of great practical value in selecting patients who are more likely to benefit from neuroprotective therapies. In this light, it should also be remarked that glutamate levels in CSF of patient with progressive strokes have been shown to have a longer time course (Davalos A et al, 1997), suggesting that worsening of the neurological deficit is paralleled by ongoing pathogenetic processes within the brain.

These studies also raise the issue of the origin of glutamate in CSF and plasma. The former case is less problematic, as raised glutamate in the CSF is most likely consequent to necrosis of glial and neuronal cells in the lesion core and to energy failure and aminoacid release in the surrounding brain tissue. The origin of plasma glutamate levels in stroke patients and the biological significance of their correlation with clinical worsening is more arduous and open to further debate and research. This point is of particular interest, as plasma samples pose much less concerns about patient safety and compliance and are then more promising as possible diagnostic tools than the CSF ones. As a matter of facts, in some of the abovementioned studies glutamate and CSF levels were found to be mutually correlated, suggesting that peripheral glutamate might also derive from the brain lesion after leakage trough the damaged blood brain barrier.

However, recent experimental evidence by our group showed that glutamate levels in stroke patients may actually be significantly higher up to two weeks after symptoms onset, when blood-brain-barrier functions are likely restored. More complex mechanisms than simple diffusion must then be involved in the sustained elevation of the plasma levels of the aminoacid (Tremolizzo et al., 2001). Since we also found reduced glutamate uptake by platelets, one possibility is that the plasma levels of the aminoacid might be increased as a result of reduced clearance from the bloodstream. Further studies on this topic are then necessary, and may highlight new interesting features of plasma glutamate as marker of disease activity in stroke patients.

Amyotrophic lateral sclerosis. Early reports (Rothstein et al., 1991) demonstrated increased glutamate levels in the CSF of patients with ALS. Further studies in different laboratories, however, failed to confirm this finding and yielded conflicting results (Camu et al., 1993; Blin et al., 1994), probably as a consequence of the abovementioned methodological issues and of the small number of patients recruited. In a recent work, in which CSF from a large number of ALS patients was studied, a significant difference in glutamate levels respect to healthy controls or patients with different pathologies was found (Spreux-Varoquax et al., 2001). Furthermore, the levels of the aminoacid seemed to be higher in patients with a spinal onset of the disease . The biological meaning of this difference is unclear, as it may either highlight different pathogenic mechanisms in subgroups of patients or more simply reflect the extent of neuronal damage, which is thought to be greater in spinal ALS. Increased glutamate levels were also reported in plasma of patients (Plaitakis et al., 1987), but the interpretation of this finding is arduous, since muscle wasting in advanced disease may lead to elevation of serum aminoacids. The main interest of these data is that they support the hypothesis of the involvement of excitotoxicity in the pathogenesis of ALS. Raised CSF levels may result from impaired transport of the aminoacid by glial cells of affected patients (Rothstein et al,), and thus they may reflect the glutamate concentration around neurons. However, this interpretation should be considered with caution, since, as previously mentioned, glutamate increase may also be an indirect marker of the intensity of cell injury and neuronal loss. The usefulness of CSF glutamate levels as diagnostic tool is unclear. High glutamate levels have been reported increased in other neurodegenerative disorders and are then not specific for ALS. Furthermore, although a statistical difference between ALS patients and healthy controls or subjects with other diseases was demonstrated, data of individual patients are widely scattered and the ranges of different groups overlap. Therefore, selecting a cut-off value for diagnostic purposes might be impossible. Other proposed biological markers of the disease, such as peripheral glutamate transport, may be more specific and more sensitive in the early phases of the disease, and may be of more use as diagnostic support.

Infectious diseases. Glutamate levels have been found increased in bacterial meningitis, and were related to the severity of cognitive impairment, as scored by the Glasgow Coma Scale (Spranger et al., 1996). Furthermore, CSF of patients was shown to have a toxic effect on cultured murine cells, which could be attenuated by the NMDA receptor antagonist MK-801 These data suggest that glutamate mediated excitotoxicity might play a role in neuronal injury during acute bacterial infections. However, the neurotoxic action of the CSF of patients may not depend on high glutamate levels only, as other compounds with NMDA affinity, such as quinolinic acid, have been shown to be raised during infections of the central nervous system (Heyes et al., 1992). Glutamate CSF levels were found normal in most meningeal infections of viral etiology. One noteworthy exception is HIV infection. Levels of the aminoacid were shown to be increased in patients with HIV related neurologic involvement, a condition named AIDS Dementia Complex (ADC), while they were within normal range in seropositive subjects without cognitive impairment (Ferrarese et al., 2001). Moreover, they were correlated with the degree of cognitive impairment as expressed by the Memorial Sloan and Kettering Scale and with the degree of brain atrophy. Serum glutamate levels were also found increased in HIV patients, but none of the abovementioned correlations could be demonstrated. Furthermore, levels in blood and CSF were not correlated with each other. Therefore, while high glutamate levels in plasma probably result from a non specific alteration of aminoacid metabolism, possibly due to liver involvement in AIDS patients, the CSF ones are increased as a result of a specific action of the HIV virus within the brain parenchyma. CSF glutamate levels may than be considered a marker of disease progression and activity within the CNS and may be clinically useful to select patients which may benefit of anti-glutamatergic treatments, such as memantine, which are currently being test in clinical trials in HIV positive patients.

3. Human platelets and fibroblasts as peripheral models for studying glutamate transport.

Several peripheral tissues have been shown to transport glutamate through the same specific glutamate transporters (EAATs) found in the central nervous system (see *Chapter 2*). Glutamate transporters mostly work to facilitate the metabolic function of glutamate in the different peripheral organs where they have been detected, even though in some cases the actual role of glutamate uptake is still unclear (see Table 1).

Table 1. Peripheral localization of human glutamate transporters.

Tissue	EAAT type	Function
Heart	EAAT1(GLAST), EAAT3(EAAC).	Metabolism of heart myocytes.
Intestine	EAAT3(EAAC).	Intestinal absorbtion of glutamate. Involved in enteric nervous system activity.
Kidney	EAAT3(EAAC), EAAT2(GLT).	Re-absorption of filtered amino acid in proximal tubules.
Placenta	EAAT1(GLAST), EAAT2(GLT), EAAT3(EAAC).	Involved in foetal brain development.
Bone	EAAT1(GLAST).	Bone-forming and resorbing.
Mammary gland	EAAT1(GLAST), EAAT2(GLT).	Linked to lactation.
Platelets and fibroblasts	EAAT1(GLAST), EAAT2(GLT), EAAT3(EAAC).	Unknown. Possibly involved in platelet aggregation.

EAAT1(GLAST) and EAAT3(EAAC) mRNA have been shown in the heart (Nakayama et al., 1996). Enhanced glutamate uptake is observed in chronic ischemic heart regions (Dinkelborg et al., 1996). EAAT3(EAAC) is expressed in enterocytes (Kanai et al., 1992), where it mediates intestinal absorption of glutamate, and modulates enteric nervous system activity (Liu et al., 1997). EAAT3(EAAC) is also present in the kidney where it plays a crucial role, as demonstrated by the fact that mutant mice deficient of EAAC develop dicarboxylic aminoaciduria (Peghini et al., 1997). In bone tissue, EAAT1(GLAST) seems to be involved in bone forming and resorbing (Mason et al., 1997; Itzstein et al., 2000). During lactation EAAT1(GLAST) and EAAT2(GLT) are expressed in mammary gland, being up or down-regulated respectively by fasting and refeeding (Martinez-Lopez et al., 1998).

High-affinity glutamate uptake system is observed also in fibroblasts and in platelets, where EAAT1(GLAST), EAAT2(GLT) and

EAAT3(EAAC) have been detected (Cooper et al., 1998; Zoia et al., 2003). The study of glutamate transporters function and expression in these latter tissues is highly valuable, since they are readily accessible after venous drawing or skin biopsy from neurological patients and healthy controls. Although probably involved in metabolic processes, the actual physiological function of glutamate transport in platelets and fibroblasts is not fully understood. Recently, some groups reported an anti-aggregating role of NMDA receptors expressed by platelets (Franconi et al., 1996), and our group recently described a glutamate release from platelets during aggregation, confirming the role of peripheral glutamate signaling in the regulation of platelet function (Longoni et al., 2001).

3.1 Expression and function of glutamate transporters.

Platelets have been extensively studied as models of synaptic apparatus, since they release and reuptake neurotransmitters such as serotonin and dopamine (Barbeau et al., 1975; Zieve et al., 1968), express serotonin receptors (Sneddon, 1973) and glutamate NMDA receptors (Franconi et al., 1996). In particular, NMDA receptor super-sensitivity, assessed by using the intracellular calcium response to glutamate, was demonstrated in schizophrenic and depressed psychotic subjects (Berk et al., 2000) and in non psychotic, unipolar major depression patients (Berk et al., 2001). In these studies, platelets have been recognized as possible peripheral markers of glutamate function in these psychiatric disorders, where hypoglutamatergic function is advocated as a pathogenetic factor (see *Chapter 17*).

A high-affinity glutamate uptake, with kinetic characteristics similar to brain synaptosomes, was also shown in platelets, which appear the major site of clearance of glutamate from blood (Mangano et al., 1981a). The three major human glutamate transporters (EAAT1-3) have been detected in human platelets by RT-PCR, immunofluorescence, western blot and immunoelectronmicroscopy, although at lower density than in brain tissue (Zoia et al., 2003; see Figure 1A). Platelets display kinetic characteristics of both a Na^+-dependent, high affinity glutamate uptake and a Na^+-independent, temperature-dependent low affinity glutamate uptake. When challenged with specific glutamate uptake inhibitors, platelets proved to be sensitive to L-serine-O-sulphate (SOS) and L(-) -threo-3-hydroxyaspartic acid (THA), but not to dihydrokainic acid (DHK). These findings indicate that glutamate transport is likely to be mostly mediated by EAAT1 and EAAT3.

Kinetic studies using radioactive glutamate demonstrated that cultured human fibroblasts isolated from embryonic muscle, skin and peripheral nerve tissues possess a Na^+-dependent uptake process, strongly inhibited by glutamate/aspartate analogues. This indicates that L-glutamate transport is of the high-affinity type, mediated by specific Na^+-dependent transporters (Balcar et al., 1994). Further, monolayers of fibroblasts cultured from human embryonic skin express moderate levels of four glutamate transporters cloned (EAAT1-4), with a relative abundance of EAAT1 and EAAT2 (Cooper et al., 1998). Our group is currently studying glutamate uptake and transporters in human fibroblasts obtained by skin biopsy performed in the interscapular region of adult subjects. Our experiments indicate that adult fibroblasts display a high-affinity sodium-dependent glutamate uptake and express the three major glutamate transporters (EAAT1-3) (Zoia et al., 2001; Figure 1B), as occur in embryonic tissues.

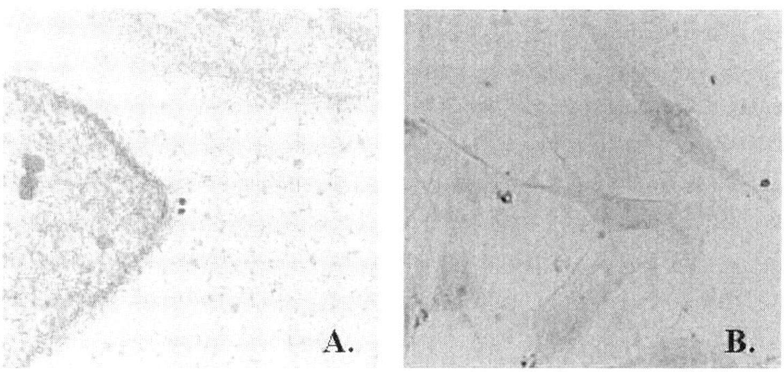

Figure 1. (A) Electron microscopy photograph showing immuno-gold localization of EAAT2(GLT) in the plasma membrane of human platelets. **(B)** Optical microscope image of human fibroblasts stained with anti-EAAT1(GLAST) antibody.

3.2 Endogenous and exogenous regulation of glutamate uptake

Expression and activity of EAATs have been shown to be amenable to fine endogenous and, potentially, pharmacological regulation in the central nervous system (for review, Beretta et al., 2003). Some of the these regulatory mechanisms have been shown to occur in peripheral tissues, such as platelets and fibroblasts.

Our group recently observed that platelets are subjected to a substrate-induced modulation of glutamate transporters, similarly to astrocytes in the CNS. Pre-incubation with glutamate causes a time and dose-dependent increase in the Vmax of radioactive glutamate uptake rate, which is associated with increased expression of EAAT1(GLAST) and enhancement of EAAT2(GLT1) activity (unpublished data).

Phorbol esters, such as phorbol myristate acetate (PMA), which acts as a protein kinase C (PKC) activator, and PKC itself, cause a sharp increase of the glutamate uptake Vmax in platelets and cultured fibroblasts (Fumagalli et al., 2002), similarly to what happens in the CNS.

Differently from neurons, experimental evidences indicate that fibroblasts cultures from healthy controls are quite resistant to oxidative stress-induced impairment of glutamate uptake (Begni et al., 2002). No information are available as far as glutamate uptake sensitivity to oxidative stress in platelets.

Biochemical and molecular studies performed in platelets suggest a physiological down-regulation of glutamate uptake and EAAT1(GLAST) transporter with aging. In fact, older subjects display decreased glutamate uptake V_{max}, without modification of transporter affinity, and a selective decrease in the expression of EAAT1(GLAST) protein and mRNA (Zoia et al., 2003).

Since glutamate transporters are modulated in the periphery as occurs in the CNS and are subjected to physiological down-regulation, their expression and activity in peripheral tissues may be also affected in CNS pathology.

4. Expression of glutamate transporters in platelets and fibroblasts from neurological patients.

Molecular studies of glutamate transporters in peripheral tissues have been perfomed in Alzheimer's disease (AD) and Down's syndrome (DS) patients.

Platelet from 10 AD patients showed a selective loss of EAAT1 (GLAST) glutamate transporter both as mRNA and as protein (Zoia et al., 2003), whereas EAAT2 (GLT1) and EAAT3 (EAAC1) were virtually unchanged. EAAT1 protein was also reduced in AD fibroblasts compared to controls, whereas EAAT1 mRNA was significantly enhanced, possibly as a result of a compensatory mechanism. Both the decrease in EAAT1 protein and the increase in EAAT1 mRNA in AD fibroblasts correlates with the severity of the disease assessed by the mini-mental state examination (MMSE) (Tagliabue et al., 2002).

Platelets from 10 DS patients with no clinical diagnosis of dementia show a not significant reduction in mRNA levels of all three glutamate transporters (EAAT1-3), among which EAAT1(GLAST) appear the more compromised (Zoia et al., 2001).

The previous findings could, at least in part, explain the reduction of glutamate uptake rate observed in the same disorders by functional studies with radioactive glutamate (see below).

The culprits for this selective down-regulation of glutamate transporter expression are only hypothetically identified. An energy defect, which is known to underlie the neurodegenerative process in AD, might be involved also at peripheral level and determine a reduced synthesis of the transporter. Otherwise, the well-documented oxidative stress occurring in the disease might structurally damage the transporters which, subsequently, undergo spontaneous internalization and proteolytic fragmentation (see *Chapter 13*). Finally, among various factors involved in the pathogenesis of AD, inflammatory-immunologic activation seems to play a major role. Inflammatory mediators, such as TNF-α, are known to decrease astroglial expression of GLAST and GLT and the same effect might occur in peripheral cells.

5. Functional studies of glutamate uptake in platelets and fibroblasts from neurological patients.

Peripheral glutamate transport has been investigated in a variety of chronic and acute neurological disorders, in most cases by assaying radioactive glutamate uptake in platelets obtained by venous drawing. Preliminary studies have also been performed in cultured fibroblasts from AD and DS patients. Major studies are summarized in Table 2.

Measurement of glutamate and aspartate high-affinity uptake in platelets from Huntington's disease (HD) did not indicate significant differences between control and HD platelets (Mangano et al., 1981b).

Our group studied high-affinity glutamate uptake in platelets from 35 AD patients, 10 multi-infarct dementia patients and 35 age-matched normal controls. A 40% decrease of this biochemical parameter was observed in AD platelets compared to controls and vascular dementia patients (Ferrarese et al., 2000). This finding supported the idea that platelet glutamate uptake might be used as peripheral marker of glutamatergic involvement in AD. A further study analyzed fibroblast cultures from 15 AD patients compared to 15 healthy age-matched controls. Basal glutamate uptake was decreased by 60% in fibroblasts from AD patients versus controls. Exposure to 4-hydroxynonenal (4-HNE) did not affect glutamate uptake in fibroblasts from controls, while reduced it by 50% in fibroblasts from AD patients. Glutathione and N-acetyl-cysteine completely blocked 4-HNE effects and also increased basal uptake in AD cells (Begni et al., 2002).

Table 2. Peripheral glutamate uptake in neurological disorders. Abbraviations: PMA = phorbol myristate acetate, HNE = 4-hydroxynonenal.

Neurological disorders	*Platelet glutamate uptake*	*Fibroblast glutamate uptake*
HUNTINGTON DISEASE	No difference versus healthy subjects.	Not investigated.
ALZHEIMER DISEASE	↓ 40% versus healthy subjects. Impaired response to PMA.	↓ 60% versus healthy subjects. No response to PMA. Further ↓ 50% after exposure to HNE, rescue by gluthathione/n-acetyl-cysteine.
DOWN SYNDROME	↓ 40% versus healthy subjects.	↓ 50% versus healthy subjects. Correlation with disease severity and mitochondrial point mutations.
PARKINSON DISEASE	↓ 48% versus healthy subjects. Correlation with disease severity.	Not investigated.
AMYOTROPHIC LATERAL SCLEROSIS	↓ 43% versus healthy subjects.	Not investigated.
ACUTE ISCHEMIC STROKE	↓ in the early weeks after stroke and persisting for 3 months.	Not investigated.

Regulatory responses proved to be altered in platelets from AD patients, which showed a much smaller increase in glutamate uptake rate after exposure to PMA, compared to control platelets. An altered regulation was also observed in AD fibroblasts, which showed virtually no response to PMA, in contrast to the sharp increase of glutamate uptake rate that occurred in controls (Fumagalli et al., 2002).

A parallel study was carried out on platelets and cultured fibroblasts from DS patients. Glutamate uptake was significantly decreased in both the peripheral tissues from DS patients compared to controls. A more severe impairment of glutamate uptake was observed in patients with free trisomy and mitochondrial point mutations, indicating a possible role of mitochondria in the genesis of excitotoxic phenomena (Begni et al., 2003).

Further, glutamate uptake was investigated in platelets from 38 sporadic PD patients, 13 patients with parkinsonian syndromes and 28 age-matched controls. A 48% reduction of glutamate uptake was observed in

PD patients with respect to control groups. The impairment of this parameter correlated with the disease severity assessed by the UPDRS and genetic studies did not show differences in polymorphisms of alpha-synuclein and ApoE genes among the three groups (Ferrarese et al., 1999; Ferrarese et al., 2001a). Even though it is in theory possible that the limited sample size could be responsible for the lack of correlation between glutamate uptake impairment and genetic abnormalities, the more likely explanation is that the biochemical defect is independent of genetic factors and represent a state marker of the disease. As occurred in DS, mitochondrial abnormalities, in particular the presence of a mild, systemic defect of complex I has been demonstrated in Parkinson's disease (PD) (Swerdlow et al., 1996).

It is well established that decreased glutamate uptake and specific loss of the astrocytic glutamate transporter EAAT2 (GLT1) occur in spinal cord and motor cortex of patients with ALS (Rothstein et al., 1992). Following the hypothesis that mechanisms of decreased glutamate uptake might be systemic in ALS, platelet glutamate uptake was assessed in 42 patients with ALS, 21 subjects with chronic neurologic diseases of vascular origin and 40 age-matched controls. A 43% reduction of high-affinity glutamate uptake rate was observed in patients with ALS compared to normal controls and cerebrovascular disease patients, suggesting a systemic impairment of glutamate uptake in ALS (Ferrarese, 2001b).

Although not disease-specific, all these studies indicate that glutamate uptake impairment appears associated with the neurodegenerative process, because similar clinical forms of vascular origin (multi-infarct dementia, vascular parkinsonism, chronic cerebrovascular and spinal cord disorders) show normal uptake values. Thus, peripheral glutamate uptake may be regarded as an indicator of glutamatergic dysfunction in neurodegenerative disorders.

As far as the mechanisms which may be responsible for the observed inhibition of glutamate uptake activity in platelets and fibroblasts, three major candidates could be theoretically identified: genetics, oxidative stress and inflammation (see Figure 2).

The impairment of glutamate transport occurring in DS, which is a genetically determined disease, is only indirectly due to trisomy 21. In fact, human glutamate transporter genes are on chromosome 5 (EAAT1), 11 (EAAT2) and 9 (EAAT3) and other pathogenetic factors, such as amyloid β-peptide hyperproduction, are more likely involved in determining excitatory injury.

Systemic oxidative stress, notably reactive oxygen species (ROS) present in the blood stream, is a well-characterized feature of neurodegenerative disease, such as AD (Mecocci et al., 2002), PD

(Martignoni et al., 1999) and ALS (Bonnefont-Rousselot et al., 2000). High-affinity glutamate transport system has been shown to be oxidant-vulnerable in primary cultures of cortical astrocytes (Sorg et al., 1997) and Xenopus oocytes expressing EAAT2(GLT) (Trotti et al., 1999). Several recent studies show that EAATs possess specific redox-sensing elements, consisting of cysteine residues, which regulate transport rate by thiol-disulphide redox inter-conversion (Trotti et al., 1997; Reis et al., 2000). Glutamate uptake presents maximal rate by the reduced state of reactive cysteine residues and minimal activity by oxidized state without any change in affinity for glutamate (Km). This regulatory mechanism probably reflects variation in transporter conformation. Lipids are specially vulnerable to oxidative stress and products of lipoperoxidation, such as acrolein and 4-hydroxy-2-nonenal (HNE), are able to reduce glutamate uptake (Lovell et al., 2000). Such vulnerability to biological oxidants, demonstrated in astroglia and in other *in vitro* models, might also occur in platelets and fibroblasts as a cause of systemic oxidative stress.

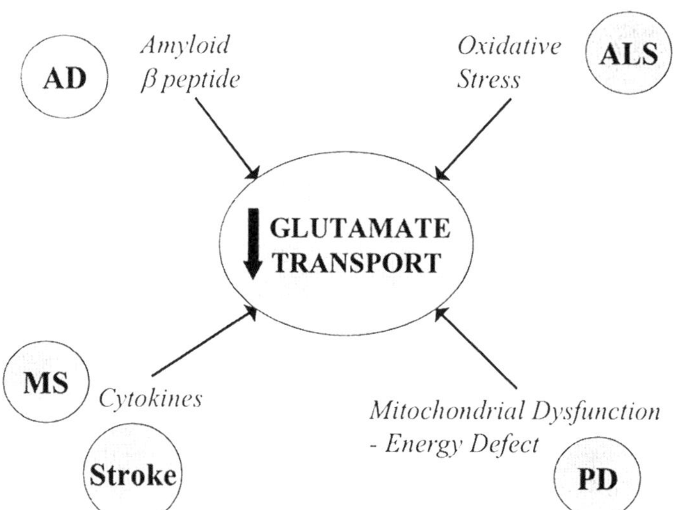

Figure 2. Systemic factors which are hypothetically responsible for peripheral glutamate uptake impairment in various neurological disorders. Abbraviations: AD = Alzheimer's disease; PD = Parkinson's disease; MS = multiple sclerosis; ALS = Amyotrophic lateral sclerosis.

Inflammatory mediators negatively influence the activity of glutamate transport, in particular interleukin-1β and interferon-γ decrease glutamate uptake in rat hippocampal cultures (Ye et al., 1996). Another cytokine, tumor necrosis factor (TNF-α), seems to produce this inhibitory effect on glutamate uptake by directly decreasing astroglial expression of EAAT1(GLAST) and EAAT2(GLT) (Szymocha et al., 2000). Several neuropathological studies showed a close association between AD neuritic plaques and local inflammatory response (Akiyama et al., 2000). The presence of Abeta plaques may keep microglia persistently activated, leading to a condition of chronic inflammation at central nervous system level. Abeta stimulates both the synthesis and the release of pro-inflammatory cytokines such as interleukin-1β (Lorton et al., 1996), interleukin-6 (Gitter et al., 1995) and TNF-alpha (Meda et al., 1995). The same cytokines have been found over-expressed in the brain of AD patients (Griffin et al., 1989; Bauer et al., 1991). Inflammatory markers were also investigated in CSF and serum of AD patients: results have been somewhat controversial, but they all showed an activation of the peripheral immune status, probably linked to the inflammatory condition present in AD brain. Such an immune response observed in AD patients could be easily hypothesized to be involved in the alteration of glutamate uptake activity observed in peripheral tissues.

As far as acute neurological disorders, alteration of platelet glutamate uptake has been demonstrated in the early weeks after ischemic stroke and a significant reduction persisted for 3 months (Tremolizzo et al., 2001). No difference between controls and patients was detected after 12 months, consistently with the previously reported data about platelet glutamate uptake in chronic cerebrovascular diseases (Ferrarese et al., 2000; Ferrarese et al., 2001b). These findings suggest that increased levels of glutamate, previously described in stroke patients (see above), could be linked to impairment of glutamate uptake, rather than to increased release from CNS. Various factor, such as proinflammatory cytokines, are elevated both in CNS and in blood after stroke and may modulate peripheral and central glutamate uptake.

6. The clinical significance of ex vivo markers for diagnosis and therapy.

All the observations in plasma, platelets and fibroblasts indicate that glutamate levels, receptors and transporters are modified also in peripheral tissues in neuropsychiatric disorders. Thus, although the targets of excitotoxicity are neurons, its players and actors are working also in cells outside the central nervous system. This fact is not surprising, since many of the mechanisms affecting glutamate transporters and receptors, i.e.

oxidative stress products, cytokines, mitochondrial impairments, appear to act systemically and are not restricted to the CNS. Indeed, glutamate transport and transporters are regulated by physiological and pathological mechanisms in similar ways in neuronal and glial cells, as in platelets and fibroblasts.

The natural consequence of these observations is that peripheral cells may be easily accessible tools to investigate such mechanisms in patients, during the course of the disease.

However, the most important question, which needs to be addressed, is the clinical relevance of these findings: are they useful for the diagnosis and the treatment of these disorders?

Since alterations in glutamate plasma levels, platelet or fibroblast glutamate uptake have been described in various neurodegenerative disorders, as well as after acute stroke, they are not disease-specific, but appear as common consequences of damage of transporters. Therefore, they are not useful for the diagnosis, which mainly stands on clinical grounds for all these disorders, but rather to analyse the possible role of excitotoxic mechanisms in such pathologies and in individual patients. Wide dispersion of data among patients also suggests that glutamate is not playing the same role in all individuals; for this reason it could be analysed as prognostic tool, as data from stroke patients may suggest. Longitudinal assessment of these markers in plasma and platelets can be used to follow the course of the disease and the effects of therapies. Moreover, fibroblast cell primary cultures may be used for pharmacological studies, to investigate mechanisms involved in alterations of glutamate transport and possible therapeutic tools.

Thus, these peripheral *ex vivo* markers from patients might bridge the gap between preclinical studies in animal models and patient care, and may help to verify the excitotoxic hypothesis at the bedside.

REFERENCES

Akiyama H., Barger S., Barnum S., Bradt B., Bauer J., Cole G.M., Cooper N.R., Eikelemboom P., Emmerling M., Fiebich B.L., Finch C.E., Frautschy S., Griffin W.S.T., Hampel H., Hull M., Landreth G., Lue L.F., Mrak R., Mackenzie I.R.; McGeer P.L., O'Banion M.K., Pachter J., Pasinetti G., Plata-Salaman, Rogers J., Rydel R., Sheng Y., Streit W., Strohmeyer R., Tooyoma I., Van Muiswinkel F.L., Veerhuis R., Walker D., Webster S., Wegrzyniak B., Wenk G., Wyss-Coray T. Inflammation and Alzheimer's disease. Neurobiol of Aging 2000; 21(3):383-421.

Balcar V.J., Shen J., Bao S., King N.J. Na(+)-dependent high affinity uptake of L-glutamate in primary cultures of human fibroblasts isolated from three different types of tissue. FEBS Lett 1994; 339(1-2):50-54.

Barbeau A., Campanella G., Butterworth R.F., Yamada K. Uptake and efflux of 14-C-dopamine in platelets: evidence for a generalized defect in Parkinson's disease. Neurology 1975; 25(1):1-9.

Bauer J., Strauss S., Schreiter-Gasser U., Ganter U., Sclegel P., Witt I., Yolk B., Berger M. Interleukin-6 and α-2-macroglobulin indicate an acute phase state in Alzheimer's disease cortex. FEBS Lett 1991; 285:111-114.

Begni B., Brighina L., Fumagalli L., Andreoni B., Ferrarese C. Excitotoxicity and oxidative stress in fibroblasts from Alzheimer disease patients. 3rd Forum of European Neuroscience, July 13-17, 2002, Paris. Abstract.

Begni B., Brighina L., Fumagalli L., Andreoni S., Castelli E., Francesconi C., Del Bo R., Bresolin N., Ferrarese C. Altered glutamate uptake in peripheral tissues from Down's sindrome patients. Neuroscience Letters 2003; in press.

Beretta S., Begni B., Ferrarese C. Pharmacological manipulation of glutamate transport. Drugs of Today 2003; in press.

Berger C., Annecke A., Aschoff A., Spranger M., Schwab S. Neurochemical monitoring of fatal middle cerebral artery infarction. Stroke 1999; 30(2):460-463.

Berger C., Schabitz W.R., Georgiadis D., Steiner T., Aschoff A., Schwab S. Effects of hypothermia on excitatory amino acids and metabolism in stroke patients: a microdialysis study. Stroke 2002; 33(2):519-524.

Berk M., Plein H., Belsham B. The specificity of platelet glutamate receptor supersensitivity in psychotic disorders. Life Sci 2000; 66(25):2427-2432.

Berk M., Plein H., Ferreira D. Platelet glutamate receptor supersensitivity in major depressive disorder. Clin Neuropharmacol 2001; 24(3):129-132.

Blin O., Samuel D., Nieoullon A., Sterratrice G., Changes in CSF aminoacid concentrations during the evolution of amyotrophic lateral sclerosis. J Neurol, Neurosurg, Psychiatry 1994;57:119-121.

Bonnefont-Rousselot D., Lacomblez L., Jaudon M., Lepage S., Salachas F., Bensimon G., Bizard C., Doppler V., Delattre J., Meininger V. Blood oxidative stress in amyotrophic lateral sclerosis. J Neurol Sci 2000; 178(1):57-62.

Camu W., Billiard M., Baldy-Moulinier M. Fasting plasma and CSF amino acid levels in amyotrophic lateral sclerosis: a subtype analysis Acta Neurol Scand 1993; 88:51-55.

Castillo J., Davalos A., Naveiro J., Noya M. Neuroexcitatory amino acids and their relation to infarct size and neurological deficit in ischemic stroke. Stroke 1996; 27(6):1060-1065.

Castillo J., Davalos A., Noya M. Progression of ischaemic stroke and excitotoxic aminoacids. Lancet 1997, 349(9045):79-83.

Cooper B., Chebib M., Shen J., King N.J., Darvey I.G., Kuchel P.W., Rothstein J.D., Balcar V.J. Structural selectivity and molecular nature of L-glutamate transport in cultured human fibroblasts. Arch Biochem Biophys 1998; 353(2):356-364.

Davalos A., Castillo J., Serena J., Noya M. Duration of glutamate release after acute ischaemic stroke Stroke, 1997,28(4):708-710.

Dinkelborg L.M., Kinne R.K., Grieshaber M.K. Transport and metabolism of L-glutamate during oxygenation, anoxia, and reoxygenation of rat cardiac myocytes. Am J Physiol 1996; 270(5 Pt 2):H1825-32.

Dirgnagl U., Iadecola C., Moskowitz M.A. Pathobiology of ischaemic stroke: an integrated view. Trends Neurosci 1999;22:391-397.

Ferrarese C., Tremolizzo L., Rigoldi M., Sala G., Begni B., Brighina L., Ricci G., Albizzati M.G., Piolti R., Crosti F., Dalpra L., Frattola L. Decreased platelet glutamate uptake and genetic risk factors in patients with Parkinson's disease. Neurol Sci 2001a; 22(1):65-66.

Ferrarese C., Sala G., Riva R., Begni B., Zoia C., Tremolizzo L., Galimberti G., Millul A., Bastone A., Mennini T., Balzarini C., Frattola L., Beghi E., The Italian ALS Study Group. Decreased platelet glutamate uptake in patients with amyotrophic lateral sclerosis. Neurology 2001b; 56(2):270-272.

Ferrarese C., Aliprandi A., Tremolizzo L., Stanzani L., De Micheli A., Dolara A., Frattola L. Increased glutamate in CSF and plasma of patients with HIV dementia Neurology 2001c,57(4):671-675.

Ferrarese C., Begni B., Canevari C., Zoia C., Piolti R., Frigo M., Appollonio I., Frattola L. Glutamate uptake is decreased in platelets from Alzheimer's disease patients. Ann Neurol 2000; 47(5):641-643.

Ferrarese C., Zoia C., Pecora N., Piolti R., Frigo M., Bianchi G., Sala G., Begni B., Riva R., Frattola L. Reduced platelet glutamate uptake in Parkinson's disease. J Neural Transm 1999; 106(7-8):685-692.

Ferrarese C., Pecora N., Frigo M., Apollonio I., Frattola L. Assessment of reliability and biological significance of glutamate levels in cerebrospinal fluids. Ann Neurol 1993, 33(3),313-369.

Fleisher G.A., Wakim K.G., Goldstein N.G. Glutamic oxalacetic transaminase and lactic dehydrogenase in serum and cerebrospinal fluid of patients with neurologic disorders. Proc Staff Meet Mayo Clin 1957;32:188-197.

Franconi F., Miceli M., De Montis M.G., Crisafi E.L., Bennardini F., Tagliamonte A. NMDA receptors play an anti-aggregating role in human platelets. Thromb Haemost 1996; 76(1):84-7.

Fumagalli L., Begni B., Brighina L., D'Orlando C., Andreoni S., Garofolo R., Ferrarese C. Impaired modulation of glutamate by PKC in fibroblasts and platelets from Alzheimer disease patients. 6th Convegno Annuale ITINAD, 18-20 June, 2002, Sorrento. Abstract.

Gitter B.D., Cox L.M., Russel R.E., May P.C. Amyloid beta peptide potentiates cytokine secretion by interleukin-1β-activated human astrocytoma cells. Proc Natl Acad Sci USA 1995; 92:10741-10783

Griffin W., Stanley L.C., Ling C., White L., Macleod V., Perrot L., White C.L., Araoz C. Brain interleukin-1 and S-100 immunoreactivity elevated in Down's syndrome and Alzheimer's disease. Proc Natl Acad Sci USA 1989; 86:7611-7615.

Hattori N., Abe K., Sakoda S., Sawada T. Proton MR spectroscopic study at 3 Tesla on glutamate/glutamine in Alzheimer's disease. Neuroreport 2002; 13(1):183-186.

Heyes M.P., Saito K., Crowley J.S., Davis L.E., Demitrack M.A., Der M., Dilling L.A., Elia J., Kruesi M.J., Lackner M.A. Quinolinic acid and kynurenine pathway metabolism in inflammatory and non inflammatory neurological disease, Brain 1992,115:1249-1273.

Hutchinson P.J., O'Connell M.T., Al-Rawi P.G., Kett-White C.R., Gupta A.K., Maskell L.B., Pickard J.D., Kirkpatrick P.J. Increases in GABA concentrations during cerebral ischaemia: a microdialysis study of extracellular amino acids. J Neurol Neurosurg Psychiatry 2002; 72(1):99-105.

Itzstein C., Espinosa L., Edelmass P.D., Chenu C. Specific antagonists of NMDA receptors prevent ostoclast sealing zone formation required for bone resorption. Biochem Biophys Res Commun 2000; 268:201-209.

Kanai Y., Hediger M.A. Primary structure and functional characterization of a high-affinity glutamate transporter. Nature 1992; 360(6403):467-471.

Kanthan R., Shuaib A., Griebel R., Miyashita H. Intracerebral human microdialysis: in vivo study of an acute focal ischemic model of the human brain. Stroke 1995;26: 870-873.

Liu M.T., Rothstein J.D., Gershon M.D., Kirchgessner A.L. Glutamatergic enteric neurons. J Neurosci 1997; 17:4764-4784.

Longoni M., Tremolizzo L., Galimberti G., Vaccaro M., Ceresa C., Aliprandi A., Stanzani L., Ferrarese C. Platelet glutamate release after aggregation in normal subjects and in stroke patients. Func Neurol 2001; 16(2):199.

Lorton D., Kocsis J.M., King L., Madden K., Brunden K.R. β-Amyloid induces increased release of interleukin-1β from lipopolysaccharide-activated human monocytes. J Neuroimmunol 1996; 67:21-29.

Lovell M.A., Xie C., Markesbery W.R. Acrolein, a product of lipid peroxidation, inhibits glucose and glutamate uptake in primary neuronal cultures. Free Radic Biol Med 2000; 29(8):714-720.

Mangano R.M., Schwarcz R. Platelet glutamate and aspartate uptake in Huntington's disease. J Neurochem 1981b; 37(4):1072-1074.

Mangano R.M., Schwarcz R. The human platelet as a model for the glutamatergic neuron: platelet uptake of L-glutamate. J Neurochem 1981a; 36(3):1067-76.

Martignoni E., Blandini F., Godi L., Desideri S., Pacchetti C., Mancini F., Nappi G. Peripheral markers of oxidative stress in Parkinson's disease. The role of L-DOPA. Free Radic Biol Med 1999; 27(3-4):428-437.

Martinez-López I., García C., Barber T., Vina J.R., Miralles V.J. The L-glutamate transporter GLAST(EAAT1) and GLT1 (EAAT2): expression and regulation in rat lactating mammary gland. Mol Membr Biol 1998; 15:237-242.

Mason D.J., Suva L.J., Genever P.G., Patton A.J., Steuckle S., Hillam R.A., Skerry T.M. Mechanically regulated expression of a neural glutamate tranporter in bone: a role for excitatory amino acids as osteotropic agents? Bone 1997; 20:199-205.

Mecocci P., Polidori M.C., Cherubini A., Ingegni T., Mattioli P., Catani M., Rinaldi P., Cecchetti R., Stahl W., Senin U., Beal M.F. Lymphocyte oxidative DNA damage and plasma antioxidants in Alzheimer disease. Arch Neurol 2002; 59(5):794-798.

Meda L., Cassatella M.A., Szendrei G.I., Otvos L. Jr, Baron P., Villalba M., Ferrari D., Rossi F. Activation of microglial cells by β-amyloid protein and interferon-γ. Nature 1995; 374:647-650.

Nakayama T., Kawakami H., Tanaka K., Nakamura S. Expression of three glutamate transporter subtype mRNAs human brain regions and peripheral tissues. Mol Brain Res 1996; 36:189-192.

Peghini P., Janzen J., Stoffel W. Glutamate transporter EAAC1 deficient mice develop dicarboxylic aminoaciduria and behavioural abnormalities but no neurodegeneration. EMBO J 1997;16:3822-3832.

Petroff O.A., Errante L.D., Rothman D.L., Kim J.H., Spencer D.D. Glutamate-glutamine cycling in the epileptic human hippocampus. Epilepsia 2002; 43(7): 703-710.

Pioro E.P., Majors A.W., Mitsumoto H., Nelson D.R., Ng T.C. 1H-MRS evidence of neurodegeneration and excess glutamate + glutamine in ALS medulla. Neurology 1999; 53(1):71-79.

Plaitakis A., Caroscio J.T., Abnormal glutamate metabolism in amyotrophic lateral sclerosis. Ann Neurol 1987;22:575-579.

Reis H.J., Gomez M.V., Kalapothakis E., Diniz C.R., Cordeiro M.N., Prado M.A., Romano-Silva M.A. Inhibition of glutamate uptake by Tx3-4 is dependent on the redox state of cysteine residues. Neuroreport 2000; 11(10):2191-2194.

Rothstein J.D., Kuncl R.W., Chaudhry V., Clawson L., Cornblath D.R., Coyle J.T., et al. Excitatory amino acids in amyotrophic lateral sclerosis: an update. Ann Neurol 1991;30:224.

Rothstein J.D., Martin L.J., Kuncl R.W. Decreased brain and spinal cord glutamate transport in amyotrophic lateral sclerosis. New Engl J Med 1992; 326:1464-1468.

Rothstein J.D., Van Kammen M., Levey A.I., Martin L.J., Kuncl R.V. Selective loss of glial glutamate transporter GLT1 in amyotrophic lateral sclerosis. Neurology 1995,38(1):73-84.

Schulz M.K., Wang L.P., Tange M., Bjerre P. Cerebral microdialysis monitoring: determination of normal and ischemic cerebral metabolisms in patients with aneurysmal subarachnoid hemorrhage. J Neurosurg 2000; 93(5):808-814.

Shulman R.G. Functional imaging studies: linking mind and basic neuroscience. Am J Psychiatry 2001; 158(1):11-20.

Sneddon J.M. Blood platelets as a model for monoamine-containing neurones. Prog Neurobiol 1973; 1(2):151-198.

Sorg O., Horn T.F., Yu N., Gruol D.L., Bloom F.E. Inhibition of astrocyte glutamate uptake by reactive oxygen species: role of antioxidant enzymes. Mol Med 1997; 3(7):431-440.

Speaux-Varoquat O., Bensimon G., Lacombez L., Salachas F., Pradat P.F., Le Forestier N., Maroun A., Michel D., Meininger V. Journal of the Neurological Sciencies 193:73-78.

Spranger M., Schwab S., Krempien S., Winterholler M., Steiner T.; Hacke W. Excess glutamate levels in the cerebrospinal fluid predict clinical outcome of bacterial meningitis. Archives of Neurology 1996;53(10):992-996.

Stahl N., Mellergard P., Hallstrom A., Ungerstedt U., Nordstrom C.H. Intracerebral microdialysis and bedside biochemical analysis in patients with fatal traumatic brain lesions. Acta Anaesthesiol Scand 200; 45(8):977-985.

Swerdlow R.H., Parks J.K., Miller S.W., Tuttle J.B., Trimmer P.A., Sheehan J.P., Bennett J.P. Jr, Davis R.E, Parker W.D Jr. Origin and functional consequences of the complex I defect in Parkinson's disease. Ann Neurol 1996; 40(4):663-671.

Szymocha R., Akaoka H., Brisson C., Beurton-Marduel P., Chalon A., Bernard A., Didier-Bazes M., Belin M.F., Giraudon P. Astrocytic alterations induced by HTLV type 1-infected T lymphocytes: a role for Tax-1 and tumor necrosis factor alpha. AIDS Res Hum Retroviruses 2000; 16(16):1723-1729.

Tagliabue E., Zoia C., Bonfanti P., Arban S., Ferrarese C. Regulation of glutamate transporters in fibroblasts from normal subjects and Alzheimer patients. Neurol Sci 2002; 23 (suppl.):S122.

Taylor-Robinson S.D., Turjanski N., Bhattacharya S., Seery J.P., Sargentoni J., Brooks D.J., Bryant D.J., Cox I.J. A proton magnetic resonance spectroscopy study of the striatum and cerebral cortex in Parkinson's disease. Metab Brain Dis 1999; 14(1):45-55.

Tremolizzo L., Longoni M., Aliprandi A., Stanzani L., Begni B., Sala G., Ferrarese C. Platelet glutamate uptake and plasma glutamate levels in stroke patients. Ann Neurol 2001; 50(3) suppl 1: s42.

Trotti D., Rizzini B.L., Rossi D., Haugeto O., Racagni G., Danbolt N.C., Volterra A. Neuronal and glial glutamate transporters possess an SH-based redox regulatory mechanism. Eur J Neurosci 1997; 9(6):1236-1243.

Trotti D., Rolfs A., Danbolt N.C., Brown R.H. Jr, Hediger M.A. SOD1 mutants linked to amyotrophic lateral sclerosis selectively inactivate a glial glutamate transporter. Nat Neurosci 1999; 2(5):427-433.

Vespa P., Prins M., Ronne-Engstrom E., Caron M., Shalmon E., Hovda D.A., Martin N.A., Becker D.P. Increase in extracellular glutamate caused by reduced cerebral perfusion pressure and seizures after human traumatic brain injury: a microdialysis study. J Neurosurg 1998; 89(6):971-982.

Ye Z.C., Sontheimer H. Cytokine modulation of glial glutamate uptake: a possible involvement of nitric oxide. Neuroreport 1996; 7(13):2181-2185.

Zieve P.D., Solomon H.M. Uptake of amino acids by the human platelet. Am J Physiol 1968; 214(1):58-61.

Zoia C., Cogliati T., Begni B. Molecular characterization of glutamate transporters in human platelets. Trends in Neurosci 1997; 20 (Suppl. 9):41.

Zoia C., Cogliati T., Tagliabue E., Cavaletti G., Sala G., Galimberti G., Rivolta I., Rossi V., Frattola L., Ferrarese C. Glutamate transporters in platelets: EAAT1 decrease in aging and in Alzheimer disease. Neurobiol Aging 2003; in press.

Zoia C., Corsini R., Begni B., Galimberti G., Castelli E., Ferrarese C. Modulation of glutamate transporters in platelets and fibroblasts from Alzheimer and Down patients. J Neurochem 2001; 77(suppl.1):22. P01-14 Abstract.

Chapter 18

ANTI-EXCITOTOXIC THERAPIES IN NEUROLOGIC DISEASES: CURRENT EVIDENCES AND NEW PERSPECTIVES

Laura Brighina, Angelo Aliprandi, Simone Beretta and Carlo Ferrarese
Department of Neurology, University of Milano-Bicocca, Ospedale S. Gerardo, Via Donizetti 106, 20052 Monza (MI), Italy.

Abstract: To counteract excitotoxic mechamisms involved in a great deal of acute and chronic neurodegenerative disorders, efforts have been addressed, over the last decade, to search for drugs that either reduce glutamate synaptic levels or block its postsysnaptic effects. Among the large amount of compounds tested, riluzole and the NMDA receptor antagonist memantine have proved to be quite efficacious in amyotrophic lateral sclerosis and in Alzheimer, disease respectively. Other clinical trials of antiexcitotoxic agents have been unsuccessful, despite the encouraging data collected in animal models, probably because of methodological or dosage problems. Moreover, some of these drugs might increase their efficacy if administred in early stages of disease and in combination with other neuroprotective agents, such as antioxidants and antiapoptotic substances.

Keywords: antiexcitotoxic therapy, cerebral ischemia, amyotrophic lateral sclerosis, Alzheimer disease, Huntington disease, riluzole, memantine.

INDEX

1. Introduction
2. Antiexcitotoxic therapy in cerebral ischemia.
3. Antiexcitotoxic therapy in amyotrophic lateral sclerosis.
4. Antiexcitotoxic therapy in Alzheimer disease.
5. Antiexcitotoxic therapy in Huntington disease.
6. Final remarks.

1. Introduction.

While several approved drugs alleviate, to some extent, the symptoms of neurodegenerative diseases (for instance, acetylcholinesterase inhibitors in Alzheimer disease and L-dopa and dopamine-agonists in Parkinson disease), none seems to stop the progression of the degenerative processes underlying them.

In keeping with this, the development of effective preventive or protective therapies has been impeded by the limitations of our knowledge of the causes and the mechanisms by which neurons die in neurodegenerative disorders.

The glutamatergic system dysfunction seems to be an early event in the pathogenesis of a large number of acute and chronic neurological disorders, in strict conjunction with other important mechanisms, such as oxidative stress and energetic failure.

In consideration of that, drug discovery efforts over the last decade have been focused on the search for drugs that either reduce glutamate synaptic levels or block its postsynaptic effects. With respect to the latter, NMDA glutamate receptor antagonists have demonstrated a therapeutic potential in several CNS disorders, ranging from acute (e.g. stroke and traumatic brain injury) to chronic neurodegeneration (e.g. Parkinson disease, Alzheimer disease, ALS, Huntington disease). In the following sections, we will discuss the main antiglutamate strategies attempted or currently under investigation for the treatment of each of these disorders.

2. Antiexcitotoxic therapies in cerebral ischemia.

The relevance of excitotoxic phenomena in the pathogenesis of neuronal damage during ischemia is now firmly enstablished, as extensively reviewed in other sections of this book (chapter 8). As a result, and given the high prevalence and social impact of stroke in western societies, the possibility of preventing neuronal death by blocking the chain of events triggered by glutamate stimulation has been the object of a huge investigational and clinical effort.

As a matter of facts, experimental results in animal models of stroke have yielded encouraging results, showing reduction of infarcted areas with administration of a number of different compounds. This type of therapeutic strategy, aimed at increasing neuronal resistance to ischemic conditions, has been termed "neuroprotection".

Unfortunately, so far all attempts to reproduce these data in clinical condition have been disappointing, and no drug was shown to be effective in phase three investigational trials. The reasons for the discrepancy between the animal models and the human patology have also been previously exposed (chapter 8); in this section, the results obtained with different classes of neuroprotective agents will be briefly summerized, and some data on clinical trials currently being performed will be reported.

Na+ -channel blockers. Drugs with affinity for Na+ channels may exert a neuroprotective action by reducing neuronal depolarisation and calcium influx, and by preventing further glutamate release. Phenytoin, a drug commonly used as an antiepileptic agent, was shown to reduce ischemic areas in animal models of stoke. The precursor of the drug, phosphenytoin, has been tested in a phase III clinical trial, which was suspended at an interim analysis for lack of efficacy (Pulsinelli, W.A. 1999)

Glutamate receptor antagonists. An obvious strategy to prevent excitotoxicity at its onset is to administer NMDA receptor antagonists. Unfortunately, most of these drugs induce serious side effects at therapeutic dosages, such as allucinations, delusions and motor hyperactivity, and were then stopped after phase I or II trials. Among the compounds tested, only selfotel and aptiganel could reach phase III trials, but were than dropped because of the risk to benefit ratio was unfavourable (Davis, S.M. 1997). Another molecule with a better tolerability, eliprodil, believed to bind to the NMDA polyamine domain, entered phase III experimentation but did not show significant clinical efficacy

GABA stimulation. Glutamate excitotoxicity could also be blocked by reducing neuronal depolarisation with the administration of drugs with GABAergic activity. One of such agents, clomethiazole, was shown to protect neurons from ischemic death in animal models. Although a phase III clinical trial failed to reach the primary endpoints investigated and no functional improvement was observed in patients (Whalgren, N.G. et al, 1999) analysis of disagreggated data suggested that the drug may be of benefit in patients with a total anterior circulation (carotid) stroke. Hence, a new trial on these type of patients, the North American Acute Stroke Study-Ischemia (CLASS-I) was started and is currently in progress.

Ca+ channel blockers. One of the most extensively studied class of drugs believed to have neuroprotective properties is calcium channel blockers. Among these flunarizine and nimodipine have both been shown to reduce infarct size in animals. Also in the case of these conpounds, however, data in clinical trials have been much less encouraging. Flunarizine was tested in the FIST trial, with negative results (Franke C.L. et al, 1996), while nimodipine, either intravenous or oral, has been investigated in several trials, all of which did not show clinical efficacy. A metanalysis of these study suggested that some benefit might be obtained if the drug was administered within 12 hours from stroke (Mohr JP et al, 1994). Another study, the VENUS trial, in which the time window permitted for patient enrolment was very short (6 hours), was than started, but no effect on functional outcome of patients could be demonstrated (Horn J et al, 1999). *Antioxidants and radical scavengers.* Another possible site at which excitotoxic damage might be counteracted is more downstream in the pathogenetic cascade, by reducing free radicals formation or buffering them with scavengers. Among the drugs acting with this putative activity, three molecules reached phase three trials: lubelozule, tirilazad and ebselen. The first, which seems to act more specifically as an inihibitor of nitric oxide syntase, has been tested in three randomised trials, all of which failed to reach a significant difference in the primary endpoints (Diener, H.C. et al, 1998; Grotta J. et al., 1997; Diener H.C., 1999). Tirilazad, a free radical scavenger, did not improve clinical outcome in two trials (Peters G.R. et al 1996) and was discontinued in a third, in which higher doses were being used, because of serious side effects (Haley E.C., 1998). Finally, the selenium-related compound ebselen appeared to be effective on clinical outcome after one month, but not after three months from stroke.(Yamaguchi T. et al, 1998).

3. Antiexcitotoxic therapies in amyotrophic lateral sclerosis.

ALS is a progressive degenerative disease of unknown etiology, whose devastating course places it at the centre of the ethical debate about end of life decision. It should not be surprising, therefore, that an impressing number of drugs have been tested in clinical trial for the treatment of the disease, mostly in small, short-lasting studies (Dib, 2003). All the drugs clinically evaluated so far, with the exception of riluzole, failed to produce a relevant effect in ALS patients (Table 1).

As discussed in Chapter 10, ALS is a multisystem disorder and can be viewed as the consequence of a complex neurodegenerative process involving neuron-glia interactions. Excitotoxicity, oxidative stress, mitochondrial dysfunction, cytoskeletal defects, neurotrophic factors and apoptosis are all putative mechanism which seems to operate in ALS and might be amenable of pharmacological intervention.

The current gap between hypothesized pathogenetic mechanisms and efficacy of therapeutics in humans is due to two methodological issues: the need of accurate biological markers, that would mean an early diagnosis and objective monitoring of the disease progression; the reliability of animal models of the disease, which do not reproduce human disease with total accuracy.

Table 1. Major clinical trials with neuroprotective drugs in ALS. For references, see the text.

Drug (year)	*phase*	*number of patients*	*duration (months)*	results
Riluzole (1994)	II	155	12	↑ survival ↑ functional status
Riluzole (1996)	III	959	18	↑ survival
Gabapentin (2001)	III	204	9	no effect
Lamotrigine (1993)	II	67	18	no effect
Dextromethorphan (1996)	II	49	12	no effect
Acetylcysteine (1995)	II	110	12	no effect
Selegiline (1998)	II	111	6	no effect
α-tocopherol (2001)	III	289	12	↑ functional status
Creatine (2003)	II	175	16	no effect

To date, riluzole is the only approved agent for the treatment of amyotrophic lateral sclerosis (ALS) (Talbot, 2002), after its efficacy on prolonging survival in ALS patients was demonstrated by multicenter clinical trials. The first randomised, double-blind, placebo-controlled trial (Bensimon et al., 1994) enrolled 155 ALS patients and showed a favourable, although limited, effect of the drug (50 mg, taken orally twice a day) on survival and functional evaluations. The effect on survival was prominent in patients with bulbar onset. The second study (Lacomblez et al., 1996) involved 959 ALS patients and tested different doses of riluzole (50, 100 and 200 mg/day). Riluzole showed a positive and dose-dependent effect on tracheostomy-free survival, while no effect on functional outcomes was reported, by contrast with the initial study.

The pharmacological profile of riluzole is complex and many effects exerted by this molecule have been shown, which could be responsible for its neuroprotective action. The anti-excitotoxic properties of riluzole, mediated by inhibition of multiple ion channels, such as glutamate-gated ion channels (Debono et al., 1993), volume-sensitive chloride channels (Bausch et al., 1996), voltage-gated sodium channels (Song et al, 1997) and voltage-gated calcium channels (Huang et al., 1997), are currently thought of as its principal mechanisms of action. However, it has been recently demonstrated that this agent attenuates oxidative cell death and inhibits protein kinase C activation (Koh et al., 1999; Noh et al., 2000).

These findings are particularly important, considering that other anti-glutamate drugs, such as gabapentin (Miller at al., 2001), lamotrigine (Eisen et al., 1993) and dextromethorphan (Blin et al., 1996) showed no evidence of beneficial effect in phase III trials. Thus, the unique efficacy of riluzole, compared to other more potent and selective anti-excitotoxic agents, could be due to other additional mechanisms, such as counteracting oxidative injury.

At the present time, the anti-epileptic drug topiramate and a dextromethorphan derivative, which should overcome the pharmacokinetic problems associated with the rapid metabolism of dextromethorphan itself, are anti-glutamate agents under investigation.

Oxidative stress is a well-established feature of ALS. Concentrations of 3-nitrotyrosine and 4-hydroxynonenal, biochemical markers of peroxynitrite-mediated oxidative damage and lipid peroxidation, respectively, were reported to be remarkably elevated in CSF from sporadic ALS patients (Tohgi et al., 1999; Smith et al., 1998). Besides, CSF from ALS patients has been shown to be neurotoxic to rat cortical neurons and its toxicity was counteracted by antioxidants (Terro et al., 1996). Nevertheless, the antioxidant agents tested so far, such as acetylcysteine (Louwerse et al., 1995), reduced glutathione (Chio et al., 1998) and selegiline (Lange et al., 1998), have yelded negative results. The only promising results were obtained for alpha tocopherol in patients taking riluzole: an higher proportion of patients receiving the double therapy remained longer in a milder ALS health state (Desnuelle et al., 2001). Much of these negative outcome might be explained by the fact that the therapeutic use of most antioxidants is limited since they do not cross the blood-brain barrier after systemic administration, as recently discussed by Gilgun-Sherki et al. (2001).

Other strategies include energy stimulators, such as creatine, which was found ineffective at the dose tested (10 mg) in a well-designed clinical trial (Groeneveld et al., 2003) and coenzyme Q, which is currently under investigation.

The hypothesis that a synergistic effect between anti-excitotoxic and antioxidant actions is the cause of the success of riluzole arises new perspectives for the future treatment of ALS, which go far beyond the glorification of this drugs, which is recognized to have a small clinical impact. Combination therapies or drugs with multiple mechanisms of actions are likely to be the only way forward for ALS therapeutics. This could be the case of minocycline. This second-generation tetracycline readily crosses the blood-brain barrier and displays multiple mechanisms of action, such as inhibition of caspases, mitochondrial cytochrome c release, nitric oxide synthase activity, microglial activation and protein tyrosine nitration (Zhu et al., 2002; Kriz et al., 2003). The pharmacodinamic properties of minocycline are somehow complementary to those of riluzole. A combination therapy with riluzole and minocycline would represent a multi-level pharmacotherapy for ALS and it is currently under clinical evaluation.

4. Antiexcitotoxic therapies in Alzheimer disease.

There is compelling evidence that glutamatergic system is impaired in AD, where Abeta and Abeta-induced oxidative stress may inhibit glial glutamate transporters (Lauderback et al, 2000) and augment neuronal vulnerability to glutamate toxicity via NMDA receptors (see Chapter 13). The prolonged mild activation of these receptors has two major consequences:
1) long term potentiation impairment (LTP), due to a decrease in signal to noise ratio;
2) induction of neural death through ROS production, lytic enzymes activation and mitochondrial derangement.

The NMDA receptor's role in excitotoxicity has driven the search for antagonists as neuroprotective agents. Its role in synaptic plasticity, on the other hand, has inspired research into drugs that modulate the activity of NMDA receptors to treat cognitive dysfunction.

The first drug tested was D-cycloserine, a partial agonist acting at the glycine B site, which unfortunately failed to demonstrate any positive cognitive effect, probably due to its narrow therapeutic window (Laake et al., 2002).

Many NMDA receptor antagonists previously evaluated in human clinical trials either prevented glutamate from binding to the NMDA receptor, or blocked the NMDA receptor channel for a longer period of time than was safe. While they protected neurons from excitotoxicity, they also interfered with physiologic learning processes. Such impairment resulted, in some cases, in hallucinations, psychosis or even coma.

The first NMDA receptor antagonist tested in humans, MK-801, produced such neuropsychiatric side effects. Another well-known NMDA receptor antagonist, phencyclidine, and certain other compounds tested in clinical trials have also shown serious side effects similar to those of MK-801.

Unlike other NMDA receptor antagonists, the low affinity, uncompetitive one memantine has been shown to reduce neurotoxic damage in vivo and in vitro at concentrations that do not inhibit induction of LTP or produce learning deficits.

This drug infact blocks NMDA receptor in the resting state, similar to the physiological blocker Mg^{2+}, and dissociates rapidly from the channel upon transient physiological activation by uM concentrations of synaptic glutamate which underlines learning and memory processes.

Nevetherless, since memantine action is less voltage-dependent than that of Mg^{++}, it does not leave NMDA channel so easily after moderate prolonged depolarisation during chronic excitotoxic insults (Parsonset al., 1999). This relatively fast on/off receptor kinetic also differs from those of high affinity antagonists that produced adverse effects in earlier trials. As a proof of that, in vivo experiments showed that memantine administration at therapeutically relevant concentrations (up to 2,5 uM plasma level), allows normal synaptic transmission and LTP (Barnes et al., 1996). In another study, memantine administration reduced cognitive impairment in rats with entorhinal cortex injury (Zajaczkowski et al., 1996) and lesions induced by NMDA injection in nucleus basalis magnocellularis (Wenk et al., 1997).

Further supporting use of this drug in AD therapy, a recent study by Hidalgo and coll. (2002) have demonstrated that memantine prevents pathological changes and learning impairment induced by beta amyloid injection in rat hippocampus. Furthermore, at a clinically relevant dose, the same drug was found to markedly increase BDNF protein levels and to induce receptor trkB in the limbic rat cortex (Marvanova et al., 2001).

Taken together, these data suggest that the neuroprotective properties of memantine could be mediated also by the increased production of neurotrophic factors in the brain. Introduced in the clinical setting as an antiviral agent more than 30 years ago, memantine has been approved for use in dementia treatment in Germany for over ten years, showing promising effects in several clinical studies.

Recently, results of a placebo-controlled study with memantine in the US were published (Reisberg et al., 2003), which revealed statistically significant benefit for up to 28 weeks treatment with 20 mg per day memantine in patients with moderate and severe AD. Significant differences in favor of the memantine treated group was observed in the Clinician's Interview-Based Impression of Change Plus Caregiver Input (CIBIC-Plus), a scale which measures overall global changes in cognition and behaviour based on patient's and care-giver's interviews, and in the Alzheimer Disease Cooperative Study activities of Daily Living Inventory (ADCS-ADL), which assesses functional ability in the activities of daily living. Furthermore, a mitigation of the burden on care-givers was observed in the treated group. In this study the tolerability of memantine was found to be excellent, while major concerns attained to the high dropout rate for the total study population, similar in the two groups, due to the severe stages of the disease in enrolled patients.

Memantine's clinical efficacy has been shown also in an etiologically mixed population of demented patients, including those with vascular dementia (VaD) (Winblad et al., 1999), and, more recently, in patients with mild-to-moderate vascular dementia (Borgogozo et al.,

2002). Given the difficulties to of diagnosing VAD in clinical practice, an optimal antidementia drug should be beneficial in both AD and VaD.

Furthermore this drug can be safely associated to acetylcholinesterase inhibitors which are the mainstay of current symptomatic treatment of Alzheimer's disease. A recent paper by Kamenetz et al. (2003) found that NMDA receptor antagonists are able to block APP-induced synaptic depression, which seems to be an early event in AD pathogenesis, thus suggesting a possible benefit of these agents also in early stages of the disease. These authors hypothesize that a negative feedback process physiologically exists wherein higher levels of neuronal activity may increase Abeta production from endogenous APP and lead to synaptic depression that could curtail excess activity. In AD, hypersensitivity and unresponsiveness to this feedback system might contribute to early cognitive dysfunction and to disease progression respectively.

Recently a competitive NMDA antagonist, CGP-43487, has been shown to stimulate neuronal proliferation in the aged rat hippocampus (Nacher et al., 2003). Since some reports indicate that basal extracellular glutamate levels are increased in the hippocampus during age and may contribute to age-related memory impairment, similar agents, able to exert at the same time neurotrophic and neuroprotective effects, would seem very promising.

Another research area consists in the development of agents that would potentiate AMPA receptor responses, since hippocampal LTP, which underlies memory and learning processes, is AMPA receptor-mediated. In particular benzoylpiperidines, such as 1-BCP, CX516 and BDP-12 (also known as ampakines), and the diazoxide derivatives, such as IDRA1, improved performances on a variety of learning and memory tasks in animal models and in aged humans (see Lynch 1998 for review). They act slowing AMPA receptors desensitisation and deactivation. Although short-term safety has been established for these compounds in several animal studies and even in small numbers of humans, a recent study indicated that CX516 increased hippocampal BDNF expression at concentrations also known to augment the frequency of spontaneous firing in cultured hippocampal slices (Lauterborn et al., 2001). From a clinical standpoint, these data warrant enhanced vigilance for adverse effects in patients suffering for conditions possibly associated with sustained AMPA receptors activation, such as ischemia or seizures.

A more comprehensive understanding of the mechanisms of toxicity and the regions most vulnerable to injury facilitated by AMPA modulators may maximize ampakines therapeutic index and allow further studies of their cognitive enhancing potential.

In recent times, a role for estrogenic replacement therapy in the prevention of dementia has been widely discussed. In vitro studies show that these hormones protect cortical neurons and glial cells from glutamate toxicity (Singer 1996; Shy, 2000) and have antioxidant properties. In vivo studies also found lower levels of glutamate and higher affinity of glutamate transporters in the brain during proestrous, which is characterized by a high estradiol level (Mitrovic et al., 1999). In a recent study, Liang and coll. (2002) suggest that one potential mechanism through which estrogens may exert a neuroprotective role in Alzheimer disease is the enhancement of glutamate transport in astrocytes, where it is pathologically decreased. These data seem to indicate glutamate transporters as emerging therapeutic targets in prevention or treatment of AD. The availability of the cloned transporters will facilitate the screening of selective compounds acting on the different excitatory aminoacid transporter subtypes.

5. Antiexcitotoxic therapies in Huntington disease.

The hypothesis that antiglutamate therapies may have an application in Huntington disease (HD) stems from many reports showing the involvement of excitotoxic mechanisms in its pathogenesis (see Chaper 12). Initial treatments for HD targeted inhibition of glutamate release, such as baclofen (Shoulson at al. 1989) or lamotrigine (Kieburtz et al., 1999), which nevetherless failed to slow or stop illness progression.

However, antiglutamate approach has been continued, since other studies demonstrated that riluzole, shown to inhibit glutamate release, has neuroprotective effects in 3-nitropropionic acid animal models of HD (Guyot et al., 1997) and in transgenic HD mice (Schiefer et al., 2002). Recent clinical trials indicate that riluzole has transient antichoreic effects and more prolonged effects on psychomotor speed and behaviour in HD patients (Seppi K et al., 2000; Rosas et al., 1999).

Another drug, remacemide, which is a non-competitive, low-affinity NMDA receptor antagonist, in combination with coenzyme $Q_{10,}$ produced a significant improvement of survival, delay of onset and disease progression in a transgenic mouse model of HD (Ferrante et al.,, 2002). Unfortunatly, a clinical trial performed over 340 patients randomized to coenzyme Q_{10} and/or remacemide for 30 months didn't show any significant effect of either agents on the primary outcome measure, which was the total functional capacity (TFC) scale of the unified Huntington's disease rating scale (Huntington Study Group, 2001), probably due to dosage inadequacy.

A recent paper indicates the activation of the NR2B-subtype of NMDA receptor as a trigger for selective neuronal degeneration in HD (Zeron et al., 2002). New piperidine derivatives have been developed with a potential effect on this subunit, and it will be interesting to evaluate their biological effects in the animal models of the disease. Development of such compounds selective for specific glutamate receptor subunits will help to find new therapeutic strategies by improving regional specificity and avoiding side effects.

Another emerging therapeutic strategy is the modulation of glutamate extracellular levels through drugs acting on adenosine (A) receptors. In particular, A1 receptors subtype stimulation can induce hyperpolarisation of the cell membrane, inhibit NMDA-receptor activation, and reduce glutamate release within the striatum in transgenic and lesional mice model of HD; the inhibition of presynaptic A2A receptors decreases extracellular glutamate concentrations, too.

There are several physiological limitations to the use of A1 agonists: the variable ability to cross the blood–brain barrier; the desensitising effect on receptor activity; several side-effects. On the other hand, at high concentrations, overinhibition of postsynaptic A2A receptors by antagonists could accelerate neurodegeneration, and worsen the hyperkinetic states which are undesirable in HD (Blum et al., 2003). Dose optimization or, alternatively, early use of these compounds in a presymptomatic stage in absence of overt hyperkinetic symptoms, could help to solve these problems.

7. Final remarks.

In recent times, several genetic and molecular advances relevant to the biology of neurodegeneration have been made, allowing the identification of selective therapeutic targets.

Promising therapeutic interventions have geared toward mitigating excitotoxicity, which is a critical, early event in the pathogenesis of a number of both acute and chronic neurodegenerative disorders.

Unfortunately, the success of these treatments in preclinical *in vitro* and animal models has not always translated into equal success in clinical trials, also due to the fact that in humans the onset of symptoms does not equate with the onset of the disease. There is in fact increasing evidence that the onset of most of the aforementioned diseases (AD, PD, ALS, HD) occurs after a long-lasting period of silent degeneration through which millions of neurons are destroyed by complex neurodegenerative mechanisms. Since available treatments, as well as expected novel compounds in the near future, are unlikely to be fully effective due to the severity of brain damage when the clinical symptoms appear, prevention should be at the forefront of our intervention strategies with the population at risk. In most instances, to date, the lack of presymptomatic markers precludes our ability to adopt this strategy.

Another issue to consider is that excitotoxicity probably occurs through different pathways in the different diseases. It's well known that glutamate toxicity acts via two distinct pathways: an excitotoxic one in which glutamate receptors are hyperstimulated, and an oxidative one in which cystine uptake is inhibited, resulting in glutathione depletion and oxidative stress. These two mechanisms might be envolved in different measure in different neurodegenerative diseases. Therefore it would be useful a therapeutic approach combining a strict antiglutamatergic effect with an antioxidant and an antiapoptotic one.

A further issue to consider is the role of second messenger systems in regulating the excitotoxic process. Since second messenger systems modulate many receptors including the NMDA ones, as well as glutamate transporters, they represent an alternative pharmacological site at which excitatory aminoacids neurotransmission might be subtly modulated. The analysis of the results of past clinical trials and of the reasons for their failure will certainly help in planning more selective and efficacious excitoprotective strategies for the future.

REFERENCES

Barnes CA, Danysz W, Parsons CG. Effects of the uncompetitive NMDA receptor antagonist memantine on hippocampal long-term potentiation, short-term exploratory modulation and spatial memory in awake, freely moving rats. Eur J Neurosci. 1996; 8(3):565-71.

Bausch, A.R., Roy, G., 1996. Volume-sensitive chloride channels blocked by neuroprotective drugs in human glial cells (U-138MG). Glia 18(1), 73-77.

Bensimon, G., Lacomblez, L., Meininger, V., 1994. A controlled trial of riluzole in amyotrophic lateral sclerosis. ALS/Riluzole Study Group. N Engl J Med 330(9), 585-91.

Blin O, Azulay JP, Desnuelle C, Bille-Turc F, Braguer D, Besse D, Branger E, Crevat A, Serratrice G, Pouget JY. A controlled one-year trial of dextromethorphan in amyotrophic lateral sclerosis. Clin Neuropharmacol. 1996;19(2):189-92.

Blum D, Hourez R, Galas MC, Popoli P, Schiffmann SN. Adenosine receptors and Huntington's disease: implications for pathogenesis and therapeutics. Lancet Neurol. 2003;2(6):366-74.

Chio A, Cucatto A, Terreni AA, Schiffer D. Reduced glutathione in amyotrophic lateral sclerosis: an open, crossover, randomized trial. Ital J Neurol Sci. 1998;19(6):363-6.

Davis SM, Albers JV, Diener HC, Lees KR, Norris J Termination of Acute Stroke Studies Involving Selfotel Treatment. ASSIST Steering Committed. Lancet. 1997, 349(9044):32.

Debono, M.W., Le Guern, J., Canton, T., Doble, A., Pradier, L., 1993. Inhibition by riluzole of electrophysiological responses mediated by rat kainate and NMDA receptors expressed in Xenopus oocytes. Eur J Pharmacol 235(2-3), 283-289.

Desnuelle, C., Dib, M., Garrel, C., Favier, A., 2001. A double-blind, placebo-controlled randomized clinical trial of alpha-tocopherol (vitamin E) in the treatment of amyotrophic lateral sclerosis. ALS riluzole-tocopherol Study Group. Amyotroph Lateral Scler Other Motor Neuron Disord 2(1), 9-18.

Dib M. Amyotrophic lateral sclerosis: progress and prospects from treatment. Drugs 2003; 63(3):289-310.

Diener H.C., Stroke, 1999; 30: 234

Diener HC, Cerebrovasc. Dis. 1999, 8, 172-181

Eisen A, Stewart H, Schulzer M, Cameron D. Anti-glutamate therapy in amyotrophic lateral sclerosis: a trial using lamotrigine. Can J Neurol Sci. 1993;20(4):297-301.

Ferrante RJ, Andreassen OA, Dedeoglu A, Ferrante KL, Jenkins BG, Hersch SM, Beal MF.

Franke CL, Palm R, Dalby M, Schoonderwaldt HC, Hantson L, Eriksson B, Lang-Jenssen L, Smakman J. Flunarizine in stroke treatment (FIST): a double-blind, placebo-controlled trial in Scandinavia and the Netherlands. Acta Neurol Scand. 1996, 93(1):56-60

Gilgun-Sherki, Y., Melamed, E., Offen, D., 2001. Oxidative stress induced-neurodegenerative diseases: the need for antioxidants that penetrate the blood brain barrier. Neuropharmacology 40(8), 959-975.

Groeneveld JG, Veldink JH, van der Tweel I, Kalmijn S, Beijer C, de Visser M, Wokke JH, Franssen H, van den Berg LH. A randomized sequential trial of creatine in amyotrophic lateral sclerosis. Ann Neurol. 2003;53(4):437-45.

Grotta J Lubeluzole treatment of acute ischemic stroke. The US and Canadian Lubeluzole Ischemic Stroke Study Group. Stroke. 1997, 28(12):2338-46.

Guyot MC, Palfi S, Stutzmann JM, Maziere M, Hantraye P, Brouillet E. Riluzole protects from motor deficits and striatal degeneration produced by systemic 3-nitropropionic acid intoxication in rats. Neuroscience. 1997;81(1):141-9.

Haley EC High-dose tirilazad for acute stroke (RANTTAS II). RANTTAS II Investigators. Stroke. 1998, 29(6):1256-7.

Horn J et al Cerebrovasc. Dis. 1999, 9 (Supp. 1), 127

Huang, C.S., Song, J.H., Nagata, K., Yeh, J.Z., Narahashi, T., 1997. Effects of the neuroprotective agent riluzole on the high voltage-activated calcium channels of rat dorsal root ganglion neurons. J Pharmacol Exp Ther 282(3), 1280-1290.

Huntington Study Group. A randomized, placebo-controlled trial of coenzyme Q10 and remacemide in Huntington's disease. Neurology. 2001;57(3):397-404.

Kamenetz F, Tomita T, Hsieh H, Seabrook G, Borchelt D, Iwatsubo T, Sisodia S, Malinow R. APP processing and synaptic function. Neuron 2003; 37(6):925-37.

Kanai Y., Hediger M.A. Primary structure and functional characterization of a high-affinity glutamate transporter. Nature 1992; 360(6403):467-471.

Kieburtz K. Antiglutamate therapies in Huntington's disease. J Neural Transm Suppl. 1999; 55:97-102.

Koh, J.Y., Kim, D.K., Hwang, J.Y., Kim, Y.H., Seo, J.H., 1999. Antioxidative and proapoptotic effects of riluzole on cultured cortical neurons. J Neurochem 72(2), 716-723.

Kriz J, Gowing G, Julien JP, Efficient three-drug cocktail for disease induced by mutant superoxide dismutase. Ann Neurol. 2003, 53(4):429-436.

Laake K, Oeksengaard AR. D-cycloserine for Alzheimer's disease. Cochrane Database Syst Rev. 2002;(2):CD003153.

Lacomblez, L., Bensimon, G., Leigh, P.N., Guillet, P., Meininger, V., 1996. Dose-ranging study of riluzole in amyotrophic lateral sclerosis. Amyotrophic Lateral Sclerosis/Riluzole Study Group II. Lancet 347(9013), 1425-1431.

Lange DJ, Murphy PL, Diamond B, Appel V, Lai EC, Younger DS, Appel SH. Selegiline is ineffective in a collaborative double-blind, placebo-controlled trial for treatment of amyotrophic lateral sclerosis. Arch Neurol. 1998;55(1):93-6.

Lauderback CM, Hackett JM, Huang FF, Keller JN, Szweda LI, Markesbery WR, Butterfield DA. The glial glutamate transporter, GLT-1, is oxidatively modified by 4-hydroxy-2-nonenal in the Alzheimer's disease brain: the role of Abeta 1-42. J Neurochem. 2001;78(2):413-6.

Lauterborn JC, Lynch G, Vanderklish P, Arai A, Gall CM. Positive modulation of AMPA receptors increases neurotrophin expression by hippocampal and cortical neurons. J Neurosci. 2000;20(1):8-21.

Louwerse ES, Weverling GJ, Bossuyt PM, Meyjes FE, de Jong JM. Randomized, double-blind, controlled trial of acetylcysteine in amyotrophic lateral sclerosis. Arch Neurol. 1995;52(6):559-64.

Lynch G. Memory and the brain: unespected chemistries and a new pharmacology. Neurobiol Learn Mem. 1998; 70(1-2):82-100.

Marvanova M, Lakso M, Pirhonen J, Nawa H, Wong G, Castren E. The neuroprotective agent memantine induces brain.derived neurotrophic factor and trkB receptor expression in rat brain. Mol Cell Neurosci 2001; 18(3):247-58.

Miguel-Hidalgo JJ, Alvarez XA, Cacabelos R, Quack G. Neuroprotection by memantine against neurodegeneration induced by beta-amyloid(1-40). Brain Res. 2002;958(1):210-21.

Miller RG, Moore DH 2nd, Gelinas DF, Dronsky V, Mendoza M, Barohn RJ, Bryan W, Ravits J, Yuen E, Neville H, Ringel S, Bromberg M, Petajan J, Amato AA, Jackson C, Johnson W, Mandler R, Bosch P, Smith B, Graves M, Ross M, Sorenson EJ, Kelkar P, Parry G, Olney R; Western ALS Study Group. Phase III randomized trial of gabapentin in patients with amyotrophic lateral sclerosis. Neurology. 2001;56(7):843-8.

Mitrovic AD, Maddison JE, Johnston GA. Influence of the oestrous cycle on L-glutamate and L-aspartate transport in rat brain synaptosomes. Neurochem Int 1999; 34(2):101-8.

Mohr JP et al Cerebrovascualr Disorders 1994, 4,197-203

Nacher J, Alonso-Llosa G, Rosell DR, McEwen BS. NMDA receptor antagonist treatment increase the production of new neurons in the aged rat hippocampus. Neurobiol Aging 2003: 24(2):273-84.

Noh, K.M., Hwang, J.Y., Shin, H.C., Koh, J.Y., 2000. A novel neuroprotective mechanism of riluzole: direct inhibition of protein kinase C. Neurobiol Dis 7(4), 375-383.

Orgogozo JM, Rigaud AS, Stoffler A, Mobius HJ, Forette F. Efficacy and safety of memantine in patients with mild to moderate vascular dementia: a randomized, placebo-controlled trial (MMM 300). Stroke 2002, 33(7):1834-9.

Parsons CG, Danysz Q, Quack G. Memantine is a clinically well tolerated N-methyl-D-aspartate (NMDA) receptor antagonist. A review of preclinical data. Neuropharmacology 1999; 38(6):735-67.

Peters GR Stroke 1996, 27: 195

Pulsinelli, W.A. et al, Neurology 1999 52 (Supp 2), A384

Reisberg B, Doody R, Stoffler A, Shmitt F, Ferris S, Mobius HJ, Memantine Study Group. Memantine in moderate-to-severe Alzheimer's disease. N Engl J Med 2003; 348(14):1333-41.

Rosas HD, Koroshetz WJ, Jenkins BG, Chen YI, Hayden DL, Beal MF, Cudkowicz ME. Riluzole therapy in Huntington's disease (HD). Mov Disord. 1999;14(2):326-30.

Schiefer J, Landwehrmeyer GB, Luesse HG, Sprunken A, Puls C, Milkereit A, Milkereit E, Kosinski CM. Riluzole prolongs survival time and alters nuclear inclusion formation in a transgenic mouse model of Huntington's disease. Mov Disord. 2002;17(4):748-57.

Seppi K, Mueller J, Bodner T, Brandauer E, Benke T, Weirich-Schwaiger H, Poewe W, Wenning GK. Riluzole in Huntington's disease (HD): an open label study with one year follow up.J Neurol. 2001;248(10):866-9

Shoulson I, Odoroff C, Oakes D, Behr J, Goldblatt D, Caine E, Kennedy J, Miller C, Bamford K, Rubin A, et al. A controlled clinical trial of baclofen as protective therapy in early Huntington's disease. Ann Neurol. 1989;25(3):252-9.

Shy H, Malaiyandi L, Timiras PS. Protective action of 17beta-estradiol and tamoxifen on glutamate toxicity in glial cells. Int Dev Neurosci 2000; 18(2-3):289-97.

Singer CA; Rogers KL. Strickland TM; Dorsa DM. Estrogen protects primary cortical neurons from glutaamate toxicity. Neuosci. Lett. 1996;212(1):13-6.

Smith, R.G., Henry, Y.K., Mattson, M.P., Appel, S.H., 1998. Presence of 4-hydroxynonenal in cerebrospinal fluid of patients with sporadic amyotrophic lateral sclerosis. Ann Neurol 44(4), 696-699.

Song, J.H., Huang, C.S., Nagata, K., Yeh, J.Z., Narahashi, T., 1997. Differential action of riluzole on tetrodotoxin-sensitive and tetrodotoxin-resistant sodium channels. J Pharmacol Exp Ther 282(2), 707-714.

Takenori Yamaguchi, Keiji Sano, Kintomo Takakura, Isamu Saito, Yukito Shinohara, Takao Asano, and Hajime Yasuhara Ebselen in Acute Ischemic Stroke : A Placebo-Controlled, Double-blind Clinical Trial Stroke 1998, 29: 12-17

Talbot, K., 2002. Motor neurone disease. Postgrad Med J 78(923), 513-519.

Terro, F., Lesort, M., Viader, F., Ludolph, A., Hugon, J., 1996. Antioxidant drugs block in vitro the neurotoxicity of CSF from patients with amyotrophic lateral sclerosis. Neuroreport 7(12), 1970-1972.

Therapeutic effects of coenzyme Q10 and remacemide in transgenic mouse models of Huntington's disease. J Neurosci. 2002;22(5):1592-9.

Tohgi, H., Abe, T., Yamazaki, K., Murata, T., Ishizaki, E., Isobe, C., 1999. Remarkable increase in cerebrospinal fluid 3-nitrotyrosine in patients with sporadic amyotrophic lateral sclerosis. Ann Neurol 46(1), 129-131.

Wahlgren NG, Ranasinha KW, Rosolacci T, Franke CL, van Erven PMM, Ashwood T, and Claesson L. Clomethiazole Acute Stroke Study (CLASS) : Results of a Randomized, Controlled Trial of Clomethiazole Versus Placebo in 1360 Acute Stroke Patients Stroke 1999 30: 21 - 28

Wenk GL, Zajaczkowski W, Danysz W. Neuroprotection of acetylcholinergic basal forebrain neurons by memantine and neurokinin B. Behav Brain Res. 1997; 83(1-2): 129-33.

Winblad B, Poritis N. Memantine in severe dementia: results of the 9M-Best Study (Benefit and efficacy in severely demented patients during treatment with memantine). Int J Geriatr Psychiatry 1999, 14(2):135-46.

Zajaczkowski W, Quack G, Danysz W. Infusion of (+)-MK-801 and memantine contrasting effects on radial maze learning in rats with entorhinal cortex lesion. Eur J Pharmacol. 1996; 296(3):239-46.

Zeron MM, Hansson O, Chen N, Wellington CL, Leavitt BR, Brundin P, Hayden MR, Raymond LA. Increased sensitivity to N-methyl-D-aspartate receptor-mediated excitotoxicity in a mouse model of Huntington's disease. Neuron. 2002;33(6):849-60

Zhu S, Stavrovskaya IG, Drozda M, Kim BY, Ona V, Li M, Sarang S, Liu AS, Hartley DM, Wu du C, Gullans S, Ferrante RJ, Przedborski S, Kristal BS, Friedlander RM. Minocycline inhibits cytochrome c release and delays progression of amyotrophic lateral sclerosis in mice. Nature. 2002, 417(6884):74-78.

INDEX

Acrolein, 254
Adeno-associated virus (AAV) vaccines, 62
AIDS neuropathology, 20–23
α-Amino-3-hydroxy-5methylisoxazoleproptionate (AMPA)
  definition of, 66
Alzheimer's disease (AD)
  definition of, 252
  diagnosis of, 257
  glutamatergic system in, 253–258
  glutamine synthetase in, 255
  neurodegeneration, 254
  oxidative stress in, 258
Amantadine, 61
  as NMDA receptor antagonist, 233
AMPA/Kainate receptors
  activation of, 19
  cell death linked to, 293
  modulation of, in traumatic brain injury, 200
AMPA receptors (AMPARS)
  effect of LTD on, 71
  exocytosis, 72–73
  4.1 N interaction with, 69
  GluR-NSF interaction with, 69–70
  increased functions of, 219
  intracellular signaling in, 71–72
  NARP interaction with, 69–70
  PDZ protein interactions of, 68 69
  phosphorylation, 70–71
  post-synaptic trafficking, 74f
  response rate of, 53
  schematic representation of, 66f
  Stargazin interaction with, 69–70
  structure of, 66–68
  subunits of, 232
  of ventral spinal cord, 219
Amyloid β-peptide (Aβ)
  APP processing, 254
  role of, in pathogenesis of Alzheimer's, 252
Amyotrophic lateral sclerosis (ALS)
  AMPA receptor expression in, 220
  effect of EAAT2 on, 221
  role of glutamate in, 218
  ventral spinal cord in, 219
  in vitro studies of, 218–221
  in vivo studies on, 221–224
Animal models
  for multiple sclerosis, 287–288
  of prion induced neurodegeneration, 268–269
  of toxic domains, 270
Anti-excitotoxic approaches, 120f
  drugs, 177–179
Antiglutamatergic drugs
  effects of, on motor neuron diseases, 224t
Antiretroviral therapy, 22
AOAA (complex I), 244
Apoptosis
  as excitotoxic mechanism, 111–112
  mitochondrial changes during, 112
  phases of, 111f
  in prion disease, 269
  triggering, 269–270
Apoptotic biochemical cascades
  activation of, 112
Astrocytes
  active roles of, 15–17
  calcium-dependent release from, 14
  compound synthesis of, 16
  effect of PrP106-126
  excitotoxicity of, 118–119
  glutamate release of, in pathology, 17
  glutamate release of, in physiology, 17
  glutamate uptake by, 15, 274, 293–295
  in hippocampus, 15
  neuron signaling, 18
  neurotransmitter receptor expression, 14
  proliferation of, 272
  PrP protein replication and, 273
  synaptic functions regulated by, 24
Astroglial cells
  glutamate transporter expression in, 39
Astrogliosis
  in prion disease, 276

BFGF receptors
activation of, 114
Blood-brain barrier (BBB)
in traumatic brain injury, 194
Bovine spongiform encephalopathy (BSE), 264
Brain cells
glutamate transporters expressed by, 29f
intracellular transporters in, 30
Brain plasma membrane
glutamate-cystine exchanger expressed by, 29

C-terminus
C1, 52
C2, 52
truncation of, 57
Calcium
cytoplasmic, 106f
enzymes activated by, 176–177
as excitotoxic mechanism, 105–107
glutamate-induced rise in, 176
loading, 58–59
neuronal homeostasis, 117
preceding cell death, 2
Calcium-calmodulin-dependent protein kinase II (CaMKII), 55
Caspase-Activated DNase (CAD), 143
Cell death
excitotoxicity and, 271–276
glutamate as cause of, 194
in hippocampal neurons, 2
mechanisms of, 268–270
Cellular mechanisms
of ecotoxicology, 2
Cerebral ischemia, 60
cell death in, 177
glutamate receptors in, 175–176
neuropathology of, 173–174
role of mGlu1 receptors in, 179–180
role of poly(ADP-ribosyl) polymerase-1 (PARP-1) in, 180–183
Cerebrospinal fluid (CSF)
in amyotrophic lateral sclerosis patients, 218
glutamate measurements in, 292
Cerestat
clinical trials of, 202
Clinical trials
future of, 178–179
phase III with glutamate receptor antagonists, 202–203
of strokes, 177–179
Copper binding
prion proteins and, 265–266
CP 101-606
clinical trials of, 203
Creutzfeldt-Jakob disease
definition of, 264
CXCR4
astrocyte-glutamate release mediated by, 23f
signaling cascade, 22
TNF shedding mediated by, 21f
Cyclooxygenase (COX) inhibitors
glutamate release blocking, 19
Cytokines
definition of, 20
as excitoprotective mechanism, 113–114

D-CPP-ene
clinical trials of, 203
Dendritic plasticity
excitotoxic mechanisms in, 112–113
Dextramethorphan
as NMDA receptor channel blocker, 196
Dextrorphan
as NMDA receptor channel blocker, 196
DNA damage
as excitotoxic mechanism, 110
DNA fragmentation
kainic acid induced, 220
Dopamine
electrophysiological changes induced by, 236

EAAT1 (GLAST)
expression of, 33, 254
immunocytochemical localization of, 34f
levels of, 33
EAAT2 (GLT)
downregulation of, 291
expression of, 34–35, 254, 255
glutamate uptake mediated by, 41
HNE binding to, 254t
immunocytochemical localization of, 34f
immunoreactivity of, 218
levels of, 35
locations of, 36
loss of, 221
neurodegeneration linked to, 39
regulation of, 40
EAAT3 (EAAC)
distribution of, 36–37
expression of, 36, 274
immunocytochemical localization of, 34f

in white matter, 294
EAAT4
distribution of, 37
EAAT transporters
attachment of, 33
definition of, 32
distribution of, 33–37
as ion channel, 31–32
mechanisms of, 31–32
physiological role of, 37–40
structure of, 32–33
Electroshock seizures (ECS)
pre-exposure to, 151f
Endoplasmic reticulum (ER)
NR2 subunits in, 57
Energy impairment
as excitotoxic mechanism, 110
Epilepsy
definition of, 138
Epileptic tolerance, 147
Excitatory amino acids (EAAs)
neurotoxicity of, 80
release of, 175
Excitoprotective mechanisms, 113–118
cytoskeleton-mediated, 117–118
neurotrophic factors, 113–114
preconditioning pathways, 116
signaling pathways of, 115f
stress response pathways, 116
Excitotoxic mechanisms
apoptosis, 111
Calcium, 105–107
in dendritic plasticity, 112–113
DNA damage, 110
energy impairment, 110
oxidative stress, 107–110
in synaptic plasticity, 112–113
Excitotoxicity
in astrocytes, 118–119
defense against, 41
definition of, 1
free radicals linked to, 3–4
in glial cells, 118–119
in Huntington's disease, 244
membrane lipid peroxidation promoting, 109
of microglia, 119
nitric oxide linked to, 4–5
NMDA-mediated, 22
NMDA receptors in, 58–59
in oligodendrocytes, 119
pharmacological approaches targeting, 198t
slow, 5–6
status epilepticus induced, 140
in transgenic mouse models of Huntington's disease, 246
in traumatic brain injury, 192–202
of white matter, 292–297
Excitotoxicity mechanisms
glutamate receptors, 175–176
glutamate release, 175
intracellular events, 176–177
in post-ischemic neurodegeneration, 174–177
Excitotrophicity
seizure-induced, 153
Experimental autoimmune encephalomyelitis (EAE)
glutamate homeostasis in, 290–292
inducing, 288
multiple sclerosis (MS) and, 287–288
NBQX treatment for, 289

Focal ischemia
neuropathology of, 173
4.1 N
AMPA receptor interaction with, 69–70
Free radicals
DNA strand breaks produced by, 182f
excitotoxicity linked to, 3–4
formation of, 181
hydroxyl, 245

GABA receptors, 83
blocking, 150
Glial cells
definition of, 14
excitotoxicity of, 118–119
Interleukins 1β (IL-1β) expression in, 21
tumor necrosis factor α (TNFα) expression in, 21
Gliosis
definition of, 271–272
in prion disease, 271
Global ischemia
neuropathology of, 173–174
GLT1
decreases in, 221–22
expression of, 222
immunoreactivity, 222f
Glucose-oxygen deprivation (OGD)
protection from, 181f
GluR-NSF
AMPA receptor interaction with, 69–70
GluR2 receptors

increased levels of, 223–224
GluR3 receptors
decreased levels of, 223–224
GluRA-D, 66
Glutamate
altered homeostasis, 290–292
astrocyte uptake of, 15
calcium-dependent release of, 14
as cause of cell death, 194
conversion of, 255, 257
excitotoxic damage to oligodendrocytes, 297
radiolaballed, 274
re-uptake of, 194–195
receptors activated by, 232
role of, 28, 230
role of, in amyotrophic lateral sclerosis, 218
role of, in synaptic transmission, 253
secretion of, in multiple sclerosis, 291
stress response pathways induced by, 116
synaptically released, 16f
traumatic brain injury and, 192–194
Glutamate catabolism
in multiple sclerosis, 291
process of, 290
Glutamate-cystine exchanger
expression of, 29
Glutamate decarboxylase (GD)
conversion to, 257
Glutamate dehydrogenase (GDH)
conversion, 290
Glutamate-glutamine cycle, 16f
Glutamate receptors
activation of, 117
in cerebral ischemia, 175–176
clinical phase III trials with, 202–203
concentration of, 112
ionotropic, 232–234
neuron death caused by, 104
role of, in motor neuron degeneration, 222–223
Glutamate release
Calcium dependent, 18–20
Calcium independent, 17–18
CXCR4-mediated, 23f
hypoosmotic, 18
in hypoxic/hypoglycemic conditions, 175
increased, 193
inhibitors of, in traumatic brain injury, 201–202
linked to neurodegeneration, 20–23
non-synaptic, 295–296
in pathology, 17
in physiology, 17
traumatic brain injury and, 192–194
in white matter, 295–296
Glutamate transport
in amyotrophic lateral sclerosis patients, 218
role of potassium in, 31
role of sodium in, 31
stimulation of, 258
stoichiometric, 31–32
Glutamate transporters
attachment of, 33
brain tissue expression of, 29f
distribution of, 33–37
EAAT1, 33–34
EAAT2, 34–36
EAAT3, 36–37
EAAT4, 37
expression of, 195
functions of, 30f
localization of, 38f
necessity of, 38–39
in oligodendrocytes, 295
properties of, 30–33
reverse uptake of, 32f
role of, in disease, 40
structure of, 32–33
uptake of, 32f
Glutamate uptake
astrocyte, 293–295
global brain, 39
kinetic profile of, 294
oligodendrocyte, 293–295
PrP proteins and, 274
Glutamatergic pathways, 234–236
in Parkinson's disease, 238
Glutamatergic system
in Alzheimer's disease, 256f
Glutaminase
role of, 257
Glutamine
conversion to, 255
Glutamine synthetase (GS)
activity of, in Alzheimer's disease, 255
conversion, 290
effect of oxidative stress on, 256
Glycine site
in traumatic brain injury, 197–200
Golgi apparatus
NMDA receptors processed at, 57

Haloperidol, 237

Hippocampus
astrocytes in, 15
HNE
binding, 254t
HU-211
as NMDA receptor channel blocker, 196
Huntington's disease (HD)
etiology of, 245–246
excitotoxic models of, 244
mitochondrial toxin models of, 244–245
NMDA receptors in, 244
transgenic mouse models, 245–246
Hypoglycemia
glutamte transporters role in, 40

Ifenprodil, 61
Immature brain
seizures in, 154–156
status epilepticus resistance of, 147–148
Integrin receptors
transduction mechanisms of, 115–116
Interferons (IFN), 20
Interleukins 1β (IL-1β)
expression of, in glial cells, 21
Intracellular events, 176–177
Ion channels
EAAT transporters as, 31
Ion homeostasis
role of oxidative stress in, 109
Ischemia
glutamte transporters role in, 40
Ischemic tolerance, 147

Kainic acid
DNA fragmentation induced by, 220
in primary motor neuron cultures, 220f
Ketamine
as NMDA receptor channel blocker, 196
Krebs cycle
glutamate linked to, 290

Levodopa, 238
for Parkinson's disease treatment, 232–234
Lipid perxoidation
measuring, 252
LTD
effect of, on AMPA receptor endocytosis, 71
LTP
effect of, on AMPA receptor exocytosis, 72–73
LY367385
role of, in glucose-oxygen deprivation, 181

Magnesium
as NMDA receptor channel blocker, 196
MAGUK proteins
NMDA receptors with, 56–57
Malonate, 244
MAP kinase/ERK pathway
in astrocyte changes, 272–273
Memantine, 61
Membrane lipid peroxidation
excitotoxicity promoted by, 109
Metabotropic receptors
modulation of, in TBI, 201
Methyl-.4-phenyl-1,2,3,6-tetrahydropyridine (MPTP)
as parkinsonian model, 231
MGlu4/7/8 receptors
activation of, 92–93
as drug targets, 92
expression of, 93
modulation function of, 94
role of, in neuroprotection, 91–94
MGlu2/3 receptors
activation of, 89f, 90
role of, in neuroprotection, 88–91
MGlu receptors
activation of, 81–82
distinguishing, 80
in Parkinson's disease, 237–238
role of group 1, in neuroprotection, 82–87
role of group II, in neuroprotection, 87–91
role of group III, in neuroprotection, 91
MGlu1 receptors
antagonists of, 181f
role of, in cerebral ischemia, 179–180
role of, in neuroprotection, 83
MGlu5 receptors
antagonists of, 87
NMDA receptor interaction with, 86f
role of, in neuroprotection, 84–87
on striatal dopaminergic terminals, 86
Microglia
activated, 296
activation of, 273
astrocyte proliferation and, 272
excitotoxicty of, 119
Microtubules
homeostasis modulated by, 117
MK-801
as NMDA receptor channel blocker, 196
Motor neurons

degeneration of, 222–223
kainate toxicity in, 220f
MPEP, 84, 87
MPTP toxicity
protection against, 88–89
Multiple sclerosis (MS)
animal model for, 287–288
definition of, 286
epidemiology of, 286
experimental autoimmune encephalomyelitis (EAE) and, 287–288
glutamate homeostasis in, 290–292
glutamate secretion in, 291–292
lesions caused by, 287
oligodendrocyte loss in, 289
pathology of, 287
research tools for, 287–288
susceptibility to, 286–287

N-terminal domain (NTD)
sequence homology of, 67–68
NARP
AMPA receptor interaction with, 69–70
NBQX
as AMPA antagonist, 223
EAE treatment with, 289
Neurodegeneration
in Alzheimer's disease, 254
animal models of, 268–269
calcium load hypothesis of, 58
glutamate release linked to, 20–23
pathogenesis of, 232
Neuron-astrocyte metabolic coupling, 16f
Neuronal death
after status epilepticus, 142
apoptic, 142–144
excitotoxic, 174
necrotic, 142–144
Neuroprotection
alternative, 179–183
clinical trials, 178
cytoskeleton-mediated, 117–118
mechanisms of, 92
role of group 1 mGlu receptors in, 82–87
role of group II mGlu receptors in, 88–91
role of group III mGlu receptors in, 91
role of mGlu 4/7/8 receptors in, 91–94
role of mGlu2/3 receptors in, 88–91
role of mGlu1 receptors in, 83
role of mGlu5 receptors in, 84–87
role of poly(ADP-ribosyl) polymerase-1 (PARP-1) in, 180–181
of status epilepticus (SE), 145–147
Neurotoxicity
studies on, 266–271
Neurotransmitter receptors
astrocyte expression of, 14
Neurotrophic factors
as excitoprotective mechanism, 113–114
in signaling pathways, 115f
Nitric oxide
excitotoxicity linked to, 4–5
NMDA receptors
activation of, 82, 195–196
amino-acid sequences for, 54f
anchorage of, 84
Ca2+-dependent inactivation of, 53
channel blockers in traumatic brain injury, 196–197
definition of, 52
as drug target, 61
drugs blocking, 233
glutamate stimulation of, 182f
glycine-dependent desensitization of, 53
Golgi apparatus process of, 57
in Huntington's disease, 244
integration of, 56
with MAGUK proteins, 56–57
mediation mechanisms of, 59–61
mGlu5 receptor interaction with, 86f
modulation of, 56–57
neuronal death linked to, 275
in postsynaptic density, 56f
regulation of, at synapse, 54–56
response rate of, 53
role of, in excitotoxicity, 58–59
role of, in hyperexcitability, 60–61
structure of, 52–54
subunits of, 232
trafficking of, 56–58
zinc modulation of, 200
NNOS
postsynaptic density binding, 55
Non-NMDA ionotropic receptors
excitotoxicity of, 118
NR1
cytoskeletal proteins linked to, 55
forms of, 52
retention of, 57
NR2, 52
cytosolic tails of, 55
neuroprotective properties of, 61
in postsynaptic density, 55
retention of, 57
NR3, 52

NR3A
expression of, 53
NR3B
expression of, 53

Oligodendrocytes
excitotoxic damage to, 292–293, 297
excitotoxicity of, 119
glutamate metabolism in, 294
glutamate transporters expressed in, 295
glutamate uptake by, 293–295
loss of, 289
pathology of, 289
1-aminoindan-1,5-dicarboxylic acid (AIDA)
role of, in glucose-oxygen deprivation, 181f
Oxidative stress, 107–110
in Alzheimer's disease, 258
cell death and, 269–270
effect of, on glutamatergic system, 256
measuring, 252
neuron vulnerability and, 109–110
role of, in excitotoxicity, 108f
role of, in ion homeostasis, 109

P-loop sequence
Re-entrant pore loop homology with, 68
Parkinson's disease (PD)
animal models of, 230–232
definition of, 230
glutamatergic pathways in, 234–236, 238
haloperidol-induced catalepsy model, 237
ionotropic glutamate receptors in, 232–234
mGlu receptors in, 237–238
MPTP model of, 89–90
reserpine-induced akinesia model, 237
striatal circuitry in, 235f
therapy for, 88
PDZ proteins
AMPA receptor interactions with, 68–69
Phosphotyrosine
mediation of, in PSD fraction, 55
Poly(ADP-ribosyl) polymerase-1 (PARP-1)
activation of, 182–183
as neuroprotective agent, 180–181
role of, in glutamate-mediated necrosis, 182f
Poly(ADP-ribosyl)ation
definition of, 180
Polyamine site
in traumatic brain injury, 197
Polypeptides
cytokines, 20
Postsynaptic density (PSD)
definition of, 54–55
expression of, 60
NMDA receptors in, 54–55
NR2 in, 55
phosophotyrosine mediation in, 55
structure of, 59
Potassium
role of, in net transport, 31
Potassium cyanide, 244
Preconditioning pathways
as excitoprotective mechanism, 116
Prion disease
animal models of, 268–269
astroglial proliferation in, 275
definition of, 263–264
gliosis in, 271
neurodegeneration mechanisms of, 271
oxidative stress in, 269–270
Prostaglandins (PGs), 14
Protein phosphorylation
role of, in AMPA receptors, 70–71
PrP106-126
astrocyte proliferation stimulated by, 275
effects of, on astrocytes, 272–273
neuronal death caused by, 275
neurotoxic mechanisms of, 270, 276
proposed mechanisms of, 266–268
transmembrane channel interaction with, 267–268
PrP proteins
copper binding and, 265–266
expression of, 264
glutamate uptake and, 274
microglia activated by, 273
prion disease and, 264
proposed mechanisms of, 266–267
replication of, 273
toxic domains in, 270–271
toxic peptide mimics of, 266

Re-entrant pore loop
P-loop sequence homology of, 68
Remacemide hydrochloride
as NMDA receptor channel blocker, 196
Reserpine, 231
Riluzole, 223, 246
RNA splicing
NR1 forms generated by, 52
ROS production, 107
Rotenone, 231
RPR1199900
as AMPA antagonist, 223

Scrapie, 264
Seizure network
  mapping, 142f
Seizures
  excitotrophicity induced by, 153–154
  in immature brain, 154–156
  recurrent, 150–152
Selfotel
  clinical trials of, 202
SIB-1757, 84, 87
SIB-1893, 84, 87
6-hydroxydopamine (6-OHDA)
  denervation, 237
  as parkinsonian model, 231
SOD1
  mutated, 222
Sodium
  role of, in glutamate binding, 31
Sodium azide (complex IV), 244
Stargazin
  AMPA receptors (AMPARS) interaction with, 69–70
Status epilepticus (SE)
  duration of, 144–145
  immature brain resistance to, 147–148
  long-term consequences of, 148–150
  neuronal death after, 142–144
  neuroprotection of, 145–147
  rate of onset of, 144–145
  uses of, 140
Stoichiometry
  of glutamate transport, 31–32
Stress response pathways
  as excitoprotective mechanisms, 116
Striatal circuitry
  in parkinsonian state, 235f
Striatal dopaminergic terminals
  mGlu5 receptors on, 86
Stroke
  clinic trials for, 177–179
  therapy for, 173–174
  World Health Organization on, 172
Structural lesions
  occurrence of, 139
Substantia nigra pars compacta, 232
Synaptic plasticity
  excitotoxic mechanisms in, 112–113

T cells
  myelin-specific, 288
Thalamocortical pathway, 235
3-nitropropionic acid (3-NP), 244
Transduction mechanisms
  of integrin receptors, 115–116
Transgenic mouse models
  excitotoxicity in, 246–247
  of Huntington's disease, 245–246
Transmembrane channels
  PrP106-126 interaction with, 267
Traumatic brain injury (TBI)
  AMPA/KA receptor modulation in, 200
  blood-brain barrier in, 194
  clinical studies on, 193
  controversies surrounding, 203–204
  definition of, 190
  experimental treatment for, 191–192
  glutamate levels and, 192–193
  glutamate-mediated excitotoxicity in, 192–202
  glutamate re-uptake in, 194–195
  glutamate release inhibitors in, 201–202
  glycine site modulation in, 197–200
  metabotropic receptor modulation in, 201
  NMDA receptor channel blockers in, 196–197
  pathophysiology of, 191–192
  pharmacological approaches to, 198t
  polyamine site modulation in, 197
Tumor necrosis factor α (TNFα), 14–15
  CXCR4-mediated shedding of, 21f
  expression of, in glial cells, 21
Tyrosine kinases
  activation of, 115f

Vesicular glutamate transporters (VGLUTs)
  role of, 30

White matter
  EAAC1 in, 294
  excitotoxicity in, 292–297, 298
  glutamate release in, 295–296
World Health Organization
  on strokes, 172

Zinc
  NMDA receptors modulated by, 200
ZK187638
  as AMPA inhibitor, 223